T0260282

WIRELESS MESH NETWORKING

WIRELESS MESH NETWORKING

Architectures, Protocols and Standards

Edited by
Yan Zhang · Jijun Luo · Honglin Hu

CRC Press
Taylor & Francis Group
Boca Raton London New York

CRC Press is an imprint of the
Taylor & Francis Group, an **informa** business
AN AUERBACH BOOK

CONTENTS

EDITORS

Yan Zhang received a Ph.D. degree from the School of Electrical & Electronics Engineering, Nanyang Technological University, Singapore. From August 2004 to May 2006, he worked with NICT Singapore, National Institute of Information and Communications Technology (NICT). From August 2006, he has worked with Simula Research Laboratory, Norway (http://www.simula.no/).

Dr. Zhang is on the editorial board of *International Journal of Network Security* and is currently serving as the series editor for the book series "Wireless Networks and Mobile Communications" (Auerbach Publications, CRC Press, Taylor & Francis Group). He also serves as co-editor for several books: *Resource, Mobility and Security Management in Wireless Networks and Mobile Communications; Wireless Mesh Networking: Architectures, Protocols and Standards; Millimeter-Wave Technology in Wireless PAN, LAN and MAN; Distributed Antenna Systems: Open Architecture for Future Wireless Communications; Security in Wireless Mesh Networks; Wireless Metropolitan Area Networks: WiMAX and Beyond; Wireless Quality-of-Service: Techniques, Standards; and Applications; Broadband Mobile Multimedia: Techniques and Applications; Internet of Things: From RFID to the Next-Generation Pervasive Networked Systems;* and *Handbook of Research on Wireless Security.*

He serves as program co-chair for IEEE PCAC'07, special track co-chair for "Mobility and Resource Management in Wireless/Mobile Networks" in ITNG 2007, special session co-organizer for "Wireless Mesh Networks" in PDCS 2006, is a member of the Technical Program Committee for IEEE AINA 2007, IEEE CCNC 2007, WASA'06, IEEE GLOBECOM'2006, IEEE WoNGeN'06, IEEE IWCMC 2006, IEEE IWCMC 2005, ITST 2006, and ITST 2005. His research interests include resource, mobility, energy and security management in wireless networks and mobile computing. He is a member of IEEE and IEEE ComSoc. Email: yanzhang@ieee.org

Jijun Luo received his master of engineering (M. Eng.) degree from Shandong University, China in 1999 and a master of science (M. Sc.) degree from Munich University of Technology, Germany in 2000. He joined Siemens in 2000 and pursued his doktor-ingenieur (Dr. Ing.) degree at RWTH Aachen University, Germany. Until now he has published more than 100 technical papers, co-authored three books and holds many patents. His technical contributions mainly cover his findings on wireless communication system design, radio protocol, system architecture, radio resource management, signal processing, coding and modulation technologies. He is active in international academy and research activities. He serves as invited reviewer of *IEEE Transactions on Vehicular Technology, IEEE Communications Magazine, IEEE Wireless Communications Magazine, EURASIP Journals, Frequenz*, etc. He has been nominated as session chair and reviewer of many high-level technical conferences organized by IEEE and European research organizations. He is leading European research projects and is active in international industrial standardization bodies. His main interests are transmission technologies, radio resource management, reconfigurability (software-defined radio) and radio system design. He is a member of IEEE. Email: jesse.luo@ieee.org

Honglin Hu received his Ph.D. degree in communications and information systems in January 2004 from the University of Science and Technology of China (USTC), Hefei, China. From July 2004 to January 2006, he was with Future Radio, Siemens AG Communications in Munich, Germany. Since January 2006, he has been with the Shanghai Research Center for Wireless Communications (SHRCWC), which is also known as the International Center for Wireless Collaborative Research (Wireless CoRe). Meanwhile, he serves as the an associate professor at the Shanghai Institute of Microsystem and Information Technology (SIMIT), Chinese Academy of Science (CAS). Dr. Hu is mainly working for international standardization and other collaborative activities. He is a member of IEEE, IEEE ComSoc, and IEEE TCPC. In addition, he serves as a member of Technical Program Committee for IEEE WirelessCom 2005, IEEE ICC 2006, IEEE IWCMC 2006, IEEE ICC 2007, IEEE/ACM Q2SWinet 2006. Since June 2006, he has served, on the editorial board of *Wireless Communications and Mobile Computing*, John Wiley & Sons. Email: hlhu@ieee.org

CONTRIBUTORS

Hamid Aghvami
Centre for Telecommunications
   Research
King's College
London, United Kingdom

Daniel Almodovar
Vodafone Group Research
   and Development
Madrid, Spain

Michael Bahr
Siemens AG
Corporate Technology
Munich, Germany

Volker Blaschke
Nachrichtentechnik
Universität Karlsruhe
Karlsruhe, Germany

Anna Calveras
Wireless Networks Group
Technical University of Catalonia
Barcelona, Spain

Marisa Catalan
Wireless Networks Group
Technical University of Catalonia
Barcelona, Spain

Chung-Ju Chang
Department of Communication
   Engineering
National Chiao-Tung University
Hsinchu, Taiwan
People's Republic of China

Tzi-Cker Chiueh
Department of Computer
   Science
Stony Brook University
Stony Brook, New York

Arindam Das
Department of Electrical
   Engineering
University of Washington,
Seattle, Washington

Juan Carlos De Martin
Computer and Control Engineering
   Department
Politecnico di Torino
Torino, Italy

Jean-Christophe Dunat
Motorola Laboratories
Gif sur Yvette, France

Rainer Falk
Siemens AG
Corporate Technology
Munich, Germany

Josep Lluis Ferrer
Wireless Networks Group
Technical University of Catalonia
Barcelona, Spain

Vasilis Friderikos
Centre for Telecommunications
   Research
King's College
London, United Kingdom

Masayuki Fujise
Wireless Communications
   Laboratory
National Institute of Information
   and Communications Technology
Singapore

Sophie Gault
Motorola Laboratories
Gif sur Yvette, France

Carles Gomez
Wireless Networks Group
Technical University of Catalonia
Barcelona, Spain

Michelle X. Gong
Cisco Systems
San Jose, California

David Grandblaise
Motorola Laboratories
Gif sur Yvette, France

Brian Hart
Cisco Systems
San Jose, California

Honglin Hu
Shanghai Research Center for
   Wireless Communications
Shanghai, People's Republic of China

Chin-Tser Huang
Department of Computer Science
   and Engineering
University of South Carolina
Columbia, South Carolina

Jane-Hwa Huang
Department of Communication
   Engineering
National Chiao-Tung University
Hsinchu, Taiwan
People's Republic of China

Holger Jaekel
Nachrichtentechnik
Universität Karlsruhe
Karlsruhe, Germany

Xiaohua Jia
Department of Computer Science
City University of Hong Kong
Kowloon, Hong Kong

Friedrich K. Jondral
Nachrichtentechnik
Universität Karlsruhe
Karlsruhe, Germany

Clemens Kloeck
Nachrichtentechnik
Universität Karlsruhe
Karlsruhe, Germany

Florian Kohlmayer
Siemens AG
Corporate Technology
Munich, Germany

Peng-Yong Kong
Institute for Infocomm
    Research
Singapore

Rupa Krishnan
Department of Computer Science
Stony Brook University
Stony Brook, New York

Jijun Luo
Future Radio, Siemens AG
Munich, Germany

Hui Ma
Department of Electrical
    Engineering
University of Washington
Seattle, Washington

B.S. Manoj
Department of Electrical and
    Computer Engineering
University of California
San Diego, California

Shiwen Mao
Department of Electrical and
    Computer Engineering
Auburn University
Auburn, Alabama

Enrico Masala
Computer and Control Engineering
    Department
Politecnico di Torino
Torino, Italy

Scott F. Midkiff
ECE Department,
Virginia polytechnic Institute and
    State University
Arlington, Virginia

Katerina Papadaki
Operational Research
    Department
London School of Economics
London, United Kingdom

Josep Paradells
Wireless Networks Group
Technical University of Catalonia
Barcelona, Spain

Pau Plans
Wireless Networks Group
Technical University
    of Catalonia
Barcelona, Spain

Marius Portmann
School of ITEE
University of Queensland
Brisbane, Australia

Shah I. Rahman
Cisco Systems
San Jose, California

Ashish Raniwala
Department of Computer
    Science
Stony Brook University
Stony Brook, New York

Ramesh R. Rao
Department of Electrical and
    Computer Engineering
University of California
San Diego, California

Sumit Roy
Department of Electrical
    Engineering
University of Washington
Seattle, Washington

Javier Rubio
Vodafone Group Research
and Development
Madrid, Spain

Antonio Servetti
Computer and Control
Engineering Department
Politecnico di Torino
Torino, Italy

Ai-Fen Sui
Siemens Ltd. China
Corporate Technology
Department Information Security
Beijing, People's Republic of China

Kean-Soon Tan
Institute for Infocomm Research
Singapore

Rajiv Vijayakumar
Department of Electrical
Engineering
University of Washington
Seattle, Washington

Jianping Wang
Department of Computer and
Information Science
University of Mississippi
University, Mississippi

Li-Chun Wang
Department of Communication
Engineering
National Chiao-Tung University
Hsinchu, Taiwan
People's Republic of China

David Wisely
Mobility Research, BT Group
Ipswich, United Kingdom

Xiaodong Zhang
Shanghai Research Center for
Wireless Communications
Shanghai, People's Republic
of China

Yan Zhang
Simula Research Laboratory
Oslo, Norway

Jun Zheng
Computer Science Department
Queens College-City University
Flushing, New York

PREFACE

Wireless Mesh Networks (WMN) are believed to be a highly promising technology and will play an increasingly important role in future generation wireless mobile networks. WMN is characterized by dynamic self-organization, self-configuration and self-healing to enable quick deployment, easy maintenance, low cost, high scalability and reliable services, as well as enhancing network capacity, connectivity and resilience. Due to these advantages, international standardization organizations are actively calling for specifications for mesh networking modes, e.g., IEEE 802.11, IEEE 802.15, IEEE 802.16 and IEEE 802.20. As a great extension to the ad hoc network, WMN is becoming an important mode complementary to the infrastructure-based wireless networks. The experiences obtained from studying and deploying WMN provide us knowledge and reference to the future networks evolution.

Wireless Mesh Networking: Architectures, Protocols and Standards provides a comprehensive technical guide covering introductory concepts, fundamental techniques, recent advances and open issues in wireless mesh networks. It focuses on concepts, effective protocols, system integration, performance analysis techniques, simulation, experiments, and future directions. It explores various key challenges, diverse scenarios and emerging standards such as those for capacity, coverage, scalability, extensibility, reliability, and cognition. This volume contains illustrative figures and complete cross-referencing on routing, security, spectrum management, medium access, cross-layer optimization, load-balancing, multimedia communication, MIMO, and smart antennas, etc. It also details information on particular techniques for efficiently improving the performance of a wireless mesh network.

This book is organized in three parts:

- Part I: Architectures
- Part II: Protocols
- Part III: Standardization and Enabling Technologies

In Part I, WMN fundamentals are briefly introduced as are various types of network architecture. Part II concentrates on the techniques necessary to enable a complete, secure and reliable wireless network, including routing, security, medium access control (MAC), scalability, load balancing, cross layer optimization, scheduling, multimedia communication, MIMO (or multiple antenna system). Part III explores standardization activities and particular mesh network specifications in the emerging standards, for instance, mesh mode in the IEEE 802.11 Wireless LAN and in the IEEE 802.16 WiMAX. In addition, the applications of mesh networks in emergency management, and public safety are exploited.

This book has the following salient features:

- Provides a comprehensive reference on state-of-the-art technologies for wireless mesh networks
- Identifies basic concepts, techniques, advanced research topics and future directions
- Contains illustrative figures that enable easy understanding of wireless mesh networks
- Allows complete cross-referencing via the broad coverage of different layers of protocol stacks
- Details particular techniques for efficiently improving the performance of wireless mesh networks

This book can serve as a useful reference for students, educators, faculties, telecom service providers, research strategists, scientists, researchers, and engineers in the fields of wireless networks and mobile communications.

We would like to acknowledge the effort and time invested by all contributors for their excellent work. They were extremely professional and cooperative. Special thanks go to Richard O'Hanley, Kim Hackett, Catherine Giacari, Glenon Butler and Jessica Vakili of Taylor & Francis Group for their support, patience and professionalism from the beginning to the final stage. We are grateful for Suryakala Arulprakasam for her great efforts during the typesetting period. Last but not least, a special thank you to our families and friends for their constant encouragement, patience and understanding throughout this project.

Yan Zhang, Jijun Luo,
and Honglin Hu

PART I

ARCHITECTURES

1

WIRELESS MESH NETWORKS: ISSUES AND SOLUTIONS

B.S. Manoj and Ramesh R. Rao

CONTENTS

1.1 INTRODUCTION

Wireless mesh network (WMN) is a radical network form of the ever-evolving wireless networks that marks the divergence from the traditional centralized wireless systems such as cellular networks and wireless local area networks (LANs). Similar to the paradigm shift, experienced in wired networks during the late 1960s and early 1970s that led to a hugely successful and distributed wired network form—the Internet—WMNs are promising directions in the future of wireless networks. The primary advantages of a WMN lie in its inherent fault tolerance against network failures, simplicity of setting up a network, and the broadband capability. Unlike cellular networks where the failure of a single base station (BS) leading to unavailability of communication services over a large geographical area, WMNs provide high fault tolerance even when a number of nodes fail. Although by definition a WMN is any wireless network having a network topology of either a partial or full mesh topology, practical WMNs are characterized by static wireless relay nodes providing a distributed infrastructure for mobile client nodes over a partial mesh topology. Due to the presence of partial mesh topology, a WMN utilize multihop relaying similar to an ad hoc wireless network. Although ad hoc wireless networks are similar to WMNs, the protocols and architectures designed for the ad hoc wireless networks perform very poorly when applied in the WMNs. In addition, the optimal design criteria are different for both these networks. These design differences are primarily originated from the application or deployment objectives and the resource constraints in these networks. For example, an ad hoc wireless network is generally designed for high mobility multihop environment; on the other hand, a WMN is designed for a static or limited mobility environment. Therefore, a protocol designed for ad hoc wireless networks may perform very poorly in WMNs. In addition, WMNs are much more resource-rich compared with ad hoc wireless networks. For example, in some WMN applications, the network may have a specific topology and hence protocols and algorithms need to be designed to benefit from such special topologies. In addition, factors such as the inefficiency of protocols, interference from external sources sharing the spectrum, and the scarcity of electromagnetic spectrum further reduce the capacity of a single-radio WMN. In order to improve the capacity of WMNs and for supporting the traffic demands raised by emerging applications for WMNs, multiradio WMNs (MR-WMNs) are under intense research. Therefore, recent advances in WMNs are mainly based on a multiradio approach. While MR-WMNs promise higher capacity compared with single-radio WMNs, they also face several challenges. This chapter focuses on the issues and challenges for both single-radio WMNs and MR-WMNs, and discusses a set of existing solutions for MR-WMNs. It begins with a comparison of WMNs with ad hoc wireless networks and proceeds to discuss the issues and challenges in MR-WMNs. The main contribution of this chapter is the detailed discussion on the issues and challenges faced by MR-WMNs, presentation with illustrations of a range of recent solutions for architectures, link layer protocols, medium access control (MAC) layer protocols, network layer protocols, and topology control solutions for MR-WMNs.

1.2 COMPARISON BETWEEN WIRELESS AD HOC AND MESH NETWORKS

Figure 1.1 shows the classification of multihop wireless networks; these constitute the category of wireless networks that primarily use multihop wireless relaying. The major categories in the multihop wireless networks are the ad hoc wireless networks, WMNs, wireless sensor networks, and hybrid wireless networks. This book mainly focuses on WMNs. Ad hoc wireless networks [12] are mainly infra-structureless networks with highly dynamic topology. Wireless sensor networks, formed by tiny sensor nodes that can gather physical parameters and transmit to a central monitoring node, can use either single-hop wireless communication or a multihop wireless relaying. Hybrid wireless networks [12] utilize both single- and multihop com-munications simultaneously within the traditionally single-hop wire-less networks such as cellular networks and wireless in local loops (WiLL). WMNs use multihop wireless relaying over a partial mesh topology for its communication.

Table 1.1 compares the wireless ad hoc networks and WMNs. The primary differences between these two types of networks are mobility of nodes and network topology. Wireless ad hoc networks are high mobility networks where the network topology changes dynamically. On the other hand, WMNs do have a relatively static network with most relay nodes fixed. Therefore, the network mobility of WMNs is very low in comparison with wireless ad hoc networks. The

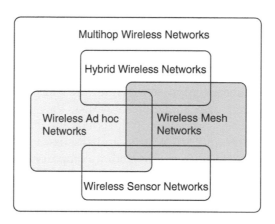

Figure 1.1   Classification of multihop wireless networks.

Table 1.1 Differences between Ad Hoc Wireless Networks and Wireless Mesh Networks

Issue	Wireless Ad Hoc Networks	Wireless Mesh Networks
Network topology	Highly dynamic	Relatively static
Mobility of relay nodes	Medium to high	Low
Energy constraint	High	Low
Application characteristics	Temporary	Semipermanent or permanent
Infrastructure requirement	Infrastructureless	Partial or fully fixed infrastructure
Relaying	Relaying by mobile nodes	Relaying by fixed nodes
Routing performance	Fully distributed on-demand routing preferred	Fully distributed or partially distributed with table-driven or hierarchical routing preferred
Deployment	Easy to deploy	Some planning required
Traffic characteristics	Typically user traffic	Typically user and sensor traffic
Popular application scenario	Tactical communication	Tactical and civilian communication

topological difference in these networks also contributes to the difference in performance in routing. For example, while the on-demand routing protocols perform better in wireless ad hoc networks, the relatively static hierarchical or table-driven routing protocols perform better in WMNs. Due to the static topology, formed by fixed relay nodes, of WMNs, most WMNs have better energy storage and power source, thus removing one of the biggest constraint in wireless ad hoc networks—the energy constraint. Finally, another important difference between these two categories of networks is the application scenario. Unlike wireless ad hoc networks, WMNs are used for both military and civilian applications. Some of the popular civilian applications of WMNs include provisioning of low-cost Internet services to shopping malls, streets, and cities.

1.3 CHALLENGES IN WIRELESS MESH NETWORKS

Traditional wireless ad hoc networks and WMNs were based on a single-channel or single-radio interface. WMNs, irrespective of its simplicity and high fault tolerance, face a significant limitation of limited network capacity. While the theoretical upper limit of the per node throughput capacity is asymptotically limited by $O(1/\sqrt{n})$, theoretically achievable capacity to every node in a random static wireless ad hoc network, with ideal global scheduling and routing, is estimated [6] as $\Theta(1/\sqrt{n}\,\log\,n)$ where n is the number of nodes in the network. Therefore, with increasing number of nodes in a network, the throughput capacity becomes unacceptably low. With the use of real MAC, routing, and transport protocols and a realistic traffic pattern, the achievable capacity in a WMN, in practice, is much less than the theoretical upper limit. It has also been found [7] through experiments using carrier sense multiple access with collision avoidance (CSMA/CA)-based MAC protocol such as IEEE 802.11 that on a string topology, the throughput degrades approximately to $1/n$ of the raw channel bandwidth. In general, the throughput capacity achievable in an arbitrary WMN is proportional to the $\Theta(W \times n^{-1/d})$ where d is the dimension of the network and W is the total bandwidth. For a two-dimensional (2D) network, the throughput can be as small as $\Theta(W \times n^{-1/2})$. One approach to improve the throughput capacity of a WMN is to use multiple radio interfaces. Although the upper limit of the capacity is unaffected by the raw bandwidth or the way the raw bandwidth is split among multiple interfaces, in practice, with realistic MAC and routing protocols, the throughput capacity can be significantly increased by the use of multiple interfaces and by fine tuning of protocols. Recently, the development of WMNs using multiple radio interfaces have taken significant process due to the availability of inexpensive and off-the-shelf IEEE 802.11-based wireless interfaces. While MR-WMNs provide several advantages such as increased network capacity, they also face several issues and challenges. This chapter primarily focuses on the issues and challenges in single-radio and MR-WMNs and proceeds to discuss some of the solutions for a multiradio wireless network. The challenges faced by WMNs are discussed in Section 1.3.1 through Section 1.3.4.

The primary challenges faced by WMNs such as throughput capacity, network scalability, and other challenges are discussed here.

Table 1.2 Throughput Degradation in a WMN with String Topology

	1 Hop	2 Hops	3 Hops	4 Hops	5 Hops	>5 Hops
Normalized throughput	1	0.47	0.32	0.23	0.15	0.14
$\dfrac{1}{\text{Hoplength}}$	1	0.5	0.33	0.25	0.2	0.16

1.3.1 Throughput Capacity

The throughput capacity achievable for WMN nodes is limited in a single-channel system compared to a multichannel system. Table 1.2 shows the throughput deviation in a string topology, as depicted in Figure 1.2, over one, two, and three hops, in a typical experimental network. From Table 1.2, it can easily be found that throughput degrades rapidly with a WMN system as the path length increases. Although there are several factors contributing to the throughput degradation, such as characteristics of MAC protocol, the exposed node problem, the hidden terminal problem, and the unpredictable and high error rate in the wireless channel, all these issues are aggravated in a single-channel system. For example, as illustrated in Figure 1.2, when node 1 transmits to node 2, especially when CSMA/CA-based MAC protocols are employed, nodes 2 and 3 cannot initiate another transmission. Node 2 is prevented from a simultaneous transmission as the wireless interface, in most WMNs is half-duplex whereas node 2 abstains from transmission because it is exposed to the ongoing transmission between nodes 1 and 2. This exposed node problem contributes to the throughput degradation in WMNs over a relayed multihop path. For example, a two-hop flow between nodes 1 and 3 has to share the bandwidth between the two and therefore, from Table 1.2, the end-to-end throughput for a two-hop path is only 47% of the single-hop throughput. In experimental arbitrary one-dimensional (ID) networks, the throughput degradation is found to be following a function of $O(1/n)$ where n is the number of hops

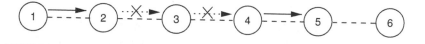

Figure 1.2 An example of string topology and exposed node problem in a wireless mesh network.

when the hop length is less than five hops and beyond five hops, the throughput remains constant albeit at a very low value.

Although there exist other factors such as the nature of routing protocol, greediness of the initial nodes and subsequent flow starvation of the latter hops, and the behavior of MAC protocols, the single most important factor contributing such a rapid degradation of throughput is the exposed node problem, aggravated by the use of a single-radio system.

1.3.2 Throughput Fairness

Another important issue in a single-radio WMN is the high throughput unfairness faced by the nodes in the system. A network is said to be exhibiting high throughput fairness if all nodes get equal throughput under similar situations of source traffic and network load. WMNs show high throughput unfairness among the contending traffic flows especially when CSMA/CA-based MAC protocols are employed for contention resolution. Figure 1.3a and Figure 1.3b show simple topologies within a WMN, causing high throughput unfairness.

Two important properties associated with CSMA/CA-based MAC protocols, when used in a WMN environment are: (i) information asymmetry depicted in Figure 1.3a, (ii) location-dependent contention depicted in Figure 1.3b, and (iii) half-duplex character of single-channel systems. In Figure 1.3a, only the the receiver of the traffic flow P is exposed to both the sender and the receiver of flow Q, and therefore, the sender of the flow P does not get any information from the channel about ongoing transmissions on other flows. On the other

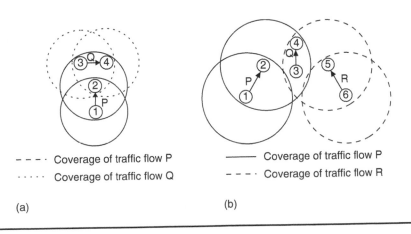

‐ ‐ ‐ ‐ Coverage of traffic flow P

· · · · · Coverage of traffic flow Q

(a)

――――― Coverage of traffic flow P

‐ ‐ ‐ ‐ Coverage of traffic flow R

(b)

Figure 1.3   Traffic flows and throughput unfairness in WMNs.

hand, the channel activity is known to be the sender of flow Q. This information asymmetry causes unfair sharing of the total throughput achieved. That is, among the flows P and Q, it is seen [1] that the flow P receives about 5% of the total throughput compared with the 95% throughput achieved by flow Q. For example, when node 1 has packets ready for transmission, upon detecting an idle channel, it may start transmission by sending request-to-send (RTS) packet to node 2. At this point, if there is an ongoing transmission between nodes 3 and 4, node 2 does not respond to the RTS, leaving node 1 to exponentially back off and retry again. This repeated back-off and several retransmission attempts lead to achieving a low throughput for flow Q. On the other hand, a similar situation can happen to node 3 with a much lower probability and that is proportional to the vulner- able period of the medium access scheme, which in this case is the propagation delay between nodes 2 and 3.

While the information asymmetry is caused by lack of information at certain nodes, having excessive information may also contribute to throughput unfairness. For example, in Figure 1.3b, flows P and R do not have information about any other flows in the network whereas flow Q has information about both the other flows. Therefore, flow Q has to set its network allocation vector (NAV) and abstain from trans- mitting, whenever it sees a transmission of control packets or data packets belonging to flows P and R. This leads the flow Q to wait for an idle channel that essentially depends on the event of both the flows P and Q simultaneously going idle. In this case, the location of the flow Q is in such a position that it experiences much more contention than the rest of the flows [1] and therefore, flow Q receives only 28% of the total throughput compared with 36% throughput share received by both the flows P and R. In fact, the throughput share of flow Q is inversely proportional to the number of neighbor flows contending for its bandwidth. Another effect of this location-dependent conten- tion is known as *perceived collision*, which may occur at flow Q. Due to the presence of contending flows, trying to access the channel, simultaneous transmission of control packets, RTS, CTS, and ACK by both the flows P and R, may result in a collision at flow Q. This results in a wrong perception of collision at flow Q that in fact may not be a collision for both the flows P and R. This perceived collision may reduce the amount of information, about the flows P and R, available at the flow Q and, therefore, leading to further degradation of throughput fairness.

In addition to the information asymmetry and the location-dependent contention, the half-duplex property of a single-interface system is

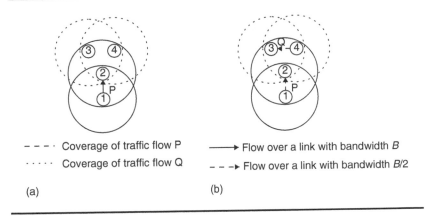

Coverage of traffic flow P

····· Coverage of traffic flow Q

⟶ Flow over a link with bandwidth B

– – –► Flow over a link with bandwidth B/2

(a)   (b)

Figure 1.4   Half-duplex radio interfaces in WMNs.

another property that causes high throughput unfairness in a single-radio WMN. Due to the half-duplex characteristics, no node can simultaneously receive and transmit. This is illustrated in Figure 1.4a in which a single half-duplex radio with a channel data rate of B bits/s is employed and therefore, only one communication could be permitted at any time. Therefore, only flow P could be transmitting while other nodes are waiting. As mentioned earlier, in certain MAC protocols such as CSMA/CA-based IEEE 802.11, there exists a strong chance of channel capture where a successful node keeps getting transmission opportunities more often than others. Such channel capturing and subsequent unfairness can be prevented by using multiple channels. In Figure 1.4b, each node uses two radio interfaces with channel data rate of $B/2$ bits/s and therefore, two simultaneous flows, P and Q, could exist. In this case, though each channel has only half the bandwidth, the throughput fairness increases as found in experimental studies [8].

1.3.3   Reliability and Robustness

Another important motivation for using WMNs and especially the MR-WMNs is to improve the reliability and robustness of communication. The partial mesh topology in a WMN provides high reliability and path diversity against node and link failures. MR-WMNs provide the most important ingredient for robustness in communication—diversity. For example, in wireless systems channel errors can be very high compared to wired networks; therefore, graceful degradation of communication quality during high channel errors is necessary. This is particularly important when the WMN system utilizes unlicensed frequency spectrum [9]. In order to achieve graceful quality degradation instead of full loss of connectivity, WMNs can employ frequency diversity, by using multiple radio interfaces, which is difficult to achieve inasingle-radio WMN system. MR-WMNs can use appropriate radio-switching modules to achieve fault tolerance in communication either by switching the radios, channels, or by using multiple radios simultaneously.

1.3.4 Resource Management

Resource management refers to the efficient management of network resources such as energy, bandwidth, interfaces, and storage. For example, the energy resources can be efficiently used in a WMN with limited energy reserve if each node in the system has a new low-power interface in addition to the regular interface. The overall power consumption, even in idle mode, depends very much on the type of interface. Therefore, in an IEEE 802.11-based WMN with limited energy reserve, an additional low-power and low-data rate interface can be used to carry out-of-band signaling information to control the high-power and high-data rate data interface. Bandwidth resources can also be managed better in a multiradio environment. For example, the load balancing across multiple interfaces could help preventing any particular channel getting heavily congested and hence becoming a bottleneck. In addition to balancing the load, bandwidth achieved through each interface can be aggregated to obtain a high effective data rate. In such a bandwidth aggregation mechanism (also known as bandwidth striping), dynamic packet scheduling can be utilized to obtain a better performance. Finally, one important advantage of using a multiradio system in a WMN is the possibility to effect provisioning quality of service through service differentiation.

1.4 DESIGN ISSUES IN WIRELESS MESH NETWORKS

There are many issues that need consideration when a WMN is designed for a particular application. These design issues can be broadly classified into architectural issues and protocol issues. The architectural design issues and protocol design issues are described in Section 1.4.1 and Section 1.4.2.

1.4.1 Network Architectural Design Issues

A WMN can be designed in three different network architectures based on the network topology: flat WMN, hierarchical WMN, and hybrid WMN. These categories are briefly discussed below.

1.4.1.1 Flat Wireless Mesh Network

In a flat WMN, the network is formed by client machines that act as both hosts and routers. Here, each node is at the same level as that of its peers. The wireless client nodes coordinate among themselves to provide routing, network configuration, service provisioning, and other application provisioning. This architecture is closest to an ad hoc wireless network and it is the simplest case among the three WMN architectures. The primary advantage of this architecture is its simplicity, and its disadvantages include lack of network scalability and high resource constraints. The primary issues in designing a flat WMN are the addressing scheme, routing, and service discovery schemes. In a flat network, the addressing is one of the issues that might become a bottleneck against scalability.

1.4.1.2 Hierarchical Wireless Mesh Network

In a hierarchical WMN, the network has multiple tiers or hierarchical levels in which the WMN client nodes form the lowest in the hierarchy. These client nodes can communicate with a WMN backbone network formed by WMN routers. In most cases, the WMN nodes are dedicated nodes that form a WMN backbone network. This means that the backbone nodes may not originate or terminate data traffic like the WMN client nodes. The responsibility to self-organize and maintain the backbone network is provided to the WMN routers, some of which in the backbone network may have external interface to the Internet and such nodes are called *gateway* nodes.

1.4.1.3 Hybrid Wireless Mesh Network

This is a special case of hierarchical WMNs where the WMN utilizes other wireless networks for communication. For example, the use of other infrastructure-based WMNs such as cellular networks, WiLL networks, WiMAX networks, or satellite networks. Examples of such hybrid WMNs include multihop cellular networks [2], throughput enhanced wireless in local loop (TWiLL) networks [3], and unified cellular ad hoc networks [4]. A practical solution for such a hybrid WMN for emergency response applications is the *CalMesh* platform [5]. These hybrid WMNs may use multiple technologies for both WMN backbone and back haul. Since the growth of WMNs depend heavily on how it works with other existing wireless networking solutions, this architecture becomes very important in the development of WMNs.

1.4.2 Network Protocol Design Issues

The design issues for the protocols can be described in a layer-wise manner starting from the physical layer to the application layer. Some of these protocol design issues are presented below.

1.4.2.1 Physical Layer Design Issues

At the physical layer, the main design issue is the choice of an appropriate radio technology. The choice of a radio technology can be based on: (i) technological considerations and (ii) economic considerations. The main technological considerations include the spectral efficiency, physical layer data rate, and the ability to operate in the presence of interference. For example, the choice of technologies such as code division multiple access (CDMA), ultra wide band (UWB), and multiple input multiple output (MIMO) are more suitable for WMN physical layer than the most popular physical layer technology, orthogonal frequency division multiplexing (OFDM) used in today's WMNs. For example, today's physical layer technology, primarily based on OFDM provides a maximum physical layer data rate of 54 Mbps. In a highly dense network with high interference, this capacity may not be sufficient. Therefore, development of new and high data rate physical layer such as UWB is a physical layer challenge. In addition to the choice of a particular physical layer technology, programable radios or cognitive radios add another dimension to the WMN physical layer design. This is emphasized by some of the applications of WMNs such as emergency response and military applications where the spectrum used for communication depends on the unused spectrum in a given locality. In such applications, a software-defined radio with cognitive capabilities would be an ideal choice. In addition to the technological considerations mentioned above, the second most important requirement is economical or social where the simplicity of the physical layer technology will lead to inexpensive devices and hence better social affordability of WMNs. An example of this is evident in the success of today's IEEE 802.11b-based WMNs where the inexpensive network interface cards contributed to the success of the proliferation of WMNs. Therefore, while choosing the physical layer technology, a network designer should look at the application and user scenario as well.

1.4.2.2 Medium Access Control Layer

The design of MAC layer protocol assumes significance in a WMN because achievable capacity depends heavily on the performance of MAC protocol. In addition to a fully distributed operation, the major issues faced by the popular CSMA/CA-based IEEE 802.11 distributed coordination function (DCF) are: (i) hidden terminal problem, (ii) exposed terminal problem, (iii) location-dependent contention, and (iv) high error probability on the channel. In order to increase the network capacity, multiple radios operating in multiple channels are used. Therefore, new MAC protocols are to be designed for operating in multichannel MR-WMN systems. MAC protocols are also to be adapted to operate in different physical layer technologies such as UWB and MIMO physical layers. Another popular research issue for better MAC performance is the use of cross-layer interaction mechanisms that enable the MAC protocol to make use of information from other layers. In traditional wireless or wired networks, each layer works with its own information making it unable to make the best use of the network-centric properties. In general, the MAC layer protocol design should include methods and solutions to provide better network scalability and throughput capacity.

1.4.2.3 Network Layer

Unlike the routing protocols for ad hoc wireless networks, the routing protocols, depending on its network scenario, face different design issues in a WMN. Since WMN is relatively a static network, the routing can make use of table-driven routing approaches such as that used in wired networks or in ad hoc wireless networks [12]. The main issues faced by routing protocol in a WMN are: (i) design of routing metric, (ii) minimal routing overhead, (iii) route robustness, (iv) effective use of support infrastructure, (v) load balancing, and (vi) route adaptability. The routing metric design plays a crucial role in achieving good performance. The best routing metric may also differ in its performance. For example, in WMNs, the routing metric design has to take the link level signal quality into account for better end-to-end performance. Routing protocols for WMNs, while providing a good end-to-end performance, should also consume minimum bandwidth for setting up paths. In addition, the use of wireless medium demands quick path reconfiguration capability in order to maintain the robustness of the path. Another important aspect is the load-balancing capability that needs to be incorporated with the routing protocol. Finally, a routing protocol for WMN must be adaptable to the network dynamics. The routing protocol can be classified into either flat routing protocol or hierarchical routing protocol based on the type of network where the routing protocol is applied.

1.4.2.4 Transport Layer

At the transport layer, the biggest challenge is the performance of transport protocols over the WMN. Since a WMN has large round-trip time (RTT) variations and these RTT variations are dependent on the number of hops in the path, the end-to-end TCP throughput degrades rapidly with throughput. The packet loss, collision, network asymmetry, and link failures can also contribute to the degradation in transport layer protocol performance. The popular transport layer for the Internet, TCP, performs very poorly in its original form over a WMN. The transport layer needs to be refined or rewritten for making it more efficient on a WMN. Some of the design issues for a transport layer protocol for WMN are: (i) end-to-end reliability, (ii) throughput, (iii) capability to handle network asymmetry, and (iv) capability to handle network dynamism.

1.4.2.5 Application Layer

The most popular application for WMNs is the Internet access service. Essentially, a WMN needs to provide Internet services for residential areas or businesses. In such a situation, though data services make primary service over a WMN, voice services such as voice over Internet protocol (VoIP) are also important. Therefore, it is very essential to provide support for both the time-sensitive and the best-effort traffics. In addition to the basic data and voice traffic support, the network provides service discovery mechanisms. Since most of the network services are in fully distributed form, static service discovery mechanisms may not be effective in a WMN. Another important requirement for the application layer protocol design is to handle the heterogeneity of networks as the data may pass through a variety of networks before being delivered to the end application.

1.4.2.6 System-Level Design Issues

The above-mentioned issues are generic to a WMN and these issues are revisited in detail for a MR-WMN system in Section 1.5. In addition to the protocol design issues, a WMN requires system-level solutions. Some examples for system-level issues are: (i) cross-layer system design, (ii) design for security and trust, (iii) network management systems, and (iv) network survivability issues.

Some of the primary challenges faced by a WMN can be alleviated by the use of an MR-WMN and therefore, subsequent sections focus on this.

1.5 DESIGN ISSUES IN MULTIRADIO WIRELESS MESH NETWORKS

The primary advantages of using an MR-WMN are the improved capacity, scalability, reliability, robustness, and architectural flexibility. Notwithstanding the advantages of using a multiradio system for WMNs, there exist many challenges for designing an efficient MR-WMN system. This section discusses the issues to be considered for designing an MR-WMN. The main issues can be classified into architectural design issues, MAC design issues, routing protocol design issues, and routing metric design issues, which are explained below.

1.5.1 Architectural Design Issues

The network architecture plays a major role in achieving the performance objectives of an MR-WMN when a network is deployed. In general, the network architecture of an MR-WMN is designed on the basis of the type of application or deployment scenario. The major architectural choices to be considered are: (a) topology-based, (b) technology-based, and (c) node-based. Based on the topology, an MR-WMN can be designed either as a flat-topology-based or as a hierarchical-topology-based. The design categories under the technology-based solution are homogeneous or heterogeneous. While the most popular form of MR-WMN system is the homogeneous type that uses only one type of radio technology such as the popular WLAN technology IEEE 802.11, it is possible to develop an MR-WMN with heterogeneous technologies that utilize a variety of communication technologies. Finally, the node-based design criteria can be classified into either host-based, infrastructure-based, or hybrid MR-WMNs. In the case of host-based MR-WMNs, the network is formed by the host nodes and is same as an ad hoc wireless network with limited or no mobility. On the other hand, in the infrastructure-based MR-WMNs, the WMN is formed by nodes placed on fixed infrastructures or buildings. An example for this architectural type is the rooftop networks formed by placing wireless mesh relay nodes on the roof of every house for building a residential communication network. Finally, a hybrid MR-WMN has both infrastructure-based backbone and wireless mesh hosts. These hosts communicate over the wireless mesh backbone. This backbone topology can be organized either as a flat topology or as a hierarchical topology as discussed in Section 1.4.1. In some application environments, the hosts are mobile and they also relay traffic on behalf of other hosts in the network. An example of such hybrid MR-WMNs is the vehicular WMNs that communicate over a wireless mesh infrastructure. Therefore, the design of an MR-WMN system must consider the type of application or deployment environment for choosing appropriate architectural solution.

1.5.2 Medium Access Control Design Issues

The MAC layer for MR-WMNs faces several challenges. The main challenges among them are the interchannel interference, interradio interference, channel allocation, and MAC protocol design. The interchannel interference refers to the interference experienced at a given channel due to the activity in neighbor channels. For example, in IEEE 802.11b, although there are a total of 11 unlicensed channels in North America (13 in Europe and 14 in Japan), only 3 of them (channels 1, 6, and 11 in North America) can be used simultaneously at any given geographical location. Therefore, the presence of multiple radios must consider the interchannel interference as the use of a new channel, at a second interface that interferes with the existing channel, will lead to significant performance degradation. In such cases, multiradio channel usage must use nonoverlapping channels. The second issue here is the interradio interference. This issue arises due to the design and implementation of radio interfaces. This type of interference is experienced at a particular radio due to the channel activity at another interface in the same WMN node. Such interferences occur even when both the interfaces use nonoverlapping channels [27]. For example, when interfaces A and B on a WMN node use channels 1 and 11, respectively, interradio interference may experience. This interference is primarily due to the design of the hardware components and the interface itself where usually a number of low-cost filters and associated RF components are used. The physical separation of interfaces may help to avoid this issue to some extent; in certain cases the separation may be difficult, especially in portable nodes. The use of certain low-cost interface cards leads to interference even when they are separated for few feet [27]. Another issue of importance to MAC is the channel allocation. This is a network-wide process where the allocation of noninterfering channels would lead to significant throughput and media access performance. The channel allocation should consider the number of channels available and the number of interfaces available. Therefore, techniques such as graph coloring are used for generating channel allocation strategies. Finally, the most important issue is the design of MAC protocols. The availability of multiple interfaces and multiple channels leads to new designs for medium access protocols that can be benefitted in the presence of multiple radios. Examples of such protocols are the multichannel carrier sense multiple access (MCSMA) [24], interleaved carrier sense multiple access (ICSMA) [8], and the two-phase time division multiple access (2P-TDMA) [26]. These protocols utilize multiple channels simultaneously and also attempt to solve the media access issue in MR-WMNs.

1.5.3   Routing Protocol Design Issues

Another important issue in designing an MR-WMN is the design of routing protocol that depends on the design of the WMN architecture and in some cases it also depends on both the network's application and the deployment scenario. The routing protocol design can be classified into several categories based on: (a) the routing topology, (b) the use of a routing backbone, and (c) the routing information maintenance approach. Based on the routing topology, routing proto-cols can be designed either as a flat routing protocol or as a hierarch-ical routing protocol. In hierarchical routing, a routing hierarchy is built among the nodes in such a way that the pathfinding responsibil-ity is delegated to higher-level nodes in the hierarchy when the lower level nodes fail to obtain a path, e.g., hierarchical state routing (HSR) [11]. On the other hand, a flat routing system does not have any inbuilt hierarchies and each node has equal responsibility to find a path to the destination and to participate in the pathfinding process of other nodes. The chosen path may include any arbitrary node in the net-work without following any particular node hierarchy. Second design category is routing based on routing backbones and is classified into tree-based backbone routing, mesh-based backboneless routing, and hybrid topology routing. Unlike a wireless ad hoc network, a WMN is relatively static or has limited mobility network; therefore, in order to increase the routing efficiency, a routing backbone can be built. An example of this routing approach is the WMN routing performed by the IP routing mechanism over a spanning tree protocol (STP)-based tree backbone [10]. In the case of STP, the link layer will form a tree topology among the WMN nodes similar to a wireless distributed system (WDS) [12] and at the network layer, routing is carried out by traditional IP-based routing method. Although this is one of the simplest approach for WMNs, it has several issues such as poor reliability and lack of network scalability. On the other hand, routing protocols designed for and implemented at the network layer may follow a backboneless mesh routing approach. A third approach is to use a network layer backbone-topology, a subset of the nodes forming a mesh-like backbone within the WMN, optimized for certain parameters such as throughput, channel quality, or network scalability can be used for aiding a backboneless mesh routing protocol. Such a routing approach that uses a dynamic backbone topology at certain specific segment of the network is called a hybrid topology routing protocol. Finally, routing protocols can be designed on the basis of the routing information maintenance approach. Examples of such routing schemes are proactive or table-driven routing protocols, reactive or on-demand routing protocols, and hybrid routing protocols. In the case of proactive or table-driven routing approach, every node exchanges its routing information periodically and maintains a routing table, which contains routing information to reach every node in the network. Examples of routing protocols that use this design approach are DSDV [13], WRP [14], and STAR [15]. On the other hand, in the reactive or on-demand routing approach, a node requests routing information and maintains the path information only when it needs to communicate with another node. Some of the routing protocols, based on this approach, are AODV [16], dynamic source routing (DSR) [17], and multiradio link quality source routing (MRLQSR) [20]. Finally, the hybrid routing protocols take benefit of both the table-driven and on-demand routing approaches. An example of such a hybrid routing approach is the zone routing protocol (ZRP) [18], which employs a table-driven routing approach within a zone and on-demand approach beyond the zone. That is, every node uses proactive approach within a k-hop routing zone and employs a reactive routing approach beyond the routing zone.

1.5.4 Routing Metric Design Issues

In addition to designing a routing protocol, another important issue is the design of a routing metric. A routing metric is the routing parameter, weight, or value that is associated with a link or path, based on which a routing decision is made. Hop count is the simplest routing metric and is an additive routing metric. Due to the special characteristics of WMNs, hop count as a routing metric performs very poorly. Therefore, the design of routing metric is very important in MR-WMNs. The routing metric plays a crucial role in the performance of a routing protocol and the design of routing metrics should take several factors such as (i) the network architecture, (ii) the network environment, (iii) the extent of network dynamism, and (iv) the basic characteristics of the routing protocol into account, in order to design an efficient routing protocol for WMNs. First, the architectural property of the network needs to be considered for designing the routing metric. For example, the network may be designed based on either a flat topology architecture or a multitiered hierarchical architecture. In addition, the architectural design may include an infrastructureless network, partially infrastructure-supported network, or an infrastructure-supported network. The routing metric to be designed should take the architectural design of the network into account. Second important factor to be considered is the network environment. For example, due to the presence of location-dependent contention, highly fluctuating and unpredictable channel conditions, and high bit error rate (BER), the characteristics of a WMN environment is radically different from that of a wired network. Therefore, the design of routing protocol and routing metric should take the specific network environment into account. Another important input for designing a WMN is the extent of the network dynamism due to the mobility experienced by the network. For a WMN designed for static nodes or for nodes with very low mobility, proactive protocols may be suitable whereas on-demand routing approach is suitable for a WMN that handles high mobility nodes. Finally, in order to design an efficient routing metric, the basic characteristics of a routing approach is important. For example, while a nonisotonic* routing protocol works well with an on-demand source-routing-based routing protocol, it may fail or perform poorly due to the formation of routing loops when used with a table-driven hop-by-hop routing protocol.

The design objectives for a routing protocol and metric are: (i) resource efficiency, (ii) throughput, (iii) freedom from routing loops, (iv) route stability, (v) quick path setup capability, and (vi) efficient route maintenance.

The challenges for designing routing protocols for MR-WMNs are: (a) interradio interference, (b) interflow interference, (c) intraflow interference, (d) hidden terminal node problem, (e) exposed terminal node problem, (f) location-dependent contention, and (g) highly dynamic channel characteristics.

1.5.5   Topology Control Design Issues

The network performance in a WMN is affected by the network topology, and by controlling the network topology the network performance can

---

* Isotonicity is the property of a routing metric that guarantees freedom from routing loops.

be improved. Topology control is defined by the network's capability to manipulate its parameters such as the location of nodes, mobility of nodes, transmission power, the properties of the antenna, and the status of the network interfaces. The topology can be controlled either as a one-time activity during the network initialization phase or as a periodic activity throughout the duration of the network lifetime. Effective use of the network topology control can help to improve the capacity. In practice, node location and mobility are not under the direct control of the network system leaving the remaining factors such as transmission power, antenna properties, and the status of the network interface cards. The objectives of topology control mechanisms are connectivity, capacity, reliability and fault tolerance, and network coverage. Section 1.9 provides a detailed description of the objectives of topology control.

1.6 LINK LAYER SOLUTIONS FOR MULTIRADIO WIRELESS MESH NETWORKS

Network scalability is the single most important problem that plagues the large-scale WMNs. The primary reasons behind the lack of network scalability in a WMN are: (i) half-duplex character of the WLAN radios, (ii) inefficient interaction between the network congestion and suboptimal congestion avoidance phase at different layers of the protocol stack, (iii) collision due to hidden terminal problem, (iv) resource wasted due to exposed terminal problem and the location-dependent contention, and (v) the difficulties in handling a multi-channel system. Some of the above-mentioned problems can be solved by an MR-WMN. However, they face several challenges such as (i) adjacent radio interference, (ii) dynamic management of spectrum resources, and (iii) efficient management of multiple radio interfaces. The adjacent radio interface problem refers to the interference caused by one radio interface to the other radio interfaces on the same node. The only way to mitigate this problem is to modify the mechanical design of the nodes to provide enough separation of the antennas or network interface cards. The dynamic spectrum management can be provided by intelligently choosing the most appropriate channel for communication between nodes. The management of multiple radio interfaces refers to the activity by which the higher layer protocols could use the desired radio interface that is appropriate for communication. There exist several link layer solutions such as multiradio unification protocol (MUP) that is discussed in detail in Section 1.6.1.

1.6.1 Multiradio Unification Protocol

The MUP [21] is a link layer solution to provide a virtual layer that controls multiple radio interfaces in order to optimize local spectrum usage in MR-WMNs. The main design goals of MUP for efficient spectrum management are: (i) to minimize hardware modifications, (ii) to avoid making changes to the higher layer protocols, (iii) to operate with legacy (non-MUP) nodes, and (iv) to not depend on the global topology information.

The MUP provides a single virtual interface to the higher layers by providing an architectural solution for concealing the multiple physical interfaces and channel selection mechanisms for choosing an appropriate channel for communication between nodes. MUP is implemented in the link layer and therefore network layer and other higher layers need not have any changes to efficiently use multiple radio interfaces. The architectural diagram of MUP is shown in Figure 1.5. One of the primary tasks that the MUP layer does is monitoring the channel quality between a node and its neighbors such that the node can choose the best possible interface for communicating with a particular neighbor. In order to virtualize multiple radio interfaces having a different MAC address, MUP uses a virtual MAC address that effectively conceals the multiple physical address. Therefore, the physical layer appears, to the higher layers, as a single interface. The complexity of choosing which radio interface to use for transmitting a particular packet to a neighbor lies with MUP. At network start-up, every node tunes its radio interfaces to orthogonal channels and this channel setting on an interface is permanent.

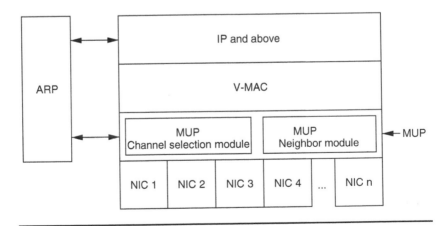

Figure 1.5   A representation of the MUP architecture.

Therefore, channel selection or switching refers to selection or switching of radio interfaces, respectively. It is, therefore, a significant requirement to have every node in the network to use the same set of orthogonal channels for its interfaces. This, however, leads to a limitation where the number of radio interfaces that can be attached to any node in the network, as per MUP, limited by the number of orthogonal channels in the system. For example, if the WMN uses an IEEE 802.11b-based system, there exist only three orthogonal channels (channels 1, 6, and 11 out of a total of 11 channels in North America). MUP employs two different schemes for the selection of radio interfaces (or channel as each interface is statically assigned a channel) that are named MUP-Random and MUP-Channel-Quality schemes. According to the MUP-Random scheme, which is the basic scheme, a node randomly chooses an interface for transmitting a packet with a destination node. Such a random selection of a radio interface has both advantages and disadvantages. Advantages include the following: (i) it is simple to implement, (ii) it does not require channel state information, and (iii) it provides a system-wide uniform distribution of traffic across the available interfaces. Some of the disadvantages are that a channel which is already congested may be chosen over other idle channels for transmission between a sender–receiver pair and thus may lead to a degraded link level throughput and performance. The MUP-Channel-Quality scheme is designed to maintain the channel state information (also referred to as channel quality metric) between nodes and choose the best possible channel based on the channel state information. The channel quality metric is derived from the neighbor table that is maintained by the MUP table (see Figure 1.5). Figure 1.5 also depicts the MUP neighbor table module and the MUP channel selection module. Probe messages are transmitted over each interface to the neighbors periodically and the round-trip delay on link is estimated. The use of probe messages enables the MUP layer to obtain the channel state information independently. For example, between two nodes A and B, the channel to be used for a transmission from node A to B may not be the channel used for transmission from node B to A. MUP comprises two major modules: (i) a neighbor module and (ii) a channel selection module. The neighbor module maintains the neighbor table and provides a classification of neighbors. On the other hand, the channel selection module makes the decision on the channel to be used for communicating with a neighbor node.

The MUP neighbor module maintains the MUP neighbor table that contains the following information for each of the node's neighbors: (i) node identifier (in most networks, the IP address) of the neighbor node, (ii) MUP status field (indicating whether the neighbor node is MUP-capable), (iii) MAC address list (list of MAC addresses associated with the neighbor node), (iv) channel quality list (channel quality values for each interface), (v) preferred channel identifier (current preferred channel with the neighbor node), (vi) selection time (the time instant at which the last channel selection is made), (vii) packet time (the time instant at which the last packet is transmitted to, or received from, the neighbor node), and (viii) probe time list (list of time instants for unacknowledged probe messages). In the MUP neighbor table, every neighbor has an entry with details on the above-mentioned fields. Initially, when an MUP-enabled node initiates communication, it assumes that every node is a legacy node and therefore, chooses address resolution protocol (ARP) for identifying the number of interfaces and the MUP status of neighbor nodes and therefore, it broadcasts an ARP request over all interfaces. Upon reception of any ARP packet, the MUP layer gleans MAC address information present, if any. Also, when the MUP layer receives a link layer packet with broadcast address from higher layers, it broadcasts the packet over all interfaces. When an ARP packet's destination node, irrespective of its status as MUP-enable or otherwise, receives an ARP packet, it originates a response packet containing the MAC address corresponds to the network address through which it receives the ARP packet. A legacy node with multiple interfaces may reply with multiple ARP packets each containing the corresponding MAC address. Upon successful ARP packet exchanges, an MUP discovery process is conducted to identify if a neighbor node is MUP enabled. During this process, an MUP-channel select (MUP-CS) packet is sent over all the interfaces to the chosen neighbor node. In response to the MUP-CS packet, an MUP-enabled node will reply with an MUP-channel select acknowledgment (MUP-CSACK) packet and a legacy node will refrain from sending any packet. Appropriate timeout mechanisms are used for MUP-CS packets to detect legacy nodes. Hence, after the initial ARP packet exchanges and MUP discovery process, the sender node detects if a neighbor node is single interface non-MUP node, multi-interface non-MUP node, or MUP-enabled node. The neighbor table information for a particular node is removed after a long period of time without any traffic between a pair of nodes. This is done to keep the neighbor table off stale entries. The neighbor discovery process is initiated before beginning further communication with such a neighbor.

When two nodes detect each other as MUP-enabled nodes, they periodically exchange channel quality information of all the channels they use. The MUP channel-selection module chooses the appropriate channel based on the MUP-Channel-Quality-based channel selection mentioned earlier. Each node chooses and maintains the channel quality information for all the interfaces by exchanging probe messages between neighbors. The round-trip delay experienced by the probe messages is used as the channel quality metric. This round-trip delay includes the delay due to MAC protocol contention, traffic load, interferences on the channel, packet collisions, queuing, and processing delay at both the end nodes. In order to reduce the queuing delay, which in general could be very high at a heavily loaded node, MUP provides high priority for probe packets either by placing the packet at the head of the que or by using priority-provisioning mechanisms defined in MAC protocols such as IEEE 802.11e. This requirement of high-priority provisioning for probe packets also leads to problems in protocols, which do not support priority in the normal operation. Popular examples of such protocols are IEEE 802.11b/a/g. In order to estimate the channel quality of a particular channel, a node periodically transmits MUP-CS packets to its neighbors. The MUP-CS and MUP-CSACK carries sequence number for detecting lost probe packets. In response to the MUP-CS packets, neighbor nodes immediately reply with MUP-CSACK packets. Upon reception of the MUP-CSACK packets, a node estimates a smoothed round-trip time (SRTT) as $SRTT = \beta \times RTT + (1 - \beta) \times SRTT$ where RTT is the round-trip time of the most recent MUP-CS–MUP-CSACK exchange. MUP also considers lost probe packets into the SRTT measurements by adding a loss penalty of three times the current SRTT value. When nodes change channels, there is a possibility for simultaneous change of multiple channels. MUP utilizes a random interval averaging about several tens of seconds to initiate the process of channel switching. Typical value used for the period of channel quality measurement is 0.5 s. One important issue to be considered while switching channels is the packets that are already in the que of the old channel. This transition from old to a new channel could lead to packet reordering and hence it may significantly affect the throughput at the application layer protocols such as TCP. In order to avoid this, MUP waits until the que of the old channel drained completely before beginning transmission onto a new interface. The MUP-Channel-Quality-based channel selection mechanism found to be providing approximately more than 50% higher throughput when compared with MUP-Random [23].

The advantages of MUP include the following: (i) it can work with legacy nodes that have either a single interface or multiple interface without MUP, (ii) it removes the higher layers in the protocol stack from knowing the complexities in handling multiple radio interfaces, and (iii) it improves the spectrum efficiency and system throughput. Some of the disadvantages include the following: (i) the channel assignment is coarse and hence MUP may not make the best use of the available orthogonal channels, (ii) the requirement of prioritized queuing for the probe packets makes MUP unusable with WMNs based on popular MAC protocols such as IEEE 802.11b, IEEE 802.11a, and IEEE 802.11g, and (iii) MUP makes a local decision on choosing a channel and that may sometimes be suboptimal as far as global resource usage is concerned. Another issue with MUP is the allocation of orthogonal channels for the nodes in the network. Since the network has multiple orthogonal channels, it becomes necessary for a new starting-up node to find out which channels are to be assigned for its interfaces in order to communicate with the rest of the network.

1.7 MEDIUM ACCESS CONTROL PROTOCOLS FOR MULTIRADIO WIRELESS MESH NETWORKS

The design of MAC protocols is important in MR-WMNs compared to the single-radio WMNs because of additional challenges faced by the former. This section presents a few of the recently proposed MAC protocols for MR-WMNs. These protocols are the MCSMA [24], the ICSMA [8], and the 2P-TDMA [25]. These protocols are explained in detail in Section 1.7.1 through Section 1.7.3.

1.7.1 Multichannel CSMA MAC

The MCSMA MAC protocol [24], is similar to an FDMA system. In this medium access scheme, the available bandwidth is divided into non-overlapping $n + 1$ channels, i.e., n data channels and a control channel. This division is independent of the number of nodes in the system. A node that has packets to be transmitted selects an appropriate data channel for its transmission. When a node is idle, i.e., not transmitting packets, it monitors all the n data channels and all the channels for which the total received signal strength (TRSS), estimated by the sum of various individual multipath components of the signal, below a sensing threshold (ST) are marked idle channels. When a channel is idle for sufficient amount of time, it is added to the free channel list.

The packet transmission mechanism of MCSMA protocol is as follows. When a potential sender node receives data packets to be transmitted, it checks in its free channel list. If the free channel list is not empty, the sender checks whether the channel on which it successfully transmitted the most recent packet is included in the list. If the most recently used channel is already present in the free channel list, the sender begins transmission on that channel. If the free channel list is empty, it waits for a channel to be idle. Upon detecting an idle channel, the sender waits for a long interframe space (LIFS) followed by a random access back-off. After the back-off period, sender checks the channel again and, if the channel is still idle, it begins transmission on that channel. In the event that the most recently used channel is not present in the free channel list, the sender randomly chooses a channel from among the idle channels. Even in such cases, the sender waits for the channel to be idle for LIFS time. During this LIFS or the back-off time, if the TRSS of the channel exceeds ST, then the back-off process is canceled and a new back-off process is begun when the TRSS of the channel goes below ST. As mentioned earlier, if a channel is used for a successful transmission, it is given priority when another transmission is intended. Therefore, when $n > N$, where n is the number of data channels and N is the number of nodes in the system, then there is a soft channel reservation. This channel reservation is implicitly provided by the free channel preference system.

The MCSMA scheme alleviates the exposed terminal problem and permits simultaneous packet transmission sessions waiting the broadcast region. This contributes to a high throughput when MCSMA is applied in WMN systems.

1.7.2 Interleaved Carrier Sense Multiple Access

The ICSMA [8] is another novel multichannel medium access protocol. This is designed to overcome one of the main disadvantages, i.e., exposed terminal problem, which is present in the single-channel carrier sense-based MAC protocol. In a topology similar to that shown in Figure 1.6, when there is an ongoing transmission between nodes 3 and 4, other nodes in the network, i.e., nodes 2 and 6 are not permitted to transmit to nodes 1 and 5, respectively. This is because of

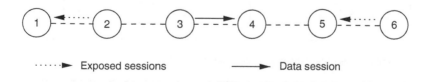

······▶ Exposed sessions ——▶ Data session

Figure 1.6 Motivation behind ICSMA protocol.

two reasons: (a) any simultaneous transmission from node 2 is prevented by its own carrier sense mechanism and (b) the acknowledgment packet received by node 3 may also be collided by the transmission from node 2. Similarly, node 6 is prevented by transmission because the acknowledgment packet originated by node 5 may collide with the data packet reception at node 4. Therefore, the nodes 2 and 6 are designated as sender-exposed and receiver-exposed [8] nodes, respectively. ICSMA is a two-channel system with similar packet exchange, RTS–CTS–DATA–ACK, as that of CSMA/CA. Compared to the CSMA/CA scheme [12], the handshaking process is interleaved between the two channels. For example, if a sender transmits RTS on channel 1 and if the receiver is willing to accept the request, it sends the corresponding CTS over channel 2. If the sender receives the CTS packet, it begins the transmission of DATA packets over channel 1. Again the receiver, if the data is successfully received, responds with ACK packet over channel 2. Figure 1.7a illustrates the

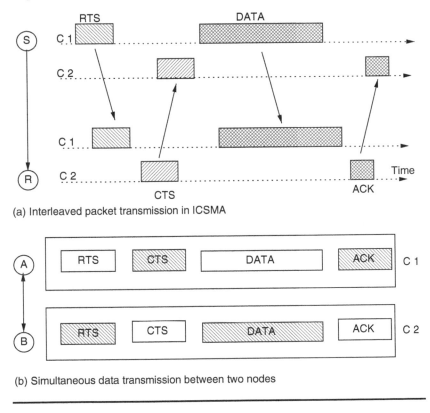

(a) Interleaved packet transmission in ICSMA

(b) Simultaneous data transmission between two nodes

Figure 1.7   Operation of ICSMA protocol.

interleaved operation of ICSMA when a sender node (S) initiates an RTS packet to a receiver node (R). Figure 1.7b shows the simultaneous transmission capability between nodes A and B. This simple mechanism of interleaving carrier sense enhances the throughput achieved by the two-channel WMNs. The ICSMA uses an extended network allocation vector (ENAV) for determining whether a particular channel is free for transmission. The ENAV is an extended form of the NAV used for CSMA/CA schemes.

This protocol improves performance by alleviating the exposed terminal problem. There are two main reasons behind the performance improvement of this protocol. (1) If a node receives RTS on channel 1, and does not receive CTS on channel 2, then it can understand its sender-exposed status. Therefore, if it needs to initiate another session with an RTS packet, it uses channel 2 to transmit the RTS packet. (2) If a node hears only CTS packet on any channel (say channel 1), and had not heard the corresponding CTS on channel 2, it realizes that it is a receiver-exposed node. Therefore, this node can now initiate a new RTS–CTS session on channel 1. Hence, interleaving the packet transmission across channels helps improve the throughput achieved by WMN nodes.

1.7.3 Two-Phase TDMA-Based Medium Access Control Scheme

The 2P-TDMA-based MAC protocol [26] is designed to provide an efficient MAC in a single channel, point-to-point, wide area WMN (WAWMN) with multiple radios and directional antennas. In WAWMNs, the main objective of using multiple radios is to improve spectrum reuse by simultaneously operating the radios connected to directional antennas. As mentioned in Section 1.7, the CSMA/CA performs extremely poor in multihop wireless networks such as ad hoc wireless networks and WMNs [25,26]. Even with the use of directional antennas with high directionality, CSMA/CA fails to provide simultaneous operation across multiple interfaces. Figure 1.8 shows an example scenario with two receivers and a central transmitter in a WAWMN. The central node (node 1 in Figure 1.8) has highly directional antennas as that of both nodes 2 and 3. Contrary to the popular notion that the two links, $1 \rightarrow 2$ and $1 \rightarrow 3$, can transmit or receive simultaneously, in practical situations it is not possible to provide error-free simultaneous communication when CSMA/CA, in its original form, is employed. Consider the case of simultaneous transmission (SynTx) where the central node 1 in Figure 1.8 attempts to simultaneously transmit to nodes 2 and 3 over interfaces 1.A and 1.B,

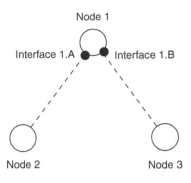

Figure 1.8 Example topology for 2P-TDMA.

respectively. There are two issues in this case: (a) the network interface that first initiates the packet (e.g., 1.A) will result in carrier sensing at the other network interface (1.B) on the same node even if they are placed several meters apart and (b) even if the interfaces 1.A and 1.B simultaneously begin transmission of their DATA packets without leading to mutual carrier sensing and subsequent back-off, the interface at node 1, which first finishes its transmission, may experience collision of its ACK packet. This is because the interface that finishes transmission expects an ACK packet and this ACK packet may experience a collision by the ongoing transmission on the other interface. Therefore, the efficient SynTx operation is not feasible. Now consider simultaneous reception (SynRx) on both interfaces at node 1. Here, both interfaces 1.A and 1.B at node 1 receive their data packets from nodes 2 and 3, respectively. However, the interface that finishes reception first (e.g., 1.A) begins transmission of its ACK packet after a time duration of interframe space (IFS) and this ACK packet may collide with the possible reception at the other interface (1.B). Therefore, the use of CSMA/CA in WAWMNs with directional antennas results in poor spectrum usage irrespective of the use of multiple radio interfaces.

The 2P-TDMA is developed to achieve simultaneous synchronous operation (SynOp) and is basically a TDMA MAC protocol without strict time synchronization requirement. The key differences, in comparison with the CSMA/CA protocol are: (a) the removal of immediate MAC-level ACK and (b) removal of carrier sensing at each interface. The higher layer protocols hold the responsibility of recovering from the packet errors. According to this protocol, every node is always in one of the two phases, SynTx or SynRx, and each node switches between these two phases. The key idea here is that when node 1, in the above example, switches over to SynTx mode, its neighbors, nodes 2 and 3, are expected to be in SynRx mode. This means that the switching process is to be synchronized for maximal performance. This demands the network topology to be bipartite graph. Even in a bipartite graph, achieving synchronization is important to operate in collision-free way. There are two main constraints in achieving a strict synchronization: (a) high global synchronization overhead in a multi-hop wireless network and (b) the high RTT involved in WAWMNs. Therefore, in order to provide a bandwidth efficient operation, 2P-TDMA utilizes loose synchronization instead of tight time synchronization. The loose synchronization is obtained purely by local decisions without involving bandwidth overhead. Every node in SynRx phase waits for the end of transmission and switches immediately to SynTx phase (starts transmission on all its interfaces) after noticing the end of transmission. In addition, every node in SynTx phase switches over to the SynRx phase after the end of transmission. These two simple steps achieve near synchronization, which may avoid two kinds of collision possibilities: (i) collision happened at a node due to the mixed operation on its interfaces (simultaneous transmission and reception on multiple interfaces, also known as mixed TxRx operation) and (ii) simultaneous SynTx phase at both the edges of a given link leading to simultaneous transmission in both directions.

It is likely that collisions or packet losses arising out of interference may lead to loss of synchronization in a 2P-TDMA system. Therefore, to avoid indefinite deadlocks, a timeout mechanism is introduced in the SynRx phase where upon timeout, if there is no signal activity from at least one of the neighbor nodes, the node in SynRx phase switches to SynTx phase. The timeout value is chosen to be a value higher than the length of the SynRx phase. Generally, the value for time-out is chosen as $1.25 \times T_{SynRx}$ where T_{SynRx} is defined as the duration or SynRx phase or the time duration for which the node receives from neighbors. Another issue that may seriously affect synchronization in 2P-TDMA is the simultaneous out of synchronization at the nodes in the network. For example, the end nodes on a link may experience exact overlap of their SynTx or SynRx phases. In order to avoid this, the timeout value is perturbed by a small random value. This perturbation in timeout results in one of the timer times out faster. For example, assume nodes 1 and 2 are now overlap asynchronously (strict overlap of SynRx and SynTx phases) and due to the perturbation in the timeout values, one node may switch faster (say nodes 1 and 2 finishes SynTx at time t_1 and t_2 where $t_2 > t_1$) and switches over to SynRx phase. Assuming a constant duration (T_{SynRx}) for SynRx phase for both nodes 1 and 2, both nodes switch over to SynTx phase at $t_1 + T_{\text{SynRx}}$ and $t_2 + T_{\text{SynRx}}$, respectively. Again, since $t_1 + T_{\text{SynRx}} < t_2 + T_{\text{SynRx}}$, there exists a short time duration after node 1 switches over to SynTx while node 2 remains in the SynRx phase. According to the synchronization rule, node 2 detects the signal from node 1 and hence remains in the SynRx phase until the transmission from node 1 ends. After the end of SynTx phase of node 1, it switches over to SynRx and at the same time node 2 detects the end of signal, which leads to switching of node 2 to SynTx and synchronization.

The primary advantages of 2P-TDMA protocol include high throughput achieved in a WAWMN and the efficiency in using multiple radios over a single channel. Disadvantages of 2P-TDMA include the inability of the protocol to operate in a general WMN network.

1.8 ROUTING PROTOCOLS FOR MULTIRADIO WIRELESS MESH NETWORKS

In addition to the architectural design and MAC protocol design, the performance of a WMN in general and an MR-WMN is affected by the design of the routing protocol and the routing metric. This section presents a number of recent routing protocols and routing metrics designed for single-radio and MR-WMNs.

1.8.1 New Routing Metrics for Multiradio Wireless Mesh Networks

Choosing the best performing routing metric in a WMN is difficult because of the three major factors present in a WMN. These factors that affect routing performance are: (i) relay-induced load, (ii) asymmetric wireless links, and (iii) high link loss. Due to the asymmetry of the links and high link loss, the shortest path routing seldom performs better. In this WMN environment, the expected transmission count (ETX) [19] routing metric is found to be a suitable routing metric to achieve high throughput. The ETX routing metric is designed to find a path based on (i) the packet delivery ratio of each link, (ii) the asymmetry of the wireless link, and (iii) minimum number of hops. The above-mentioned objectives add to advantages such as energy savings and spectrum usage.

The ETX routing metric helps an underlying routing protocol such as DSR [17] and DSDV [13] to find a path that provides a much better throughput performance. This ETX metric has the additive property and hence can be incorporated into other, traditionally shortest path-based, on-demand or table-driven routing schemes with minimal changes to the routing protocol. The ETX of a link is estimated as the expected number of transmissions required for successfully transmitting a packet over that link. Since ETX is an additive metric, the ETX of an end-to-end path is defined as the sum of the ETX of each of the links in that path. The ETX of a link is estimated by the following equation:

$$ETX = \frac{1}{FDR \times RDR} \qquad (1.1)$$

where FDR and RDR are the forward delivery ratio and reverse delivery ratio, respectively. The FDR is the estimated value of the probability of a data packet successfully received at a receiver over a given link. Similarly, the RDR is the estimation of the probability of the ACK packet successfully received at the sender of the data packet over a given link. In Equation 1.1, the denominator term of the RHS, containing the product of FDR and RDR, represents the expected probability of a successful data packet transmission and the ACK packet transmission. The ETX value of a given link provides the average number of transmission attempts to be made for sending a packet successfully over a given link. Each node periodically broadcasts short probe packets once in every T seconds. Every receiver node collects these probe packets for a period P_{window}. An empirical method is used for choosing the P_{window} for optimization and experiments [19] found that a value of $P_{window} = 10 \times T$ performs well.

The packet delivery rate is obtained by $\frac{Probe\ count(P_{window})}{P_{window}/T}$ where probe count (x) refers to the number of probes received during window x. On a link formed between nodes A and B, $(A \rightarrow B)$, node A estimates the RDR and node B estimates the FDR. Node B includes its measured value of FDR from the last P_{window} duration, estimated for each of its neighbors, in its periodic probe packets. Upon receiving the probe packets, node A estimates both FDR and RDR and proceeds with the calculation of ETX for the A → B link. When a particular node receives a probe packet from its neighbor, it estimates both FDR and RDR. The periodicity of probe packets could lead to the large-scale collision of probe packets when the transmission times are synchronized. Therefore, to avoid the possible synchronization of probe packet transmission schedule, a random jitter is added to the periodicity of the probe packets. A typical value for the jitter is taken as $\pm 0.1 \times T$.

Some of the advantages of ETX routing metric include its high throughput and efficiency. The ETX routing metric saves resources such as spectrum and energy. Some of the disadvantages of ETX metric include the following: (1) the ETX may not work efficiently when the traffic load is high. When traffic load is very high, the probe packets may either be lost or queued. (2) Adding a separate queue for the probe packets may prevent this protocol from being used with popular MAC and routing protocols. (3) When nodes are mobile, ETX calculation may not be correct for a specific duration and ETX routing metric depends on the underlying routing protocol to quickly reconfigure path and or communicate the correct ETX value to each neighbor.

1.8.2 Multiradio Link Quality Source Routing

The multiradio link quality source routing (MRLQSR) [20], an extension of the DSR protocol, is designed to work with MRWMNs. The main contribution of MRLQSR is the use of a new routing metric called weighted cumulative expected transmission time (WCETT). The WCETT tries to avoid shortest path routing in an MRWMN environment. The major modules in the MRLQSR protocol are: (i) a neighbor discovery module, (ii) link weight assignment module, (iii) link weight information propagation module, and (iv) pathfinding module. The neighbor discovery and link weight information propagation modules are similar to the DSR protocol whereas the link weight assignment and pathfinding modules differ from DSR. While DSR assigns equal weights to all links, MRLQSR assigns link weights in a better way to improve performance. The link weight assigned by the MRLQSR is proportional to the expected amount of time necessary to successfully transmit a packet through that link. This expected transmission time essentially depends on the link data rate and the packet loss rate. In addition, while DSR utilizes a shortest path routing based on an additive hop count-based routing metric, MRLQSR uses WCETT as the routing metric.

The main design philosophy behind the WCETT routing metric is to obtain a link cost metric that represents the following properties: (i) loss rate and the bandwidth of a link, (ii) a nonnegative link cost, and (iii) consideration of the cochannel interference. The WCETT routing metric for a path is estimated by the following equation:

$$\text{WCETT} = (1 - \alpha) \times \sum_{i=1}^{L} \text{ETT}_i + \alpha \times \max_{1 \le j \le k} T_j \qquad (1.2)$$

where ETT_i is the expected transmission time (ETT) of link i in a path of length L, α is a tunable parameter that ranges from 0 to 1, and T_j is the sum of the transmission times on a particular channel j. The value of ETT is taken as $ETT = ETX \times \frac{S}{B}$ where ETX is the expected transmission time, S denotes packet length, and B refers to the bandwidth of the link. The value of ETX is obtained using the packet-probing mechanism. Equation 1.2 has mainly two parts. The first is provided by the sum of ETTs of every link on the path and is representing the end-to-end delay factor. This end-to-end delay factor essentially provides the approximate value of the end-to-end delay that a particular packet may face. The second part is the channel diversity factor that reduces the chance of a link, which uses a heavily used channel from being included in the path. The value of T_j can be obtained as follows:

$$T_j = \forall_{1 \le j \le k} \sum_{\text{Link } i \in L \text{ uses channel } j} ETT_i \tag{1.3}$$

where k is the number of channels in the system and L is the path length. Generally, the path bandwidth is upper-bounded by the bottleneck link, which has the lowest bandwidth. The channel diversity factor reduces the chance of choosing a path that has bottleneck links. The tunable parameter α provides a balance between the factor that reduces the end-to-end delay and the second factor, which reduces the bottleneck links. Figure 1.9 illustrates an example of the operation of the MRLQSR protocol. In this example, a 15-node network in which each node has two radio interfaces. Each radio interface can operate in either channel 1 (Ch 1) or channel 2 (Ch 2) as marked in the figure. When the source node (node 1) decides to find a path to the destination node (node 12), it initiates a pathfinding process by transmitting a route request (RREQ) packet. This pathfinding process is similar to DSR protocol. When an intermediate node forwards a RREQ packet, it attaches the ETT value and channel information for the link through which the packet is received and forwards the packet again to the neighbors. When the RREQ packet reaches the destination, node 12, it contains the ETT values and channel informations for all the links on that path. The destination node upon receiving the RREQ replies using a route reply (RREP) over the path through which the RREQ is received. The RREP contains all informations received through the RREQ. Figure 1.9 shows that the path between nodes 1 and 12 has three choices. These path choices are 1–2–3–7–8–12, 1–5–4–12, and 1–6–10–12. When Node 1 receives information about

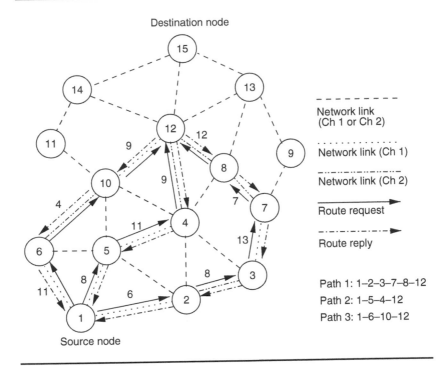

Figure 1.9   An example of MRLQSR protocol operation.

these three paths, it chooses an appropriate path based on Equation 1.2. Figure 1.10 shows different path choices and the corresponding ETT values. The estimated values for WCETT for each path choice can be calculated as shown in Table 1.3.

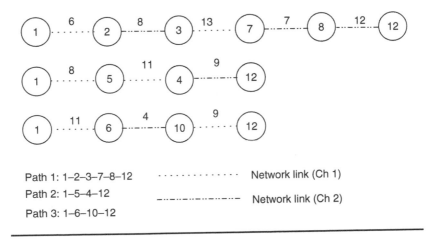

Figure 1.10   The path choices between nodes 1 and 12 and the ETT values.

Table 1.3 Estimated Values of WCETT

No.	Path	ΣETT_i	Max T_j	WCETT ($\alpha = 0$)	WCETT ($\alpha = 0.2$)	WCETT ($\alpha = 0.8$)	WCETT ($\alpha = 1$)
1	1–2–3–7–8–12	46	27	46	42.2	30.8	27
2	1–5–4–12	28	19	28	26.2	20.8	19
3	1–6–10–12	24	24	24	24.2	24.8	24

Path 1 has a path length of five hops, an end-to-end delay factor of 45 units, and a channel diversity factor of 27 units. Paths 2 and 3 have only three hops each and their end-to-end delay factor turns out to be 28 and 24, respectively. Since paths 2 and 3 have got different channel diversity factors, the WCETT may differ based on the value of α. For example, if α is chosen as 0.2, the WCETT will be computed for paths 2 and 3 as 26.2 and 24.2, respectively, and the chosen route will be path 3. When the value of $\alpha = 0.8$, the WCETT becomes 20.8 and 24.8 for paths 2 and 3, respectively, and in such a case path 2 will be preferred. The choice of α is not fixed by the protocol and could either be decided by the network operator as a fixed parameter or be made variable in a dynamic way based on the network load. Researchers found that at high network load, lower value of α provides better network throughput [20].

The main advantage of MRLQSR is the improved throughput performance compared with the throughput achieved by other multiradio routing metrics such as ETX. This throughput advantage results from the fact that MRLQSR considers a trade-off between end-to-end delay and the path throughput for the candidate paths. According to the experiments conducted in an experimental static two-radio-limited WMN test bed [20], it was found that WCETT outperforms ETX routing metric by about 80%. One of the significant disadvantages is that MRLQSR does not consider the important aspect of channel interference on neighboring links. For example, the estimation of channel diversity factor does not consider the relative positions of the usage of the same channel. In addition, the use of multiple radios on a single node may consume additional power and hence the routing metric should effectively look into energy-efficient routes when used in mobile WMNs. Therefore, MRLQSR may underperform when used in a WMN with limited mobility. In addition to the above disadvantages, the WCETT can cause loop formation within a network. Although WCETT may not form routing loops when used with on-demand routing protocols such as DSR, when used along with a distance vector routing protocol it may, in some cases, cause loop formation.

1.8.3 Load-Aware Interference Balanced Routing Protocol

The metric of interference and channel-switching (MIC) [22] is designed as the routing metric for load and interference-balanced routing algorithm (LIBRA) [23]. In comparison with the WCETT routing metric, MIC also attempts to consider the intraflow and interflow interferences while making routing decisions. The first part of the MIC metric consists of a channel switching cost (CSC), a measure of intraflow interference, which for a given path p is represented as C_{cs}^p. The C_{cs}^p is estimated at jth hop on path p as

$$\begin{aligned} C_{cs}^p &= w_a \text{ if } CH_{j-1} \neq CH_j \\ &= w_b \text{ if } CH_{j-1} = CH_j \end{aligned} \tag{1.4}$$

where $0 \leq w_a < w_b$. The CSC essentially takes into account the intraflow interference arising out of the reuse of the same channel over neighboring links and therefore, prefers to give a lower weight for paths, which utilizes multiple channels. The second factor that is part of MIC metric, interference-aware resource usage (IRU) factor, is for the interflow interference. The IRU factor considers the amount bandwidth resource consumed by the transmission over a link under consideration. This is estimated for a given link k as $IRU_k = ETT_k \times N_k$ where ETT_k, the expected transmission time, is estimated either by the same mechanism used for WCETT routing metric or by the data rate provided by the interface and N_k is the number of nodes affected by the transmission over the link k. For example, the IRU factor not only considers the expected transmission time for the transmitter of the link, but also takes into account the neighbor nodes which when using CSMA-based MAC protocols, wait for the transmission to be completed. In order to obtain a route from among a set of path choices, the MIC routing metric is estimated for each path. For a given path k, the MIC routing metric is calculated as follows:

$$MIC_k = \sum_{\text{node } j \in k} C_{cs}^j + IFF \times \sum_{\text{link } i \in k} IRU_i \tag{1.5}$$

where IEF refers to the interflow interference normalization factor for a network having N_T number of nodes and is estimated as

$IFF = \dfrac{1}{N_T \times \text{MIN(ETT)}}$. Therefore, the MIC metric provides a routing metric that can balance the effects of both the intra- and the interflow interferences.

The primary advantages of MIC metric include the following: (i) it provides better throughput and delay performance, (ii) it considers intra- and interflow interference, and (iii) it considers the channel diversity. The first disadvantage is that the MIC metric does not guarantee isotonicity and therefore, when used with hop-by-hop routing protocols, routing loops are likely to be formed. Another major disadvantage is the high overhead that is associated with obtaining the total number of nodes in the network. The total number of nodes in the network is essential for estimation of routing metric, which in large networks may become very expensive. Finally, the scalability of routing metric is affected by the requirement that every node needs to estimate the minimum value of ETT in the network. Therefore, the fluctuating nature of ETT affects the scalability of the routing scheme.

1.9 TOPOLOGY CONTROL SCHEMES FOR MULTIRADIO WIRELESS MESH NETWORKS

In WMNs, topology control refers to the alteration of network topology by modifying one or more parameters such as mobility, location, transmission power, directionality of antennas, and the status of network interfaces. This section presents the objectives of topology control and the existing solutions for topology control in WMNs.

1.9.1 Objectives of Topology Control Protocols

The network capacity is influenced by the network topology and therefore, altering the topology can increase the capacity and scalability. For example, reducing the transmission power reduces the neighbor node density which further influences number of nodes affected by a single transmission, thus leading to increase in spectrum reuse. In addition to the spectrum reuse, the number of nodes contending for a given spectrum will be reduced with decreased transmission range, which will further reduce the collision and congestion on the channel. Decreasing the transmission power below the critical transmission power may leave the network disconnected. Even if the transmission power is above the critical transmission power, the number of neighbors may not be optimal [28,29] to achieve maximum capacity. It is important to design topology control algorithms that maintain the network topology in such a way that the network capacity and scalability are maintained near optimal values. One of the novel techniques for the backbone topology management for WMNs is the backbone topology synthesis (BTS) [30] approach, which is explained in detail below.

1.9.2 The Backbone Topology Synthesis Algorithm

The BTS [30] approach provides a dynamic mechanism to manage a backbone network in a WMN. WMNs generally employ multiple WLAN access points that have wireless relaying capability to achieve an extended service set (ESS). This ESS may use a nonmesh network topology, e.g., the WDS that uses an STP in order to form a basic communication backbone within the WMN. According to BTS, there are three types of nodes in the network: (a) backbone nodes (BNs) that form the communication backbone, (b) the backbone capable nodes (BCNs), and (c) regular nodes (RNs). The BNs are those, which take part in the current network backbone and perform backbone services such as control and data packet relaying for the rest of the network. The BCNs are nodes that are capable of acting as BCs, but may act as BCs on a case-to-case basis. The RNs are neither BCs nor BCNs and they may not even have multiple interfaces. The RNs can be considered as client devices. The BNs are dynamically elected from a set of BCNs. Here every node in the WMN has two timers: (a) short timeout timer and (b) long timeout timer. The short timeout timer is used for beaconing, i.e., the periodic transmission of a beacon packet that contains the node identifier, status of the node, nodal weight, BN → BCN indicator, associated BN's identifier, identifier of the predecessor node, and the BN-neighbor list. The nodal weight is a measure based on node identifier, neighbor degree, resource capability, or any other stability metric. The purpose of the beacon packets is to gather one-hop neighbor information and two-hop BN topology information. In addition, though this algorithm is originally designed to work with the BCNs and BNs having two radios, one high power radio for inter-BN relaying and another low power radio for BN–RN communication, the algorithm works for MR-WMNs with two interfaces where each of them uses the same transmission power. These beacon packets are sent over all the interfaces of the MR-WMN node. The long timeout timer is used for topology reconfiguration and therefore, when the long timeout timer expires, every node executes a set of steps. For example, the BN node executes the BN → BCN switching algorithm (refers to the algorithm that helps in making the decision to switch from a BN node to BCN node), BCN node executes the association and BCN $\rightarrow$ BN switching algorithm whereas the RN executes only the association algorithm. The BCN nodes are expected to have two interfaces, one for backbone connectivity among BCNs and BNs, and the other for access connectivity for RNs. The backbone interface may be a high power interface while the access interface is a low power interface, though in practice both may be of the same technology, e.g., IEEE 802.11a/b/g.

The BN $\rightarrow$ BCN conversion algorithm provides the logical decision-making framework for a BN node A to switch from BN to BCN. There are three main steps in this algorithm: (a) every client associated to the BN A must have at least more than one BN neighbor, (b) any two neighbors of the BN A (B and C) either are directly connected to each other and BN A does not have the highest weight among the three BNs (A, B, and C) or either BNs B or C indicate their unwillingness to switch to a BCN node or have at least one additional shared BN neighbor (BN P) and BN P is unable to switch its status or the BN P has higher node weight compared to BN A, and (c) any one of the BN A's neighbor BNs (e.g., BN K) and one of neighbor BCNs (BCN M) either are directly connected to each other and BN K indicates its inability to switch or has a higher nodal weight than BN A or have at least one shared BN neighbor (BN R) that indicates its inability to switch or has higher nodal weight than BN A. The primary requirement of the switching algorithm is to keep the network connectivity intact as a result of switching and therefore, if the above conditions are met only partially (i.e., meeting (a) and not satisfying (b) or (c) as a result of the absence of any alternate path between any pairs of neighbors of BN A, then BN A indicates its inability to switch through a BN $\rightarrow$ BCN indicator set to zero. This indicates network partitioning upon the switching of BN A to BCN. Alternatively, the conditions (b) or (c) are not satisfied due to the differences in the nodal weights, and the BN A sets it BN $\rightarrow$ BCN indicator field to 1. Therefore, the switching from BN to BCN is done to improve the performance while maintaining the network connectivity.

The BCN $\rightarrow$ BN switching algorithm helps a BCN node, BCN A, to make a decision to switch to BN. Here, BCN A decides to switch if two main rules satisfy their conditions and at least one of the remaining three supporting rules. The first main rule that must be satisfied for switching is that a BCN must only convert to a BN if the number of its BN neighbors is less than or equal to a BN neighbor threshold. This condition controls the number of BCN to BN conversions because there are enough number of existing BN neighbors and therefore avoiding a new BCN to BN switching may not lead to network partitioning. The second main rule is that a BCN must only convert to a BN if there are no new BN neighbors joined its neighborhood within the last short timeout period. This rule prevents multiple BCNs switching to become BNs within a short period. The supporting rules, at least one of them, that must satisfy for switching a BCN to BN are the following: (a) the BCN A has the highest nodal weight compared with its nonassociated BCN neighbors or the BCN A has received at least one association request during the previous cycle, (b) at least one pair of its BN neighbors (BNs X and Y) are not connected by a path with less than or equal to two hops and it has the highest weight among all its BCN neighbors, and (c) at least one of the BN neighbors (e.g., BN K) and one of BN K's neighbor, BN M, should not be connected through a common neighbor or directly and should satisfy the following two conditions: the first is that the BN K should be the node with the highest nodal weight among all BCN neighbors and the second is that none of BCN neighbors of the node should directly connect to BN K or to at least one of the neighbors. Therefore, this algorithm efficiently decides on a BCN node becoming a BN.

The regular nodes execute the association algorithm during every long timeout period and there are two main parts for this association algorithms: (a) association over the backbone channel and (b) association over the access channel. The association process over the backbone channel is basically a single-hop association process over the backbone channel (high power channel in some networks) by finding the appropriate BN among its BNs. The choice of BN is made based on the nodal weight where a tie is broken in favor of the lowest node identifier. In the absence of any appropriate BNs, the node associates with a BCN. In the event the association with a BCN node, and then the RN inserts the BCN's identifier and the associated status in its periodic beacons. The association process over the access channel (referred to as low power channel) consists of identifying a BN by BCNs and RNs and the BN with highest nodal weight is chosen to associate with. In the event that no BNs are detected, then an appropriate BCN node is detected over multiple hops, generally the nearest one is chosen, and it then broadcasts the associated BCN in its beacon packets.

The advantages of BTS algorithm include that it is a dynamic, scalable and fully distributed solution to manage the backbone topology in an MR-WMNs. It dynamically chooses the BNs and thereby maintain the network topology to obtain a better performance.

1.10 OPEN ISSUES

The open research issues in the physical layer of WMNs include the development of new radio technologies that can provide high data rates such as UWB physical layer. In addition, MIMO-based physical layers have not yet been designed for fully distributed WMNs. Further development of new radio technologies can provide high physical layer capacity, capability to operate in extreme interference levels, and the ability to operate in flexible radio bands.

The open issues in MAC layer include the design of efficient MAC protocols for both single-radio WMN and MRWMN in order to achieve high throughput. In addition, new MAC protocols are required for new physical layers such as UWB and MIMO as mentioned earlier. More than the capacity provided by the physical layer, the network scalability is also affected by the MAC protocol design, therefore, new research solutions are necessary to achieve these objectives.

At the network layer, the most important aspect as far as routing is concerned is the design of new routing metrics that can provide high performance and network scalability. New research is required to provide routing metrics that can provide inherent load balancing and fault tolerance. Another important aspect of network layer design is the routing optimization in an unplanned WMN, i.e., a WMN with an arbitrary topology.

At the transport layer, in WMNs, in order to improve the performance of TCP over WMN, we need new solutions that can maximize the throughput while minimizing the deviation from TCP. Such schemes may demand use of mechanisms such as explicit link failure notification (ELFN). Thorough investigation is necessary to answer the question on whether to develop entirely new transport layer protocols or to modify the existing TCP for improving the performance of transport layer protocol for WMNs.

In the application layer of the WMNs, the open issues are about provisioning a variety of application layer services and service discovery mechanisms. QoS provisioning is one such application layer service, which is necessary for primarily providing voice services over WMNs. In addition, other services, distributed over the entire WMN, may need additional utilities in order to provide quick discovery of services. Moreover, these services may be provided over a variety of heterogeneous networks such as WMN, the Internet, and WAWMNs. In such cases, provisioning an efficient service discovery mechanism is essential.

In addition to the above-mentioned open issues, there exist several issues that need research effort. Some of the system-level issues are: (i) network management, (ii) cross-layer design, (iii) security, and (iv) pricing and billing schemes.

1.11   SUMMARY

This chapter presents the issues and challenges in WMNs in general and MR-WMNs in particular. WMNs face several challenges such as architectural design issues and network protocol design issues. We presented the major design issues for both the above-mentioned issues. The capacity of WMNs is very limited as a result of the limited bandwidth available and the use of multihop wireless relaying. One important direction for improving the capacity of WMNs is to use multiple radio interfaces and multiple channels simultaneously. Therefore, the MR-WMNs are becoming important. In addition to all the issues that arise as a part of the wireless spectrum, the MR-WMNs face many more challenges, as discussed in the later part of this chapter. The primary challenges can be categorized as architectural, MAC, networking and routing, and topology control. This chapter presents a brief discussion on each of these challenges and explains a set of existing solutions for solving them. The chapter concludes with a discussion of a set of open issues.

REFERENCES

1. V. Kanodia, A. Sabharwal, B. Sadeghi, and E. Knightly, "Ordered Packet Scheduling in Wireless Ad hoc Networks: Mechanisms and Performance Analysis," *Proceedings of ACM MobiHoc 2002*, pp. 58–70, June 2002.
2. B.S. Manoj, R. Ananthapadmanabha, and C. Siva Ram Murthy, "Multi-hop Cellular Networks: The Architectures and Protocols for Best-Effort and Real-Time Communications," *Journal of Parallel and Distributed Computing*, January 2006.
3. B.S. Manoj, D.C. Frank, and C. Siva Ram Murthy, "Throughput Enhanced Wireless in Local Loop (TWiLL)—The Architecture, Protocols, and Pricing Schemes," *ACM Mobile Computing and Communications Review*, vol. 7, no. 1, pp. 95–116, January 2003.
4. H. Luo, R. Ramjee, P. Sinha, L. Li, and S. Lu, "UCAN: A Unified Cellular and Ad Hoc Network Architecture," *Proceedings of ACM MobiCom 2003*, pp. 353–367, September 2003.
5. Available at: http://calmesh.calit2.net.
6. P. Gupta and P.R. Kumar, "The Capacity of Wireless Networks." *IEEE Transactions on Information Theory*, vol. 46, no. 2, pp. 388–404, March 2000.

7. J. Li, C. Blake, D.S.J. De Couto, H.I. Lee, and R. Morris, "Capacity of Ad Hoc Wireless Networks," *Proceedings of ACM Mobicom 2001*, pp. 61–69, July 2001.

8. S. Jagadeesan, B.S. Manoj, and C. Siva Ram Murthy, "Interleaved Carrier Sense Multiple Access: An Efficient MAC Protocol for Ad hoc Wireless Networks," *Proceedings of IEEE ICC 2003*, pp. 1124–1128, May 2003.

9. P. Bahl, A. Adya, J. Padhye, and A. Wolman, "Reconsidering Wireless Systems with Multiple Radios," *ACM Computer Communications Review*, vol. 34, no. 5, pp. 39–46, October 2004.

10. C. Cheng, I.A. Cimet, and S.P.R. Kumar, "A Protocol to Maintain a Minimum Spanning Tree in a Dynamic Topology" *Proceedings of ACM SIGCOMM 1988*, August 1988.

11. A. Iwata, C.C. Chiang, G. Pei, M. Gerla, and T.W. Chen, "Scalable Routing Strategies for Ad hoc Wireless Networks," *IEEE Journal of Selected Areas in Communications*, vol. 17, no. 8, pp. 1369–1379, August 1999.

12. C. Siva Ram Murthy and B.S. Manoj, *Ad hoc Wireless Networks: Architectures and Protocols*, Prentice Hall PTR, New Jersey, May 2004.

13. C.E. Perkins and P. Bhagawat, "Highly Dynamic Destination-Sequenced Distance-Vector Routing (DSDV) for Mobile Computers," *Proceedings of ACM SIGCOMM 1994*, pp. 234–244, August 1994.

14. S. Murthy and J.J. Garcia-Luna-Aceves, "An Efficient Routing Protocol for Wireless Networks," *ACM Mobile Networks and Applications Journal, Special Issue on Routing in Mobile Communication Networks*, vol. 1, no. 2, pp. 183–197, October 1996.

15. J.J. Garcia-Luna-Aceves and M. Spohn, "Source-Tree Routing in Wireless Networks," *Proceedings of IEEE ICNP 1999*, pp. 273–282, October 1999.

16. C.E. Perkins and E.M. Royer, "Ad Hoc On-Demand Distance Vector Routing," *Proceedings of IEEE Workshop on Mobile Computing Systems and Applications 1999*, pp. 90–100, February 1999.

17. D.B. Johnson and D.A. Maltz, "Dynamic Source Routing in Ad Hoc Wireless Networks," *Mobile Computing*, Kluwer Academic, vol. 353, pp. 153–181, 1996.

18. Z.J. Haas, "The Routing Algorithm for the Reconfigurable Wireless Networks," *Proceedings of ICUPC 1997*, vol. 2, pp. 562–566, October 1997.

19. D.S.J. De Couto, D. Aguayo, J. Bicket, and R. Morris, "A High-throughput Path Metric for Multi-hop Wireless Routing ," *Proceedings of ACM Mobicom 2003*, 134–146, September 2003.

20. R. Draves, J. Padhye, and B. Zill, "Routing in Multi-radio, Multi-hop Wireless Mesh Networks," *Proceedings of ACM Mobicom 2004*, pp. 133–144, September 2004.

21. A. Adya, P. Bahl, J. Padhye, A. Wolman, and L. Zhou, "A Multi-radio Unification Protocol for IEEE 802.11 Wireless Networks," *Proceedings of IEEE Broadnets 2004*, pp. 344–354, October 2004.

22. Y. Yang, J. Wang, and R. Kravets, "Designing Routing Metrics for Mesh Networks," *Proceedings of IEEE WiMesh 2005*, September 2005.

23. Y. Yang, J. Wang, and R. Kravets, "Interference-aware Load Balancing for Multi-hop Wireless Networks," *Technical Report UIUCDCS-R-2005-2526*, Department of Computer Science, University of Illinois at Urbana-Champaign, 2005.

24. A. Nasipuri, J. Zhuang, and S.R. Das, "A Multi-Channel CSMA MAC Protocol for Multi-Hop Wireless Networks," *Proceedings of IEEE WCNC 1999*, vol. 1, pp. 1402–1406, September 1999.

25. S. Xue and T. Saadawi, "Revealing the Problems with 802.11 Medium Access Control Protocol in Multi-Hop Wireless Ad Hoc Networks," *Computer Networks*, vol. 38, 2002.

26. B. Raman and K. Chebrolu, "Design and Evaluation of a New MAC Protocol for Long-distance 802.11 Mesh Network," *Proceedings of ACM MobiCom 2005*, pp. 156–169, September 2005.

27. R. Draves, J. Padhye, and B. Zill, "Routing in Multi-Radio Multi-Hop Wireless Mesh Networks," *Proceedings of ACM MobiCom 2004*, pp. 114–128, September–October 2004.

28. L. Kleinrock and J. Silvester, "Optimum Transmission Radii for Packet Radio Networks or Why Six is a Magic Number," *Proceedings of IEEE National Telecommunications Conference 1978*, pp. 431–435, 1978.

29. E.M. Royer, P.M. Mellier-Smith, and L.E. Moser, "An Analysis of the Optimum Node Density for Ad Hoc Mobile Networks," *Proceedings of IEEE ICC 2001*, pp. 857–861, June 2001.

30. H.J. Ju and I. Rubin, "Backbone Topology Synthesis for Mesh Wireless LANs," *Proceedings of IEEE Infocom 2006*, April 2006.

2

MULTIRADIO MULTICHANNEL MESH NETWORKS

Rajiv Vijayakumar, Arindam Das,
Sumit Roy, and Hui Ma

CONTENTS

2.1 INTRODUCTION

The emergence of cost-effective wireless access technologies such as IEEE 802.11 to mobile end users has changed communications and computing in significant ways. Its success to date has been largely due to its deployment in the home and small enterprise segments and various "hot spot" scenarios where it has limited coverage and serves only a few users simultaneously.* Currently, there is considerable interest in expanding IEEE 802.11 networks to large-scale enterprise scenarios to provide wide-area wideband access to a significant number of users. This requires a proliferation of access points (APs) over the desired coverage area.

Wide-area coverage using IEEE 802.11 basic service sets (BSSs) should naturally look to principles of cellular systems engineering for successful scaling. The key to one-hop capacity scaling (e.g., between a client and AP) is based on frequency reuse spatially. Given any number of orthogonal channels, neighboring APs are assigned the available orthogonal set in a systematic manner (e.g., the familiar frequency reuse patterns in cellular networks). The resultant signal-to-interference + noise ratio (SINR) at the edge of the "cell" (or BSS) along with the inherent properties of the IEEE 802.11 distributed coordination function (DCF) protocol then essentially determine the throughput per cell obtainable.

Scaling the *aggregate* network throughput over the coverage area is directly related to reducing the reuse distance between cochannel APs without degrading the SINR at the cell edge (equivalently, increasing the spatial reuse factor). This can be achieved by a combination of approaches (notably, among others, the use of directional antennas for beamforming at the APs, which we will not consider)—the most obvious being the availability of increased system bandwidth (equivalently, more orthogonal channels). Currently, only a very limited number of such orthogonal channels are available: 3 in IEEE 802.11b (2.4 GHz) and 12 in 802.11a (5 GHz). Although greater worldwide allocation is anticipated for unlicensed use in the future,

---

* In this work, we confine ourselves exclusively to 802.11 networks composed of multiple BSSs in infrastructure mode. Recall that an infrastructure BSS consists of an AP and its associated users at any given time.

it is clear that relying primarily on increased available bandwidth for scaling is not a feasible option. *Accordingly, for any given system bandwidth, optimizing network performance necessarily requires improving the entire protocol stack.*

A promising option for scaling the capacity of a wireless access network is to configure a layer-2 mesh that is currently planned within the IEEE 802.11s task group. This implies a direct wirelessly interconnected set of mesh nodes to form a multihop network. These nodes comprise of APs that allow direct client access as well as "routers" which only relay packets between other mesh elements. The ad hoc (but static) nature of mesh node deployments results in a significant spatial variability of the multiaccess interference (MAI) seen at any node location that leads to variable location-dependent node throughput. Hence, effective topology modification mechanisms including power control, node clustering, and channel assignments (CAs) are anticipated to be important design degrees of freedom.

Traditional multihop wireless networks (historically called packet radio networks) have almost exclusively comprised of single-radio elements. For various reasons as clarified by subsequent discussions on CA for a single-radio mesh, such networks are unable to effectively scale to exploit the increasing system bandwidth available. Consequently, the use of multiple radio nodes in a mesh network appears to provide one of the most promising avenues to network scaling. Multiple radios greatly increase the potential for enhanced channel selection and route formation while the mesh allows more fine-grained interference management and topology (power) control.

In this chapter, we address several issues pertinent to multiradio multichannel (MRMC) networks. We begin with an overview of typical mesh architectures in Section 2.1.1, starting with single-radio meshes to establish a baseline for comparison of the advantages of multiradio mesh networks. In Section 2.1.2, we provide a qualitative description of the gains achievable through the use of multiple radios and channels. These gains are quantified by simulations in Section 2.1.3.

Key to these gains are intelligent CA and routing in MRMC networks. In Section 2.2, we discuss choices of "radio usage policy," whereby the nodes bind their radios to a particular channel. For instance, nodes may use a static CA, in which each radio is bound to a specified channel for a long period of time. Alternatively, nodes may use a dynamic policy where the radio–channel binding changes with time, in principle as frequently as with every packet transmission.

The choice of radio usage policy influences the options available for CA which determines which of several available channels should be assigned to a particular radio. In addition to CA, another key factor in determining the performance of a mesh network is the choice of routing protocol. In an MRMC network, performance gains can be achieved if the routing protocol is "channel-aware," i.e., the metric used for route formation/selection takes into account the channels used on each link along the route. In general, CA and routing are closely coupled and we discuss both these topics in Section 2.3. Finally, we conclude the chapter in Section 2.4 with a discussion of several open issues.

2.1.1 802.11 Mesh Architecture

The increasing availability of multimode radios, integrated 802.11b/g/a cards, in client and infrastructure devices will enable new mesh architectures. For example, Tier-1 client–AP connectivity may use the 802.11b/g radio while the Tier-2 backhaul AP mesh can use the 802.11a radio, thereby separating the different kinds of traffic (client–AP vs. inter-AP) and simultaneously utilizing the potential of multiband radios. We first briefly discuss the performance of a single-radio mesh (one radio per node) as a prelude to showcasing the advantages of a multiradio mesh (multiple radios per node). The nodes in a Tier-2 mesh backhaul or access network consists of two types of nodes as shown in Figure 2.1 and Figure 2.2—a predominant lightweight subset (pure

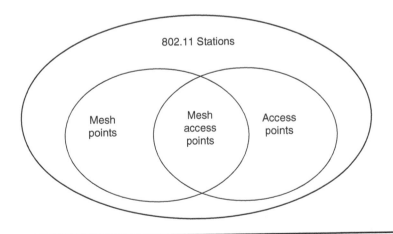

Figure 2.1 Schematic representation of the two types of mesh nodes: APs and mesh points.

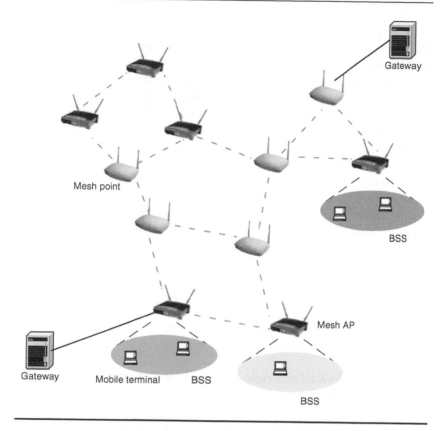

Figure 2.2 **A two-tier mesh network architecture. Tier-2 nodes may be either APs or mesh points. Each AP serves multiple Tier-1 clients which form a BSS.**

mesh points) whose only function is to route packets wirelessly to neighboring nodes and another subset of mesh AP nodes that allow direct client connectivity. A small fraction of these mesh AP nodes will be connected to the wired backbone and serve as gateways for traffic ingress/egress.

2.1.2 Capacity Scaling

Single-radio multihop wireless networks are not new; they have been studied since the 1970s under the nomenclature of packet radio networks. The end-to-end throughput in such single-radio networks reduces with the number of hops. The primary reason for this is that a single-radio wireless transceiver operates in a half-duplex mode; i.e., it cannot transmit and receive data simultaneously and an incoming frame must be received fully before the node switches from receiver to transmit mode.

Enhancing end-to-end throughput is naturally related to increasing one-hop aggregate throughput, which in turn depends critically on the number of simultaneous transmissions per channel (equivalently, minimizing reuse distance for cochannel cells) that can be achieved in a given network area. Achieving this is a complex function of many factors including the network topology and various attributes of layers 1–3 of the protocol stack. Layer 1 attributes include the type of radio, SINR requirements at the receiver for reliable detection, and the signal propagation environment. Layer 2 attributes include medium access control (MAC) attributes for interference management, and layer 3 attributes include the choice of the routing metric for path determination. Thus, the overall network optimization requires a multidimensional, cross-layer approach. In order to appear to the IP layer as a single local area network, a mesh network may implement its own routing functionality and other services at layer "2.5"; i.e., as an intermediate layer between the standard IEEE 802.11 MAC (or lower MAC) and the IP layer. Figure 2.3 illustrates this for a node with two radios.

2.1.2.1 Single-Radio, Single-Channel Mesh Networks: A Baseline

Substantial simulation evidence suggests that the per-node share of the aggregate throughput of a single-channel multihop IEEE 802.11

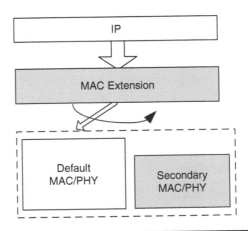

Figure 2.3 Illustrating the use of an intermediate "2.5" layer to enable mesh functionality.

network typically behaves as $1/n^{\alpha}$, where n is the number of nodes, and the exponent α is influenced by topology and traffic characteristics [1,2]. Analytical guidance is provided by an upper bound for large networks [2] that suggests $\alpha = 0.5$ for a purely ad hoc topology and random choice of source–destination pairs. Experimental evaluation [1] suggests a far more pessimistic value of $\alpha = 1.68$. Further insight can be obtained for finite networks with special topologies like a single-channel n node linear chain, for which the per-node throughput is $O\,(1/n)$ implying $\alpha = 1$.* The aggregate throughput in this case is essentially constant (independent of the number of nodes or hops) since only a single transmission can occur at any time [3] with a CSMA/CA-type MAC. The above scaling laws follow an important but *pessimistic* assumption that all nodes in the network interfere with each other, and that any pair of nodes (irrespective of their separation) communicate with equal probability. This is true only in small networks; in larger networks, traffic is more "localized" (i.e., nearby nodes communicate much more frequently). This implies that spatial reuse of channels is possible, leading to enhanced aggregate throughput. The role of spatial reuse in enhancing aggregate network throughput, facilitated by multiple (orthogonal) channels as well as multiple radios per node, is discussed in Section 2.1.2.2 and Section 2.1.2.3.

2.1.2.2 Single-Radio, Multichannel Mesh Networks

Any end-to-end path in a multihop network should utilize all the available orthogonal channels† (say C) in a manner that maximizes spatial reuse, i.e., maximizes the number of simultaneous transmissions in the network area. Unfortunately, a key limitation of commodity single-radio wireless devices is that they operate in half-duplex mode, and therefore cannot transmit and receive, simultaneously even if multiple noninterfering channels are available. A possible (but naive) approach to multihop route formation is for all nodes to use the same channel, even if multiple channels are available, at the cost of sacrificing spatial reuse. This approach does however avoid the serious drawback of large end-to-end delay when adjacent node pairs

---

* This is true for a "small" chain, or equivalently, a large carrier-sensing range that prevents any spatial reuse of the single channel in the network. Under the same assumptions, a chain with C channels will achieve an aggregate throughput of $O(C)$ that also does not scale with the number of nodes n.
† While very limited spatial reuse can be achieved even with $C = 1$, meaningful network scaling is only possible with increasing C.

use *different* channels to communicate. The latter necessitates channel scanning, selection, and switching a radio such that two adjacent nodes share a common channel; this switching delay (per node) grows with C. For example, the switching delay for present 802.11 hardware ranges from a few milliseconds to a few hundred microseconds [4]. Such frequent channel switching may be viewed as effective route lengthening because the switching delay manifests itself as virtual hops along the route [5]. Hence, exploiting the multiple orthogonal channels clearly enhances aggregate one-hop throughput vis-a-vis the single channel scenario, but at the cost of enhancing the end-to-end delay.

For all the above reasons, multiradio meshes, which introduce several new degrees of freedom that fundamentally address the key limitation of commodity single-radio wireless devices, are expected to be a key component in achieving both network scalability and adaptivity in practice (as in software-defined radios) for future wireless networks.

2.1.2.3 Multiradio Mesh

Multiple radio nodes are effectively full duplex, i.e., they can receive on channel c_1 on one interface while simultaneously transmitting on channel c_2 on another interface, thereby doubling the node throughput (in principle). For example, consider the path $1 \rightarrow 2 \rightarrow 3$ in Figure 2.4. Let R denote the maximum possible transmit rate over one hop (i.e., $1 \rightarrow 2$). With one radio, node 2 spends roughly half the time receiving from node 1 and the other half transmitting to node 3. Consequently, if the source (node 1) rate is R bps, the average receive rate at node 3 is approximately $R/2$ bps. With two radios at node 2 and two orthogonal channels, radio 1 can be tuned to channel 1 and radio 2 can be tuned to channel 2, in which case the receive rate at node 3 will be theoretically equal to R bps. Now, consider the case when node 2 has only one radio and there is a concurrent transmission on the route $4 \rightarrow 2 \rightarrow 5$. In this case, node 2 has to spend a quarter of its time receiving from nodes 1 and 4 and transmitting to nodes 3 and 5. The average receive rate at nodes 3 and 5 in this case is $R/4$ bps. Again, having multiple nonoverlapping channels does not help in this specific scenario since the limiting factor is the availability of only one radio at node 2. Finally, consider the case when node 2 is equipped with two radios and there are two available orthogonal channels. In this case, radios 1 and 2 can be tuned to channels 1 and 2, respectively. If radios 1 and 2 are used on a half-duplex mode to support the routes $1 \rightarrow 2 \rightarrow 3/4 \rightarrow 2 \rightarrow 5$, respectively, the average

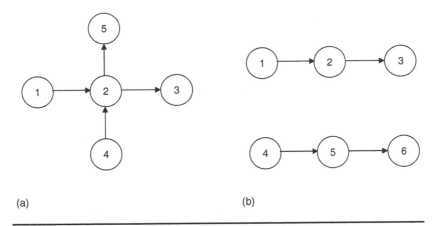

Figure 2.4 An example motivating the improvement in throughput that can be obtained with multiple radios and/or multiple channels. (a) With one radio at node 2, each of the two flows, $1 \rightarrow 2 \rightarrow 3$ and $4 \rightarrow 2 \rightarrow 5$, receive an end-to-end throughput of $R/2$ bps (where R is the source rate) if they are scheduled at different times. However, if the two flows are simultaneous, the receive rate for both flows drops to $R/4$ bps. With two radios and availability of two orthogonal channels, the receive rate for both flows increases to $R/2$ bps, the same as each flow would have received if they were scheduled at different times. (b) An illustration of a scenario when having multiple orthogonal channels is helpful even with one radio. For example, if two channels are available, one each can be used for the two transmissions. The receive throughput for each flow in this case is $R/2$ bps.

receiver throughput for each flow doubles to $R/2$ bps, the same as each flow would have received if they were scheduled at different times.

The allocation of channels to interfaces/radios will greatly influence end-to-end throughput, as will the choice of metric for route formation—we discuss these issues subsequently. In summary, we suggest that with proper design of layers 1–3, the performance of multiradio mesh scales as the size of the network increases.

2.1.3 Example

In this section, we use simulations (using the commercial simulation package OPNET) to quantify the throughout gains achievable by using multiple radios and channels. The network we consider is a 10×10 uniform 2D grid with a node separation of 10 m in both horizontal and vertical directions. We assume that any given node sends packets only to its immediate neighbors at a distance of 10 m. The carrier sense range of each node is set to 29 m* and the use of RTS/CTS is disabled. All 100 nodes have two radios each, and we study the impact of increasing the number of available channels.

The CA algorithm used is a simple heuristic. For convenience, we only consider the case when the number of available channels is even; here half the available channels are assigned to "horizontal" links and the other half to "vertical" links. All links on a given row use the same channel, as do all the links on a given column; it is easy to verify that the resulting assignment is feasible for the two-radio case that we are considering. The available "horizontal" (vertical) channels are cycled through as channels are assigned to each row (column).

The traffic model that we consider is as follows. A one-hop traffic flow is set up on each edge of the grid in both directions, for a total of 360 flows. Each flow consists of a Poisson stream of constant-length (1500 byte) packets generated directly at the IP layer. Each node is set up with a static routing table which was built using shortest-path routing; no routing protocol is used at runtime.

For each CA, the offered rate of all flows is simultaneously increased until the fraction of offered packets network-wide which is dropped reaches 10%. A packet can be dropped for two reasons: either because the MAC layer buffer (which has a capacity of 21 packets) is full and cannot accept another packet from the IP layer, or because the number of retransmission attempts for the packet exceeds the retransmission limit of 7. We denote the highest offered rate (per flow) for which the packet drop rate stays below 10% by T_{max}. Since T_{max} is the per-flow throughput and there are 360 flows in all, this means that when the offered traffic per flow is T_{max}, the carried traffic network-wide is $0.9 \times 360 \times T_{max}$.

T_{max} is an upper bound to the maximum traffic that can be carried simultaneously on each link while maintaining an acceptably low packet loss rate. We use T_{max} as our metric for comparison of network performance across various channel assignments. Since we increase rates on all links simultaneously, we are essentially considering a "fair" scenario in which all links are used equally. Our interest lies in exploring the variation of the throughput metric with the number of channels. The choice of packet loss rate of 10% is somewhat arbitrary; it was chosen such that the metric be measured reliably without excessively long simulation runs. We also do not use "saturated"

---

* The value of 29 m was found to be optimal for the single-radio single-channel case in terms of maximizing the throughput.

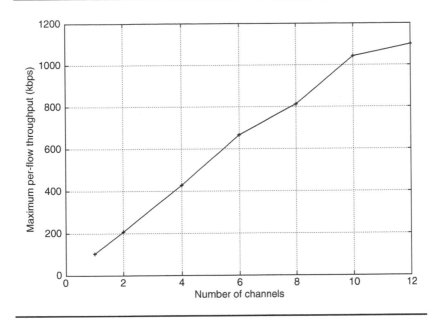

Figure 2.5 Maximum per-flow throughput as a function of the number of available channels in a 10 × 10 grid network with two-radio nodes.

sources, since saturated sources are known to cause unfairness in topologies where all nodes can not sense each other [6].

Figure 2.5 shows T_{max} for the various CAs. We see that as the number of channels C increases, T_{max} initially grows linearly with C, but the rate of growth slows as C increases beyond 6. Much further gain is not obtained by increasing the number of channels from 10 to 12. Figure 2.6 shows the network topology and the carrier sense range around a single transmitter. With only one channel, the S–D pair indicated must contend with all the flows within CS range for access to the channel; there are 25 nodes in all (including S) within CS range, and each node generates 4 flows. However, with $C = 2$ there are only half as many flows on the same channel within CS range, which leads to a doubling in the throughput T_{max}. However, as the number of channels grows further, the increases in throughput diminish, limited by having only two radios, and by the particular choice of CA algo-rithm. With six channels, the only cochannel links within CS range are those on the same "row" as S–D; this suggests that increasing the number of channels to eight will not yield additional throughput increase. However, the CS range definition is idealized: it only

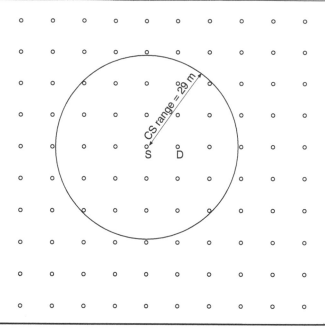

Figure 2.6 **A source–destination (S–D) pair and the carrier sense range around the source.**

indicates the distance at which a *single* simultaneous transmitter can be detected. In practice, actual CS based on aggregate energy from multiple transmitters may detect secondary sources from outside the CS range as defined (and hence the reference S–D pair must contend with all such flows). This explains the additional gain achievable from increasing C beyond 6.

A naive way to obtain linear capacity scaling with the number of channels is to increase the number of radios per node so that the number of radios equals the number of channels. This will result in there effectively being C parallel networks (one on each channel), thereby scaling the capacity by C. The key conclusion to be drawn from the simulations in this section is that, by using a careful CA, the overall network capacity can be made to scale linearly with the number of channels C (for low values of C) while using only two radios per node. This gain was achieved by using a simple heuristic for CAs. In the rest of this chapter we investigate more sophisticated approaches to the design of CA and routing algorithms.

2.2 RADIO USAGE POLICIES

We begin with an examination of possible radio usage policies—that determine which radio a node uses to transmit to a particular neighbor and when to bind the radio to a particular channel—in MRMC networks. The issue of which particular channel to use for transmission and reception will be considered separately in Section 2.3.

The simplest approach to binding channels to interfaces is to use *static* binding. In this approach, each interface is assigned to a channel when the system is initialized, and remains permanently tuned to that channel. In practice, the assignment could change occasionally; we assume that this change is slow relative to packet transmission duration. A common architecture in modern networks is to use relatively "lightweight" (or thin) APs in combination with an intelligent switch that controls multiple APs. This switch takes on the task of assigning channels to the interfaces on each node. The key advantage of using static binding is that it requires no change to the existing IEEE 802.11 standard. More complex schemes require some level of coordination among nodes, usually in the form of a modified MAC protocol.

In MRMC networks, the issue of which radio to use to communicate with a neighbor becomes interesting in the case when two neighboring nodes have more than one channel in common, i.e., they can communicate over more than one interface. One approach to using the multiple available interfaces is to use them in a round-robin manner on a packet-by-packet basis called "striping." This approach can result in packets arriving out of order at the receiver which can lead to a severe reduction in higher layer throughput. A different approach is proposed by Adya et al. [7]. Their multiradio unification protocol (MUP) combines multiple available interfaces into a single logical interface as seen by the higher layer. The MUP then transmits data over only one of the available interfaces; it selects the interface with the lowest (smoothed) round-trip time as measured by probe packets. It is shown [7] that in cases where reordering of packets can cause a significant throughput loss when using striping, MUP succeeds in maintaining a high throughput.

We now turn to more general radio usage policies than the static policy considered so far. In hybrid (or mixed) approaches, one interface is tuned to a fixed channel, while the other interface is dynamically switched to other channels. One such approach described [5] has a fixed interface at each node tuned to any of the available channels, and the choice of channel is communicated to neighbors by a separate

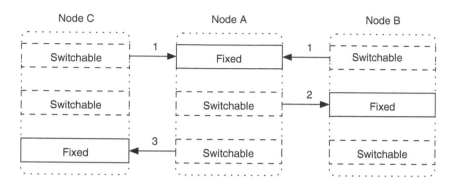

Figure 2.7 A hybrid radio usage policy in which each node uses a fixed interface for packet reception. (From Kyasanur, P. and Vaidya, N.H, *Wireless Communications and Networking Conference, 2005 IEEE,* 2005. With permission.)

higher layer protocol. When a node wishes to send a packet to its neighbor, it switches one of its dynamic interfaces to the fixed channel of the neighbor and transmits the packet. Thus, the fixed channel at each node represents the desired channel for reception at that node. This policy is illustrated in Figure 2.7 for the case of three interfaces per node.

If a node is simultaneously communicating with several neighbors who may have different reception channels, then there will be a significant penalty incurred in switching channels for each packet transmission. Another potential drawback of this scheme is that current 802.11 hardware does not allow for switching on a per-packet basis. This is because channel-switching directives and packet deliveries occur over two distinct (logical) interfaces with the 802.11 MAC/PHY hardware; there is no prescribed method in the standard for a desired channel to be associated with a packet that is delivered to the MAC from the higher layer. One way to implement the proposed switching scheme would be to install a new intermediate layer between the IP and MAC layers; this layer will ensure that only one packet at a time is delivered to the MAC layer, and that the PHY is switched to the appropriate channel before the packet is delivered to the MAC layer.

An alternate hybrid approach is for all nodes to share a common "control" channel, and to tune their fixed interface to this common channel. In this case, communication on the common channel is used to determine which channel to use for data transmission. Examples of

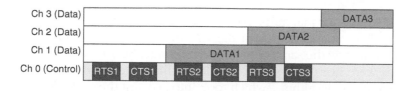

Figure 2.8 **Illustration of the common control channel radio usage policy.** *(From Mo, J., So, H.-S. W., and Walrand, J., in MSWiM '05: Proceedings of the 8th ACM international symposium on modeling, analysis and simulation of wireless and mobile systems, ACM Press, New York, 2005. With permission.)*

this approach include the work of Wu et al. [8] and Benveniste [9]; these approaches require modifications to the 802.11 MAC. An illustration of this approach is shown in Figure 2.8. Nodes exchange special Mesh-RTS and Mesh-CTS frames on the common control channel that include information on which data channel they will use for transfer of the next packet. Then they switch one of their data radios to that channel. Since the control channel is constantly being monitored by a dedicated radio, all nodes are always aware of the current data channels. The bottleneck in this approach is ultimately the control channel itself; as the network scales, the control channel becomes a scarce resource. Further, the overhead from the use of the control channel renders it inefficient when the data packets to be transmitted are short. Since the data channels may be shared by other nonmesh nodes, contention-based access would still be required on the data channel by the standard DCF of EDCA mechanisms. For a performance evaluation of the above protocols (along with a discussion of protocols which use multiple channels with single-radio nodes), see Mo et al. [10].

After discussing the various approaches to radio usage policies, we now turn to the question of how to pick which channel to assign to a particular radio.

2.3 CHANNEL ASSIGNMENT AND ROUTING

The simulation results presented in Section 2.1.3 illustrated the performance benefits that can be obtained by a careful choice of CAs. In this section, we explore the twin topics of CA and routing, beginning with a discussion of the importance of considering these two topics jointly.

For the simulations in Section 2.1.3, we used a traffic pattern that placed an identical load on each link, and the CA heuristic we used worked well for that traffic pattern. Due to the uniformity of both, the traffic load across links and the CA, the load on all *channels* was approximately the same. This would no longer be the case if the traffic loads were not uniform across links; in that case we might expect a different CA algorithm—which took into account the traffic loads—to outperform the simple heuristic that we used.

In general, it will not be the case when all links are loaded equally. In a typical topology, some of the mesh points serve as gateways, and the traffic to and from these gateways may be much higher than traffic elsewhere in the network. The traffic loads on each link are affected by the choice of routing protocol and the associated routing metric. Therefore, the optimal choice of CAs for a given set of traffic demands and a given routing protocol may be quite different from that for a different set of demands or a different routing protocol.

Conversely, if we are given a specific CA and asked to choose routes optimally, the choice of routes would depend on the CA. In a wireless mesh network (WMN), when a flow is routed over a particular link, it not only reduces the available capacity of that link, but also the available capacity on other cochannel links within carrier sense range. This is because all the cochannel links within CS range share the same total bandwidth for their transmissions. This key difference between wired and wireless networks makes it important to consider CAS when choosing routes in WMNs.

Figure 2.9 illustrates the importance of accounting for the CA in route selection. In this scenario there are two available routes between a source S and a destination D. One route is S–A–D, and involves two hops both of which use the same channel (channel 1). The other involves three hops, and uses a different channel on each hop. We consider two traffic scenarios. In the first scenario, suppose all channels are very lightly loaded, and that the traffic between S and D is also

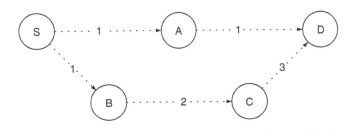

Figure 2.9   Two paths between a source S and a destination D. Each hop on the path is labeled with its assigned channel.

small as compared to the link capacities. In that case the end-to-end delay between S and D is dominated by the transmission time for the packet on each hop, which is given by L/B, where L is the length of the packet (including headers) and B is the link data rate. If all links have the same data rate B, then the best route from the viewpoint of minimizing delay is the two hop route S–A–D. The S–A–D route also uses fewer network resources since it utilizes only two links instead of three.

Now consider a different scenario in which we are interested in maximizing the throughput that the flow between S and D can obtain. Such a scenario might arise if the flow is a TCP stream, e.g., a file transfer between S and D. In this case, the end-to-end throughput available between S and D is only $B/2$ if the traffic is routed over S–A–D since both the S–A and A–D links use the same channel and hence share the same bandwidth. However, if the traffic is routed over S–B–C–D, the available end-to-end throughput is B, since each link can be active simultaneously because all three links use a different channel.

The above discussion illustrates that there is a close relationship between CA and routing in mesh networks. Therefore, in order to maximize performance, the two problems should be treated jointly rather than separately. However, in practice the joint problem is generally too hard to solve optimally. One approach to the joint problem is to solve the CA and routing problems separately and iterate over the two phases to improve the overall performance. It is therefore useful to first consider the two problems in isolation which we do in Section 2.3.1 through Section 2.3.4. We will discuss some work on treating the joint problem in Section 2.3.5.

2.3.1 Channel Assignment Basics

Our discussion of assigning channels to radio interfaces will be mostly restricted to the case of static radio usage policies. We will also assume that all nodes in the network use the same transmission power; the formulations and results are easily generalized to the case where different nodes use different transmission powers.

The CA problem is usually stated in graph-theoretic terms. Let $G = (V, E)$ be a graph in which the node set V represents the mesh nodes and the edge set E represents direct communication links. In general, a communication link will exist between every pair of nodes (v_1, v_2) such that the Euclidean distance between them $D(v_1, v_2)$ is less than some threshold D_{max}; equivalently, since all nodes use the same power, the received power at v_2 when v_1 transmits is greater than some threshold P_{min}. The threshold P_{min} is chosen to ensure that there is sufficient link margin at the receiver to allow for successful packet decoding to occur with high probability even in the presence of external sources of interference. The graph G is called the reachability graph of the network.

The basic CA problem can be posed in terms of assigning channels to either the vertices in V or the edges in E. In the vertex formulation, the goal is to assign a channel to every radio at every node $v \in V$ while ensuring that any two neighboring nodes (i.e., nodes connected by an edge) have at least one channel in common (which ensures that neighboring nodes can, in fact, communicate). In the edge formulation, channels are assigned to every edge $e \in E$ while ensuring that the number of distinct channels assigned to edges incident on a node v is not greater than the number of radios at v; this restriction is necessary because we are only considering static radio usage policies which do not allow for switching a radio among different channels. Any solution to the vertex formulation can be translated to a solution for the edge formulation and vice versa, so the two formulations are equivalent. Figure 2.10 illustrates the radio- vs. edge-based CA schemes. With the radio-based scheme, if radio $A1$ chooses frequency F_1 and radio $C1$ chooses frequency F_2, the two radios of B must be tuned to F_1 and F_2 (F_3 cannot be a choice) to ensure that it can communicate with nodes A and C. In addition, radio $D1$'s choice is also limited between F_1 and F_2 as otherwise a communication link would not exist between nodes B and D. With the edge-based assignment scheme, if the edge $A \leftrightarrow B$ chooses F_1 and $B \leftrightarrow C$ chooses F_2, $B \leftrightarrow D$ must choose between F_1 and F_2 so as not to violate the two-radio limitation on node B.

Any solution to the problem as posed above is considered a feasible CA. The feasible set is clearly nonempty since the trivial solution of assigning the same channel to every edge (or radio) is a feasible solution. At the same time, a random assignment of channels to radios or edges may not be feasible. Consider, e.g., a network in which all nodes have two radios and we assign one of four channels to each radio randomly. If some node is assigned channels 1 and 2, and its neighbor is assigned channels 3 and 4, then the two neighbors do not have a channel in common, and the assignment is not feasible.

The goal of various CA algorithms is to pick a feasible CA that optimizes some suitably chosen performance metric. Later in this section, we consider the forms that a metric can take. First, however, we consider the impact of the choice of CA on network performance, for which we need to introduce the concept of "interfering edges."

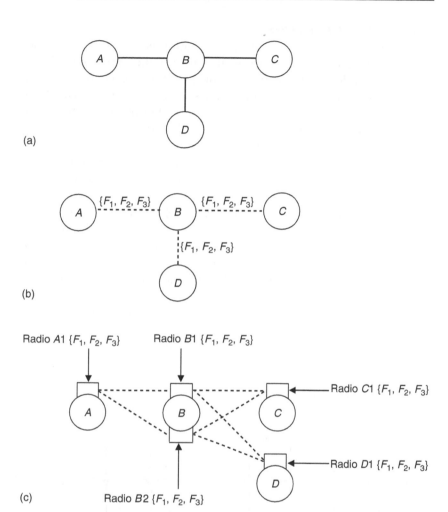

**Figure 2.10 Illustrating radio-based vs. edge-based channel assignment schemes.
(a) Desired reachability graph. Nodes *A*, *C* and *D* have a single radio each while
node *B* has two radios. The radios are numbered *A1*, *B1*, *B2*, *C1*, and *D1*. Assume
that three orthogonal channels are available for data communication. (b) In a
radio-based assignment scheme, channels are assigned to individual radios. In
order to ensure radio connectivity between nodes *A* and *B*, they must share a
common channel. Similarly, (*B* and *C*) and (*B* and *D*) must share a common
channel to ensure radio connectivity. (c) In an edge-based assignment scheme,
channels are assigned to the edges of the reachability graph. In this example, since
node *B* has only two radios, only two of the three outgoing edges from *B* can have
distinct channels.**

2.3.1.1 Interfering Edges

Consider a network whose reachability graph is a 5×6 grid network as shown in Figure 2.11a. Suppose all links use the same channel, and all nodes employ virtual carrier sensing (RTS/CTS). Then, when the link $a \leftrightarrow b$ is active (i.e., either a is transmitting to b or receiving from b), all nodes adjacent to either a or b can neither transmit nor receive packets (since they cannot transmit either RTS or CTS frames). Therefore, all edges incident on neighbors of either a or b must be inactive when $a \leftrightarrow b$ is active. We call this as the set of "interfering edges"; for the edge $a \leftrightarrow b$ in Figure 2.11a, this set consists of 22 other edges, and is shown in Figure 2.11b.

In general, for the MRMC case, the set of interfering edges for an edge e is denoted IE(e) and consists of those edges in the network that are on the same channel as e and, by virtue of either physical or virtual carrier sensing, cannot be simultaneously active with e. The concept of interfering edges is the key to understanding the impact of the choice of CA. In essence, the size of IE(e) represents the cost to the network of having edge e active. Also, by symmetry, IE(e) represents the set of edges f for which $e \in$ IE(f); this yields another viewpoint that is also useful, i.e., the size of IE(e) corresponds to the number of

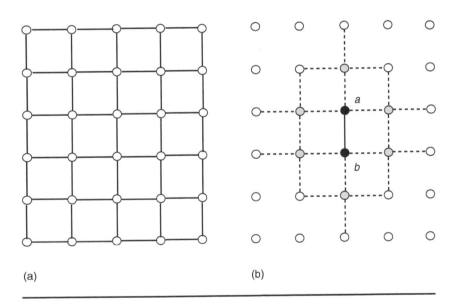

(a)   (b)

Figure 2.11   (a) Reachability graph of a 5×6 grid. (b) Set of 22 interfering links (shown dotted) for the edge $a \leftrightarrow b$. The lightly shaded nodes are neighbors of either node a or node b and must remain silent when the link $a \leftrightarrow b$ is active.

edges with which the edge e must contend for access to the channel. If any of the edges in IE(e) are active, then e must be silent.

Interfering edges are closely related to the concept of a "conflict graph" introduced [11]. The conflict graph consists of one vertex corresponding to each edge in the reachability graph. If $a \leftrightarrow b$ and $c \leftrightarrow d$ are two edges in the reachability graph, then there is an edge between the corresponding nodes v_{ab} and v_{cd} in the conflict graph if and only if $c \leftrightarrow d \in$ IE($a \leftrightarrow b$).

2.3.2   Formulations and Algorithms

We now consider several different approaches in the literature to solving the CA problem.

2.3.2.1   Integer Linear Programing

One approach to obtaining an optimal CA (for a suitable objective function) is to pose the problem as an integer linear program (ILP). This was adopted by Das et al. [12] where the objective was to maximize the number of simultaneous transmissions. This can also be reformulated in terms of finding the largest possible independent set in the conflict graph [11]. The resulting solution effectively maximizes the achievable instantaneous throughput. While an ILP-based formulation can yield optimal solutions using standard ILP software, the required run-time to find the optimal solution can be very high for even modest-sized networks. However, by terminating the search after a certain time, or by using noninteger relaxations, useful bounds to the objective function can be computed.

2.3.2.2   Graph-Theoretic Approaches

The CA problem can be viewed as a coloring problem on the reachability graph. Given a finite number of colors (i.e., channels), the goal is to assign the colors to edges while satisfying the constraint that the number of distinct colors assigned to edges incident on a node is not more than the number of radios at that node. Furthermore, we may seek to find a coloring that optimizes some objective function. In general, posing the CA problem as a coloring problem reveals the computational hardness of the problem; several authors have developed formulations which they have shown to be NP-hard [13,14].

The graph-theoretic approach can be useful for considering questions such as the minimum number of channels required to achieve maximum capacity or the maximum achievable capacity with a given number of channels. The first question is closely related to the "strong edge-coloring" problem [14,15]. The question of maximum achievable capacity is addressed [11], where the authors derive upper and lower bounds for the capacity.

Marina and Das [13] pose the CA problem as a coloring problem with the objective of minimizing the maximum size of IE(e) over all edges $e \in E$. Since the problem is NP-complete, they also provide a heuristic approach. Their algorithm proceeds in a single pass over all the nodes in the network and assigns channels while ensuring that the connectivity constraint is met. When there are multiple available choices for the channel to be assigned to a link, the choice is made based on minimizing the size of IE.

The discussion [13] reveals some interesting behavior with regard to the performance improvements achievable by increasing the number of radios per node. Following the terminology [13], the size of the largest set IE(e) (over all edges e in the network) is called the maximum "link conflict weight," denoted by W_{max}. For the case of 12 available channels (as in 802.11a), as the number of radios increases from one to three, the value of W_{max} drops sharply as expected. However, as the number of radios increases beyond three W_{max} begins to increase. This counterintuitive behavior occurs because the increase in radios leads to multiple common channels between neighbors, which in turn *increases* the number of edges in a typical set IE(e). This is because when nodes have more than one channel in common, they are allowed to communicate over all the available channels. Thus, while increasing the number of radios can increase the available capacity between neighbors, it can also have a detrimental effect on the network-wide capacity since there are now more links contending for access to the same channel. The maximum instantaneous throughput continues to increase as the number of radios increases beyond three, however the increase is far more gradual than the rapid increase in instantaneous capacity as the number of radios is increased from 1 to 3.

The formulation considered [13] can also be posed as an ILP problem along the lines of [12]. Solving the ILP for a 4×4 grid in which each node has two radios and four channels yields the CA shown in Figure 2.12. In this case, IE(e) consists of two edges for all edges e. We will discuss this CA approach in Section 2.4.

2.3.3 Limitations

The various CA approaches we have outlined assume the "protocol" model of interference [2], i.e., a transmission fails if there exists a

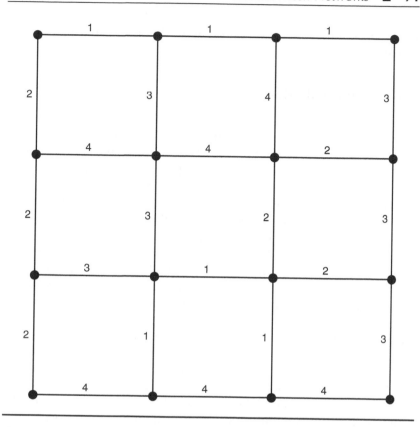

Figure 2.12 Channel assignment for a 4 × 4 grid with the optimization goal of minimizing IE(e) over all edges. Each node has two radios, and there are four available channels. Each edge is labeled with the assigned channel.

simultaneous cochannel transmitter within a certain range of the receiver. This model only considers the effect of a single interferer, and does not consider the effect of *aggregate* interference from several simultaneous transmitters. In practice, a node always sees the combined interference from all cochannel transmitters, as described by the "physical" interference model [2]. Similarly, when a node does carrier sensing by an energy-detecting mechanism, it detects the combined energy from all transmitters, not from just one transmitter at a time. The effect of aggregating energy from multiple transmitters can be significant, as shown in the simulations in Section 2.1.3. Therefore, in order to fully capture the effect of reusing a channel in a mesh network, we need to consider not only the distances of the transmitters from each other, but also the number of such transmitters that can be simultaneously transmitted. This is an area of ongoing research.

2.3.4 Routing Metrics

We now turn our attention to the problem of routing in mesh networks for a given CA. We do not discuss the various options for routing protocols themselves; usually one of the well-known routing protocols developed for ad hoc networks will suffice. Instead, we focus on the choice of *routing metric* used by the routing protocol.

The simplest routing metric, the hop count (shortest path), has been widely used in existing ad hoc routing protocols. However, it will generally fare poorly in mesh networks because it does not account for any of the issues raised in Section 2.3.3. An improved metric is the "expected transmission count" (ETX) metric proposed by De Couto et al. [16]. The ETX metric assigns a weight to each link that corresponds to the expected number of transmissions required by the 802.11 MAC to successfully transmit a packet over the link. Thus, ETX assigns a higher weight to (i.e., penalizes) links that are subject to higher packet losses. This is an improvement over the hop count approach since it captures, to a certain extent, the impact of varying levels of interference on different links. However, it does not account for the fact that different links may use different bandwidths (i.e., the same packet on different links require different transmission durations), or that reusing the same channel along a path reduces the available throughput. The "expected transmission time" (ETT) metric introduced by Draves et al. [17] addresses the former problem by multiplying the ETX by the time required for each transmission (L/B, using the notation of the Section 2.3). Another new metric introduced [17] is the "weighted cumulative expected transmission time" (WCETT) metric which attempts to address the issue of channel reuse along a path. Since this is one of the key issues in route selection in mesh networks, we consider this metric in some detail below.

Consider a path P and denote the ETT for link $i \in P$ by ETT_i and the channel assigned to i by CH_i. Suppose there are C channels available in all, and for each channel c, define

$$X_c = \sum_{i \in P; \, CH_i = c} ETT_i, \quad 1 \le c \le C. \tag{2.1}$$

X_c then is a measure of the usage of channel c along the given path. The WCETT metric for the path P is defined by

$$\text{WCETT} = (1 - \beta) \sum_{i=1}^{n} \text{ETT}_i + \beta \max_{1 \leq c \leq C} X_c, \tag{2.2}$$

where β is a tunable parameter, $0 \leq \beta \leq 1$.

There are two useful ways of interpreting the WCETT metric [17]. The first is that the two terms represent a trade-off between "global good" and "selfishness." Reducing the first term reduces global resource utilization while reducing the second term increases the throughput achievable along the path by reducing bottlenecks due to the same channel being reused along the path. A second interpretation is that the two terms in Equation 2.2 represent a trade-off between simultaneously achieving low delay and high throughput; reducing the first term reduces delay, while reducing the second term increases throughput. In either interpretation, β provides a means to adjust the relative importance of the two objectives; for instance, with the second interpretation, low values of β emphasize delay minimization while high values of β emphasize throughput maximization.

It is instructive to consider the value of the WCETT metric for the example of Figure 2.9. Let us suppose that the value of ETT is the same on all links; we will take it equal to 1 without loss of generality. Now the value of WCETT for the S–A–D path is $(1 - \beta) \times 2 + \beta \times 2 = 2$, regardless of the choice of β. The value of WCETT for the S–B–C–D path is $(1 - \beta) \times 3 + \beta \times 1 = 3 - 2\beta$. Thus, for values of $\beta < 1/2$, the WCETT metric would choose the S–A–D path, while for $\beta > 1/2$ it would pick the S–B–C–D path. This is in keeping with the intuition that for lower values of β, the path with lower delay is preferred, while for high values of β, the path which maximizes throughput is preferred.

Despite the above intuitively appealing features of the WCETT metric, it does suffer from some drawbacks. The second term in the definition, which penalizes bottleneck links, does not distinguish between cochannel links that are separated by a short distance from those that are separated by a long distance. If two cochannel links are outside carrier sense range of each other, they do not share the same bandwidth and hence should not incur the same penalty as two links that are within CS range of each other. The WCETT metric also does not explicitly account for congestion caused by links that are outside the given path. This is only considered indirectly through the impact on the ETT which does not account for increased queueing delays due to congestion. A third issue with the WCETT metric is that it is not possible to efficiently compute the value of the metric along a path using a procedure such as Dijkstra's algorithm [18]. This is because the dynamic programing principle that lies at the heart of Dijkstra's algorithm can no longer be applied: if an optimal path between S and D goes through some intermediate node A, the portion of that path between A and D may not itself be the optimal path between nodes A and D. This behavior is attributable to the second term in Equation 2.2: the contribution to that term of a given link depends not only on that link alone, but also on other links along the overall path.

A different metric that aims to improve upon WCETT is the "metric of interface and channel switching" (MIC) proposed by Yang et al. [19,20]. The MIC metric for a path P is defined in a similar vein to WCETT as follows:

$$\text{MIC} = \frac{1}{N \times \min{(\text{ETT})}} \sum_{i \in P} \text{IRU}_i + \sum_{i \in P} \text{CSC}_i, \qquad (2.3)$$

where N is the total number of nodes in the network and min(ETT) is computed over all links. The two terms, "interference-aware resource usage" (IRU) and "channel switching cost" (CSC$_i$) are defined as follows:

$$\text{IRU}_i = \text{ETT}_i \times N_i, \qquad (2.4)$$

$$\text{CSC}_i = \begin{cases} w_1 & \text{if } \text{CH}_{i-1} = \text{CH}_i, \\ w_2 & \text{if } \text{CH}_{i-1} \neq \text{CH}_i, \quad 0 \leq w_1 < w_2, \end{cases} \qquad (2.5)$$

where N_i is the number of edges within CS range of the edge i.

The IRU term in the definition of MIC incorporates both the delay along a path (as does WCETT) and the impact on the network as a whole in terms of resource usage.

The CSC term penalizes repeated use of a channel, but only locally, i.e., only on successive links. This is in a sense the opposite end of the spectrum from WCETT which penalizes channel reuse on all links along the path. While the MIC metric, like WCETT, is not directly amenable to efficient computation by the Dijkstra algorithm, the authors propose a procedure of introducing "virtual nodes" [19] which does allow for the use of efficient algorithms. Simulation results [20] illustrate the improvements possible over WCETT by using the MIC metric.

2.3.5 Joint Approaches

The CA approaches described earlier effectively consider a system in which all links carry an equal traffic load. While this can form an useful baseline, if the traffic demands in the network are known, then they can be incorporated into both CA and route selection procedures in order to improve overall network performance. Improvements can be in the form of increased network capacity, or reduced delays for a specified network load.

Since the case of uniform traffic loads on each link can be viewed as a special case of the general load and routing aware approach, it is clear that obtaining optimal solutions in the general case will be hard. Consequently, various heuristic approaches have been proposed in the literature, usually involving the CA and the routing problems separately and iterating over the two phases to find improved solutions.

One such iterative approach is proposed by Raniwala et al. [4]. CA is initially done by a greedy algorithm; flows which then exceed the available capacity for the CA are rerouted using either shortest-path or randomized multipath routing. These two phases are then iterated over multiple times. Simulation results show that a much larger portion of the offered traffic can be carried by adopting this methodology than with a CA that does not take into account the traffic demands. The algorithm [4] is centralized, and the authors also propose a distributed form of the algorithm [21].

Another approach to joint CA and routing is proposed by Alicherry et al. [22]. Their work assumes an MAC that operates synchronously and in time-slotted mode; the resulting throughput is an upper bound to the performance of 802.11. The problem is initially formulated as an ILP, and an LP relaxation of the problem is solved optimally. This is followed by several adjustment steps to obtain a valid CA and a link-scheduling policy that eliminates interference.

Kyasanur and Vaidya [23] propose a dynamic CA algorithm (used in conjunction with the hybrid radio usage policy [5]) in which nodes periodically broadcast "Hello" packets on every channel indicating that they are using for reception. If a large number of other nodes within a specified distance of a given node are also using the same channel for reception, then that node changes its reception channel with some probability p.

2.4 OPEN ISSUES

There are many promising open research questions on the subject of MRMC mesh networks. We summarize some of them in this section.

Several radio usage policies were outlined in Section 2.2. The relative merits of these approaches have not, to the best of our knowledge, been closely studied. In particular, the various schemes that employ dynamic channel switching have different associated costs (channel switching time and bottleneck of a common control channel), and a comprehensive study of the best approach to utilizing multiple interfaces remains an open topic.

Most of the CA algorithms in the literature to date have been centralized off-line algorithms. In practice, a CA procedure that was distributed and adjusted dynamically to channel conditions would be preferable (one such algorithm is described [21]). Currently the 802.11 standard does not provide much information to the higher layers with which to make CA decisions; the primary channel quality measure that is available is the "received signal strength indicator" (RSSI), whose values are allowed to be vendor-dependent. However Task Group K within the IEEE 802.11 is preparing a standard (in draft stage as of early 2006) which will enable higher layers to obtain far more detailed information about channel conditions from the MAC and PHY layers. Available measurements will include a standardized signal strength measurement as well as a "neighbor report" that includes information on neighboring nodes that have been detected. Utilizing these measures to develop more sophisticated CA schemes is an open area at present.

In our discussion of routing metrics in Section 2.3.4, we touched upon the trade-offs involved in making routing decisions in mesh networks. One potential area of interest is also to incorporate future 802.11k measurements in determining the link conditions to use as a link metric. Estimating transmission times on a link by probes is currently unreliable (e.g., see the discussion [7]), and improvements in this area could result in improvements in path metric computation, and hence in route selection.

There are also several other degrees of freedom available in the design of mesh networks that we have not considered here. These include transmission power control, carrier sense threshold selection, receive sensitivity setting, and the choice of transmission data rate. Joint optimization over these various criteria provides a rich area for future investigations.

REFERENCES

1. Piyush Gupta, Robert Gray, and P.R. Kumar, "An experimental scaling law for ad hoc networks," University of Illinois at Urbana-Champaign, May 2001.
2. Piyush Gupta and P.R. Kumar, "The capacity of wireless networks," *IEEE Trans. Inform. Theory*, vol. 46, no. 2, pp. 388–404, 2000.

3. Jinyang Li, Charles Blake, Douglas S.J. De Couto, Hu I. Lee, and Robert Morris, "Capacity of ad hoc wireless networks," in *MobiCom '01: Proceedings of the 7th Annual International Conference on Mobile Computing and Networking*, ACM Press, New York, 2001, pp. 61–69.

4. Ashish Raniwala, Kartik Gopalan, and Tzi-Cker Chiueh, "Centralized channel assignment and routing algorithms for multi-channel wireless mesh networks," *SIGMOBILE Mob. Comput. Commun. Rev.*, vol. 8, no. 2, pp. 50–65, April 2004.

5. P. Kyasanur and N.H. Vaidya, "Routing and interface assignment in multi-channel multi-interface wireless networks," in *Wireless Communications and Networking Conference, 2005 IEEE*, 2005, vol. 4, pp. 2051–2056.

6. Michele Garetto, Theodoros Salonidis, and Edward W. Knightly, "Modeling per-flow throughput and capturing starvation in CSMA multi-hop wireless networks," Proc. IEEE INFOCOM 2006.

7. Atul Adya, Paramvir Bahl, Jitendra Padhye, Alec Wolman, and Lidong Zhou, "A multiradio unification protocol for IEEE 802.11 wireless networks," in *BROADNETS '04: Proceedings of the 1st International Conference on Broadband Networks (BROADNETS'04)*, IEEE Computer Society, Washington, DC, 2004, pp. 344–354.

8. Shih-Lin Wu, Chih-Yu Lin, Yu-Chee Tseng, and Jang-Ping Sheu, "A new multi-channel MAC protocol with on-demand channel assignment for multi-hop mobile ad hoc networks," in *Proceedings of the International Symposium on Parallel Architectures, Algorithms and Networks, I-SPAN*, 2000, pp. 232–237.

9. Mathilde Benveniste, "The CCC mesh MAC protocol," IEEE 802.11-05/0610rl, 2005.

10. Jeonghoon Mo, Hoi-Sheung W. So, and Jean Walrand, "Comparison of multi-channel mac protocols," in *MSWiM'05: Proceedings of the 8th ACM International Symposium on Modeling, Analysis and Simulation of Wireless and Mobile Systems*, ACM Press, New York: 2005, pp. 209–218.

11. Kamal Jain, Jitendra Padhye, Venkata N. Padmanabhan, and Lili Qiu, "Impact of interference on multi-hop wireless network performance," in *MobiCom '03: Proceedings of the 9th Annual International Conference on Mobile Computing and Networking*, ACM Press, New York, 2003, pp. 66–80.

12. Arindam K. Das, Hamed M.K. Alazemi, Rajiv Vijayakumar, and Sumit Roy, "Optimization models for fixed channel assignment in wireless mesh networks with multiple radios," in Proc. *2nd Annual IEEE Communications Society Conference on Sensor and Ad Hoc Communications and Networks (SECON)*, Santa Clara, CA, September 2005, pp. 463–474.

13. Mahesh K. Marina and Samir R. Das, "A topology control approach to channel assignment in multi-radio wireless mesh networks," in *2nd International Conference on Broadband Networks (Broadnets)*, Boston, MA, October 2005.

14. Mohammad Mahdian, "On the computational complexity of strong edge coloring," *Discrete Applied Mathematics*, vol. 118, no. 3, pp. 239–248, 2002.

15. F. Tasaki, H. Tamura, M. Sengoku, and S. Shinoda, "A new channel assignment strategy towards the wireless mesh networks," *Proceedings 10th Asia-Pacific Conference on Communications and the 5th International*

Symposium on Multi-Dimensional Mobile Communications, 2004, vol. 1, pp. 71–75.

16. Douglas S.J. De Couto, Daniel Aguayo, John Bicket, and Robert Morris, "A high-throughput path metric for multi-hop wireless routing," in *MobiCom '03: Proceedings of the 9th Annual International Conference on Mobile Computing and Networking*, ACM Press. New York, 2003, pp. 134–146.

17. Richard Draves, Jitendra Padhye, and Brian Zill, "Routing in multi-radio, multi-hop wireless mesh networks," in *MobiCom '04: Proceedings of the 10th Annual International Conference on Mobile Computing and Networking*, ACM Press, New York, 2004, pp. 114–128.

18. Dimitri Bertsekas and Robert Gallager, *Data Networks*, 2nd ed., Prentice-Hall, Englewood Chiffs, NJ, 1992.

19. Yaling Yang, Jun Wang, and Robin Kravets, "Interference-aware load balancing for multihop wireless networks," Technical Reports UIUCDCS-R-2005-2526, Department of Computer Science, University of Illinois at Urbana-Champaign, 2005.

20. Yaling Yang, Jun Wang, and Robin Kravets, "Designing routing metrics for mesh networks," in *1st IEEE Workshop on Mesh Networks*, Santa Clara, CA, Sep. 2005.

21. A. Raniwala and Tzi-Cker Chiueh, "Architecture and algorithms for an IEEE 802.11-based multi-channel wireless mesh network," in *INFOCOM 2005. 24th Annual Joint Conference of the IEEE Computer and Communications Societies. Proceedings IEEE*, 2005, vol. 3, pp. 2223–2234.

22. Mansoor Alicherry, Randeep Bhatia, and Li (Erran) Li, "Joint channel assignment and routing for throughput optimization in multi-radio wireless mesh networks," in *MobiCom '05: Proceedings of the 11th Annual International Conference on Mobile Computing and Networking*, ACM Press, New York, 2005, pp. 58–72.

23. Pradeep Kyasanur and Nitin H. Vaidya, "Routing and link-layer protocols for multichannel multi-interface ad hoc wireless networks," SIGMOBILE Mobile Computing and Communications Review, November 2005.

3

IEEE 802.11-BASED
WIRELESS MESH NETWORKS

*Ashish Raniwala, Rupa Krishnan,
and Tzi-cker Chiueh*

CONTENTS

3.1   INTRODUCTION

IEEE 802.11 has become the de facto standard for home and enterprise deployment of wireless local area networks (LANs). Most of these deployments operate in the infrastructure mode, where a set of access points (APs) serve as communication hubs for mobile stations and provide entry points to the Internet. The current role of IEEE 802.11 is limited to mobile client to AP communication. Economies of scale make IEEE 802.11 a desirable alternative even to interconnect these APs by forming a wireless mesh network (WMN) as shown in Figure 3.1.

To enable such applications, IEEE 802.11 supports two additional modes of operation: the ad hoc mode for forming a single-hop ad hoc network where nodes communicate with each other directly without the use of an AP; and the wireless distribution system (WDS) mode for forming point-to-point AP relay links where each AP acts not only as a base station, but also as a wireless relay node. However, before IEEE

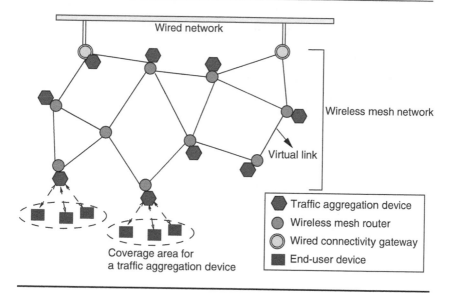

Figure 3.1 A wireless mesh network (WMN) core, which is connected to a wired network through a set of wired connectivity gateways. Each WMN node has a radio interface that is used to communicate with other WMN nodes over wireless links as shown. A WMN node is equipped with a traffic aggregation device (similar to an 802.11 access point) that interacts with individual mobile stations. The WMN relays mobile stations' aggregated data traffic to/from the wired network.

802.11 can be used to form an effective WMN, several performance, security, and management issues need to be addressed. From performance standpoint, low end-to-end network capacity and its unfair distribution among flows are two frequently mentioned problems with IEEE 802.11-based WMNs. In this chapter, we focus our attention to these performance issues and present state-of-the-art techniques proposed in literature to solve them.

The rest of the chapter is organized as follows. Section 3.2 discusses the capacity and fairness issues arising in IEEE 802.11-based WMNs. Section 3.3 presents various techniques developed to discover and utilize high-quality routes. Section 3.4 elaborates on state-of-the-art architectures and algorithms to enable use of multiple channels within a WMN. Section 3.5 focuses on solutions developed to achieve fairness on top of inherently unfair IEEE 802.11 MAC layer. Section 3.6 presents other performance related issues, while section 3.7 discusses the open research issues in use of IEEE 802.11 as platform for wireless mesh networking. Section 3.8 concludes the chapter with a discussion of IEEE 802.11s standardization activities.

3.2 PERFORMANCE ISSUES AND THEIR CAUSES

3.2.1 Limited Capacity

3.2.1.1 Protocol Overheads

Despite many advances in wireless physical-layer technologies, limited capacity remains a pressing issue even for single-hop wireless LANs: The advertised 54 Mbps bandwidth for IEEE 802.11a/g-based hardware is the peak link-level data rate. When all the overheads—medium access control (MAC) contention, 802.11 headers, 802.11 ACK and packet errors—are accounted for, the actual goodput available to applications is almost halved. In addition, the maximum link-layer data rate falls quickly with increasing distance between the transmitter and the receiver.

3.2.1.2 Interflow and Intraflow Interference

The bandwidth issue is even more severe for multihop WMNs where in order to keep the network connected all nodes operate over the same radio channel. This results in substantial interference between transmissions from adjacent nodes on the same path as well as neighboring paths reducing the end-to-end capacity of the network [1,2]. Figure 3.2 depicts an example of such interference.

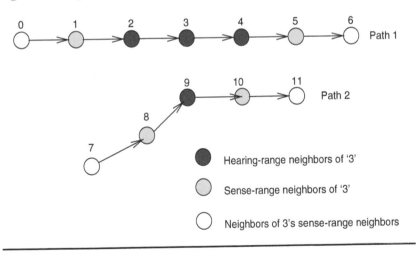

Figure 3.2   Intrapath and interpath interference in a single-channel multihop ad hoc network. Nodes 1, 2, 4, and 5 are in the interference range of node 3, and hence can not transmit/receive when node 3 is active. Nodes 8, 9, and 10 belonging to another node-disjoint path also fall in the interference range of node 3. Thus none of the wireless links shown in the figure can simultaneously operate when node 3 is transmitting to node 4.

3.2.1.3 Ineffective Route Selection

The simplest routing metric for WMNs is the hop-count metric. However, using the hop-count metric leads to suboptimal path selection. First, small hop count translates into longer, and hence more error-prone, individual hops [4]. Second, use of minimum hop count does not do anything to load-balance the traffic across the network [21]. This reduces the effective capacity of the WMN.

3.2.1.4 TCP's Control Overhead

The limited capacity problem is further compounded by transmission control protocol (TCP) that fails to effectively utilize the available bandwidth. First, TCP's reliance on ACK clocking requires it to send an acknowledgment every packet or every other packet. These end-to-end acknowledgments consume substantial network bandwidth (up to 20%) because of high fixed per-packet overhead in IEEE 802.11 wireless networks. Second, when a packet is lost on an intermediate hop, TCP's end-to-end strategy requires the retransmission to traverse the entire path all over again. This leads to wastage of bandwidth on all preceding hops where the prior transmissions of the very packet being retransmitted were successful. Table 3.1 estimates the overhead imposed by TCP ACKs by comparing TCP throughput with optimal user datagram protocol (UDP) throughput. Table 3.1 was obtained by conducting experiments on a one-hop IEEE 802.11a-based network.

Table 3.1 Relative Overhead of MAC Contention, PLCP Header, and Link-Layer ACK Increases as More Sophisticated Link-Layer Encoding Is Used. This Substantially Increases the Relative Overhead of TCP ACKs. Beyond 18 Mbps, TCP's DelACK Mechanism Kicks in, and Hence the TCP's ACK Overhead Does Not Increase Further

Link-Rate (Mbps)	TCP Thruput (Mbps)	Optimal UDP Thruput (Mbps)	ACK Overhead %
6	4.6	5.3	13.2
12	8.4	10.3	18.4
18	12.0	14.9	19.4
24	16.5	19.3	14.5
36	22.6	26.8	15.7
48	26.9	32.9	18.2

Table 3.2 TCP's Congestion Control Performance Degrades Significantly in Case of Channel Error-Induced Packet Drops. This Is Because TCP's Congestion Control Mistakes Channel Error-Induced Packet Drops as Sign of Network Congestion and Slows Down the Sender. The Channel Conditions Were Controlled by Changing the Transmission Power of the NICs

Channel Condition	TCP Thruput (Mbps)	Optimal UDP Thruput (Mbps)	TCP Underutilization (%)
Very bad	0.08	0.87	90.8
Bad	3.37	6.07	44.5
Average	14.5	18.6	22.0
Very good	26.9	32.9	18.2

3.2.1.5 Ineffective Congestion Control

TCP's congestion control relies on packet drops to detect network congestion. In wireless networks, however, packets are also dropped because of bit errors. TCP fails to distinguish between these frequent bit errors and true congestion, and inadvertently reduces its sending rate even in case of bit errors. Depending on the channel error conditions, this can lead to substantial underutilization of the network [3]. Table 3.2 shows the difference between bandwidth achieved by a TCP flow and an optimal flow running on an IEEE 802.11a-based one-hop network under various channel conditions.

3.2.2 Flow Unfairness

3.2.2.1 Hidden Terminal Problem

It is well-known that IEEE 802.11 MAC layer exhibits the hidden node problem [5] that causes one wireless link's transmission to be inhibited by another link. While the request-to-send/clear-to-send (RTS/CTS) messages in 802.11's MAC protocol effectively stop a hidden node from interfering with an ongoing communication transaction, they cannot prevent the hidden node from initiating its RTS/CTS sequence at inopportune times and subsequently suffering from long back-off delays. TCP exacerbates this unfairness problem because TCP senders further back off when their packets take a long time to get through the inhibited links. As a result, TCP flows traversing on an inhibited link could be completely suppressed in the worst case. Figure 3.3a depicts this scenario, and Figure 3.4 presents the results

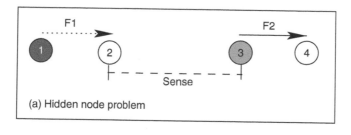

(a) Hidden node problem

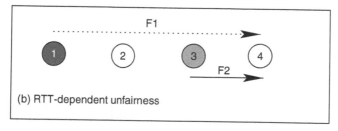

(b) RTT-dependent unfairness

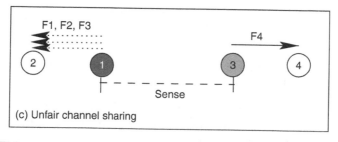

(c) Unfair channel sharing

Figure 3.3 **Three scenarios in which significant unfairness among flows arises. The wireless nodes getting a lesser than fair share of bandwidth are numbered 1 and 2, whereas the ones getting a larger share are numbered 3 and 4. (a) Node 1 lacks informations about node 3's transmissions, attempts its communication at inopportune times, and eventually backs off unnecessarily. (b) Flow F1 traverses more hops than flow F2. Some transport protocols, such as TCP, give more bandwidth to flow F2. (c) Flow F1, F2, F3, and F4 all share the same channel, but most transport protocols allocate more bandwidth to F4 than to others.**

from the corresponding experiment conducted on an IEEE 802.11a-based 4-node testbed.

3.2.2.2 Channel Sharing Problem

Existing transport protocols at best attempt to allocate a radio channel's bandwidth fairly among flows from a single node, rather than among all flows from all nodes that share the radio channel. As a result, a flow emanating from a node with fewer flows tends to get

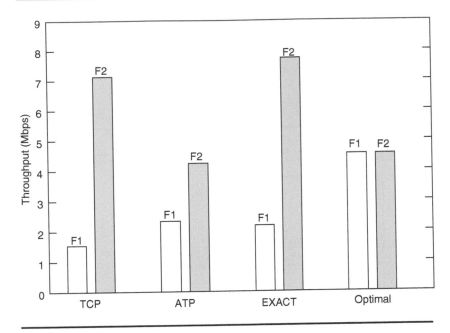

Figure 3.4 Amplification of the hidden node problem (Figure 3.3a) in existing wireless transport protocols.

a larger than fair share of channel bandwidth. Figure 3.3c depicts an example of this case, while Figure 3.5 presents the empirical results from the corresponding experiment.

3.2.2.3 RTT-Dependent Unfairness

TCP's fairness depends strongly on the round trip time (RTT) of the flows involved. Specifically, when two multihop TCP flows share the same wireless link, the flow traversing a fewer number of hops tends to acquire a higher share of bandwidth (Figure 3.3b). While this is true even for TCP operations on the wired Internet, the problem is much more frequent in a WMN. In a WMN, most of the traffic is directed to/from gateway nodes that connect the WMN to the wired Internet. As different WMN nodes are bound to be different hops away from these gateway nodes, an RTT-independent fairness model is essential for effective mesh operations (Figure 3.3b).

3.2.2.4 Bad Fish Problem

Even when two interfering links are not hidden from each other and have equal number of flows traversing them, IEEE 802.11 MAC allocates equal number of packet transmissions to each of them. However, these interfering links could be operating at vastly different link rates,

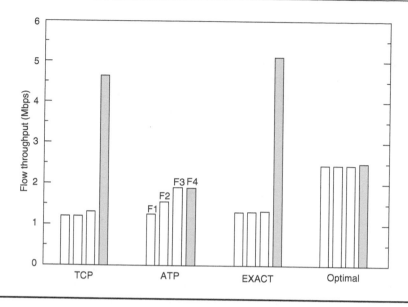

Figure 3.5 While existing transport protocols work effectively when the participating flows share some common intermediate node, they fail to allocate bandwidth fairly when flows share common radio channels (Figure 3.3c). The optimal bandwidth allocation was achieved by exhaustively trying different sending rates.

e.g., 1 and 11 Mbps. In such a case, the effective throughput of 11 Mbps link becomes limited by that of 1 Mbps link. We call this problem the bad fish problem. If instead the MAC layer allocates equal channel time to the two links, the 11 Mbps link would no longer be limited because of the interfering link operating at 1 Mbps link rate.

The above performance issues (Section 3.2.1.1 through Section 3.2.2.4) need to be tackled at many different fronts. First, to improve the overall capacity of the network, multiple channels need to be utilized within each WMN. Second, the routing layer needs to choose paths that effectively utilize this capacity by considering both the link qualities and the amount of traffic introduced by each of the flows. Finally, to make this network-layer bandwidth available to applications, the transport layer needs to perform fair and efficient allocation of bandwidth across competing flows. In the following sections, we discuss related works geared to improve network performance from each of these directions.

3.3 HIGH-PERFORMANCE ROUTING

Routing governs the flow of packets through the WMN. While shortest path routing minimizes the amount of network bandwidth used to transfer packets, it does not consider important factors such as link errors, link criticality, or channel diversity during path selection. Intelligent selection of paths based on these factors can not only improve the quality of path chosen for current stream of packets, but also potentially enable network to admit more packet load in future. In this section, we discuss several representative routing techniques proposed in the literature that aim to improve the end-to-end network throughput.

3.3.1 Link Quality-Aware Routing

In a wired network, it is usually safe to assume that a link either works well or does not work at all. The assumption does not hold for wireless networks, where a majority of the links have intermediate loss ratios [6]. Simply minimizing the hop count maximizes the distance traveled by each hop; this in turn minimizes the signal strength and thus maximizes the loss ratio of each hop.

Expected transmission count (ETX) is one of the first metrics that explicitly accounts for link quality during path selection [6]. The ETX of an individual link is defined as the expected number of transmissions (including retransmissions) it takes for a single packet to be successfully transmitted over that link. It is computed by the following equation:

$$\text{ETX} = \frac{1}{d_f \times d_r} \tag{3.1}$$

where d_f is the forward delivery ratio, i.e., the probability of successful reception of a single packet sent by the transmitter, and d_r is the reverse delivery ratio. The ETX of a route is defined as the sum of ETX of individual links composing the path. The ETX metric not only avoids low-quality (high ETX) links, but also prefers shorter (low aggregated ETX) paths to minimize network resource usage. The delivery ratios d_f and d_r are estimated through periodic exchange of HELLO packets among neighbors.

ETX ignores the fact that different links in the network could operate over different link rates and hence consume different amounts of channel time. Draves et al. [7] overcomes this limitation by using expected transmission time (ETT), which is computed as

$$\text{ETT} = \text{ETX} \times \frac{S}{B} \tag{3.2}$$

where S is the average size of the packet and B is the raw bandwidth of the link. While the true transmission time of a packet depends upon several factors such as back-off delay, the authors found that it is sufficient to use the raw bandwidth of the link to compute the expected transmission time.

3.3.2 Interference-Aware Routing

Neither ETX nor ETT captures the channel load along the path. These metrics could therefore end up choosing congested paths even if a lightly loaded path is available. Various routing protocols have been proposed that consider the traffic load during the path selection. Load-balanced ad hoc routing (LBAR) is one such protocol [9]. Here, the route discovery is done in a reactive manner, where the destination chooses the least loaded path. The path load is captured by aggregating the degree of nodal activity on the nodes composing the path. The degree of nodal activity for a node is in turn defined as the sum of number of active paths passing through the node and its interfering neighbors. This is given by the expression:

$$\text{Pathcost}_p = \sum_{i \in p} \left(A_i + \sum_{j \in N(i)} A_j \right) \tag{3.3}$$

where A_i is the number of active flows going through node i and $N(i)$ is the set of neighbors of i.

Dynamic load-aware routing (DLAR) similarly uses the number of packets queued in a node's interface queue to estimate its load [10]. The interface queue size in some sense measures how overloaded the channel of the outgoing link is. Hyacinth [21] approximates the channel load observed by a link by explicitly summing the traffic load (bytes per second or packets per second) imposed on that channel by all its interfering links.

Unlike topology-based routing, traffic-based routing may choose different paths between the same pair of nodes at different points in time. This characteristic of traffic-based routing can easily lead to route oscillations, where several sender nodes simultaneously detect and switch to an underloaded path. The latter path therefore becomes overloaded requiring all the senders to switch back. Route oscillation is a well-studied topic in wired networks and solutions there could be applied in WMN context as well.

3.3.3 Multipath Routing

Most routing protocols only utilize a single path to route packets between any communicating node pair. Another possibility is to discover and utilize multiple paths between a source and a destination node. The availability of redundant paths provides a degree of fault tolerance in case a node on the primary route fails. Further, the traffic can be distributed across multiple noninterfering paths to achieve better network load balancing or higher aggregate throughput.

Much of the work on ad hoc multipath routing has been done with the goal of improving fault tolerance. For instance, Lee and Gerla [8] perform explicit selection of multiple node-disjoint paths at the receiver of route request. Specifically, the receiver compares the path stamped on different route requests before selecting the ones to be advertised back to the sender. Similarly, Marina and Das [51] propose a multipath variant of AODV called AOMDV. Here, multiple disjoint paths are maintained for each destination. When the primary path fails an alternate path is available instantaneously. This eliminates the route rediscovery overhead. Mosko and Garcia-Luna-Aceves [13] explore the use of meshed or nondisjoint paths to increase path reliability. Their claim is that exploiting mesh connectivity increases path reliability since adding a loop-free path can never decrease reliability. Although node disjointness of routes provides tolerance against node failures, but does not necessarily ensure that the paths do not interfere. This limits the amount of spatial load balancing such schemes can achieve in the context of WMNs.

Roy et al. [12] argue that it may not be possible to find multiple paths that do not interfere with each other if the nodes are using omnidirectional antennas. They evaluate the impact of multipath routing on end-to-end throughput in two cases—one when the nodes are equipped with omnidirectional antennas, and another when they are equipped with directional antennas. Their simulation results substantiate their argument.

3.3.4 Diversity-Aware Routing

In context of multichannel WMNs, channel diversity of the paths becomes an important concern during the path selection. Specifically, to minimize intraflow interference (Figure 3.2), adjacent hops of a route should operate over different channels. One such metric that accounts for channel diversity in path selection is weighted cumulative ETT (WCETT) [7]. Formally, WCETT is defined as

$$\text{WCETT} = (1 - \beta) \times \sum_{i=1}^{n} \text{ETT}_i + \beta \times \max_{j=1}^{k} X_j \qquad (3.4)$$

where ETT_i is the expected transmission time of a packet over ith hop, X_j is the sum of transmission times over hops operating on channel j, and β is a tunable parameter between 0 and 1. The first term is used to limit the end-to-end delay incurred or the total network bandwidth consumed by each packet. The second term limits the number of hops that operate over the same channel. The authors' experiments suggest that WCETT is able to select channel-diverse routes.

WCETT suffers from two limitations. First, it does not take inter-flow interference into account. Paths whose nodes have more interfering neighbors would have worse performance. Second, WCETT is nonisotonic and that makes it unsuitable for use in link-state algorithms. A metric is said to be isotonic if it preserves the relative weight of two paths when both are prepended by a common path. Yang et al. [14] propose metric of interference and channel switching (MIC), an isotonic metric which also considers interflow interference. MIC is computed as

$$\text{MIC} = \frac{1}{N \times \min{(\text{ETT})}} \sum_{\text{link } l \in p} \text{IRU}_l + \sum_{\text{node } i \in p} \text{CSC}_i \qquad (3.5)$$

where

$$\text{IRU}_l = \text{ETT}_l \times N_l \qquad (3.6)$$

and

$$\text{CSC}_i = \begin{Bmatrix} w_1 \text{ if CH } (\text{prev}(i)) \neq \text{CH}(i) \\ w_2 \text{ if CH}(\text{prev}(i)) = \text{CH}(i) \end{Bmatrix} \qquad (3.7)$$

$0 < w_1 < w_2$, N is the total number of nodes and min(ETT) is the smallest ETT is the network. ETT_l is the ETT of a link, and N_l is the set of neighboring nodes interfering with link l. $\text{CH}(i)$ is the channel in which link l transmits, and $\text{prev}(i)$ is the channel of the previous hop along this path. IRU_l accounts for interflow interference and CSC_i represents self-interference when adjacent hops of a path operate in the same channel. However, this metric only accounts for intraflow interference among successive hops of the flow; in a realistic situation intraflow interference could extend to several hops. It is not clear if extending it to multiple prior hops would retain isotonicity of the metric.

3.3.5 Opportunistic Routing

Traditional routing protocols decide a sequence of nodes between the source and the destination and route every packet through that sequence. This routing framework fails to leverage the broadcast nature of wireless transmissions. Specifically, when an intermediate receiver fails to receive a packet, the packet has to be retransmitted by the immediate transmitter even if another neighbor of the receiver successfully received the packet. ExOR is an opportunistic routing mechanism that makes delayed forwarding decisions [11]. Specifically, ExOR broadcasts each packet on each hop, determines the set of nodes that actually received the packet, and then chooses the best receiver (that is closest to the final destination) to forward the packet. Since choosing the best receiver incurs communication overhead, delayed decisions are made for batch of packets. The advantage of delayed forwarding decisions is that ExOR can try multiple long but radio lossy links concurrently, resulting in high expected progress per transmission.

ExOR attempts to decrease the total number of transmissions, but does not explicitly leverage the multirate option at the physical layer that leads to transient variations in transmission rates. ROMER [15] another routing protocol based on this framework leverages such transient variations to select the highest throughput path instead of the closest receiver to the destination. Specifically, ROMER forms an opportunistic, forwarding mesh that is centered around the long-term stable, minimum-cost path (e.g., the shortest path or long-term minimum-delay path), but opportunistically expands or shrinks at the runtime to exploit the highest-quality, best-rate links.

3.4 MULTICHANNEL WIRELESS MESH NETWORKS

The IEEE 802.11b/g standards and IEEE 802.11a standard provide 3 and 23 nonoverlapped frequency channels, respectively, which could be used simultaneously within a neighborhood. Ability to utilize multiple channels substantially increases the effective bandwidth available to wireless network nodes. However, a conventional WMN architecture equips each node with a single interface, which is always tuned to a network-wide unique channel in order to preserve connectivity. To utilize multiple channels within the same network, either each node needs channel-switching capability [16–19] or it needs multiple interfaces each tuned to operate in a different channel [20,21]. Channel-switching requires fine-grained synchronization among nodes as to when any node will transmit/receive over a particular channel. One possible scheme is to have all the nodes switching between all the available channels in some predetermined order [50]. Here an interface switches between available channels in different slots, based on a random seed. Nodes wishing to communicate wait for a slot where their interfaces are in the same channel. These sequences are not fixed and can be altered if need arises. The advantage of this scheme is that traffic is load balanced across all available channels reducing overall interference. However such fine-grained synchronization is difficult to achieve without modifying the 802.11 MAC layer. Therefore, using multiradio wireless mesh routers is a more promising approach to form multichannel IEEE 802.11 WMNs.

The assignment of channels to radio interfaces plays an important role in harnessing the raw bandwidth capability of this multiradio architecture. For instance, an identical channel assignment to all nodes [7] would artificially limit the throughput improvement possible over single-radio architecture. Intuitively, the goal of channel assignment is to reduce interference by utilizing as many channels as possible within each neighborhood while maintaining the necessary connectivity between nodes. In this section, we discuss various techniques proposed to perform intelligent channel assignment.

3.4.1 Topology-Based Channel Assignment

Channel assignment can be done purely based on the network topology with the goal of minimizing the interference on any link. Channel assignment in this case reduces to a constrained graph coloring problem where links are colored by the channels constrained by the number of interfaces on a node. The problem is known to be computationally hard; therefore the proposed solutions are approximate. Marina and Das [30] take this topology-control approach to solve the channel assignment problem and propose a greedy solution to it. The algorithm, termed connected low interference channel assignment (CLICA), visits all the nodes in the order of their channel constraints; more constrained nodes are visited first. Upon visiting a node, the channels are chosen in a greedy fashion: the node picks locally optimal channels for each of its communication edges. The optimization goal is to minimize the maximum interference faced by any link. It should be noted that the visiting order changes dynamically to account for changing constraints on the nodes. It can however be shown that once a link is assigned a certain channel, it does not need to be readjusted in later steps.

Subramanian et al. [31] propose a search-based approximation solution, where the solution space is searched for a better solution until none is found for some number of iterations. It first uses a TABU search-based approximation algorithm to color each edge in the network graph with one of the K available colors such that the total number of conflicting edge pairs are minimized. In the next phase, edge-connected components are remerged as needed to satisfy the interface constraint on each node. Tang et al. [32] solve the problem of finding a channel assignment, such that if the original network forms a K-connected graph then the resulting network topology is also K-connected. Their objective is to minimize co-channel interference for any channel.

Das et al. [33] solve the channel assignment problem by formulating it as a linear program. Their objective is to find a channel assignment that maximizes the number of links that can be active simultaneously. The constraint of this linear program includes the number of interfaces on the node, and the fact that an interface cannot be assigned multiple channels. This problem turns out to be an integer linear program, since all the variables (channels, interfaces, interference) are all integer valued. Integer linear programs are exponential in complexity, hence the authors propose greedy heuristic algorithms to solve the problem.

3.4.2   Traffic-Aware Channel Assignment

A topology-based channel assignment is based on the premise that all network links are equally loaded. This premise does not hold true for a general traffic distribution, as some links are bound to carry more traffic than others. Intuitively, the goal of channel assignment in a multichannel WMN should be to bind each network interface to a radio channel in such a way that the available bandwidth on each virtual link is proportional to the load it needs to carry. As the link load is determined by routing algorithm, in an ideal solution the channel assignment is also performed in conjunction with the routing.

Even with complete information about network topology and traffic matrix, the channel assignment problem is NP-hard. We have proved this hardness property [20] by reducing minimum subset sum problem to the channel assignment problem. We propose a traffic-engineering solution that uses a greedy heuristic to select channels for individual network links. Specifically, given a network topology and expected link loads (based on routing), the heuristic visits all the links in decreasing order of their expected loads. Upon visiting an edge, the heuristic greedily assigns a locally optimal channel that minimizes the neighborhood interference. In each step, the heuristic attempts to maintain the previous channel assignment as constraints. The output of each run of channel assignment is fed to the routing algorithm that translates the channel assignment into bandwidth assignment and uses it to come up with more realistic routes. The routing algorithm is in turn used by the next iteration of channel assignment to come up with more informed selection of channels for each link. We experimentally prove that even with two interfaces on each node, this traffic-engineering solution achieves upto seven times improvement in network cross-section capacity when compared with a single-radio architecture.

Most theoretical work on traffic-aware channel assignment require use of omnipotent scheduler to achieve optimal MAC scheduling. For instance, Alicherry et al. [34] solve the joint problem of channel assignment and routing and devise a constant-factor approximation of the optimal. The authors follow a multistep procedure: The algorithm begins with an LP formulation for routing problem with inter-ference-free schedule based on the packing lemma. The channel assignment algorithm runs over this solution to achieve bandwidth equal to factor k-times required load at all the nodes. Finally, an LP is formulated to reduce the interference over the channel assignment and maximize the flow. It can be proven that this algorithm is within a constant factor of the optimal. The proof, however, relies on the assumption that the scheduling can be done by an omnipotent entity. The impact of using IEEE 802.11 scheduling in place of this optimal scheduler is not discussed.

3.4.3 Dynamic Channel Assignment

Unlike routing, limited research has been done to devise distributed schemes for channel assignment. A fundamental issue with distributed channel assignment is the dependency between the channels used by different nodes. In the general case, a single-channel assignment change can lead to a series of channel changes cascading through the entire network. This effect is demonstrated in Figure 3.6.

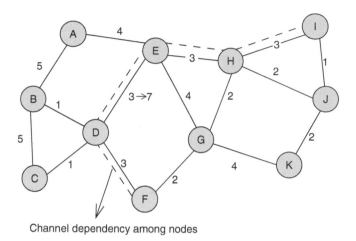

Channel dependency among nodes

Figure 3.6   This example shows how a change in channel assignment could lead to a series of channel reassignments across the network because of the channel dependency problem. In this example, when link D-E is changed from channel 3 to channel 7, link D-F, E-H, and H-I also need to change channels in order to maintain the constraint on number of interfaces per node.

We have proposed a distributed mechanism to assign channels [21]. Each node periodically exchanges channel usage information with its interfering neighbors, and constructs a channel-usage map of the neighborhood. Based on this map, the node assigns the least-used channel to its interfaces while coordinating the assignment with its immediate neighbors. To break the channel dependency, we utilize the tree structure of the network and ensure that the set of channels used by a node to communicate with the parent node is nonintersecting with the set of channels used to communicate with child nodes (Figure 3.7). Based on this restriction, each node independently assigns the channels to its downlinks, while its uplinks get assigned the same channels as its parent's downlinks.

Kyasanur and Vaidya [35] present another distributed channel assignment mechanism where a subset of each node's interfaces are assigned to fixed channels, while the remaining interfaces are switched across different channels based on immediate communication requirements. The fixed interfaces of a node are assigned the least-used channels in the neighborhood and this information is communicated to the node's neighbors. When a node needs to communicate with another, it temporarily switches one of its interface to the fixed channel of the other node.

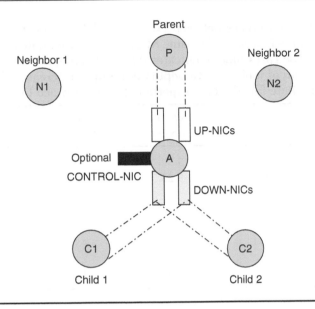

Figure 3.7 Eliminating the channel dependency problem by separating the set of NICs used in each WMN node into UP-NICs and DOWN-NICs so that any channel assignment change in a WMN node's DOWN-NICs does not affect its UP-NICs.

3.4.4 Inter-Channel Interference

Ideally, there should be no interference between nonoverlapping channels. This assumption, however, is not entirely true in practice. In our experiments with real 802.11 hardware, we observed substantial interference between two cards placed on the same machine despite operating on nonoverlapping channels. The extent of interference depends on the relative positions of the cards. Placing cards right on top of each other lead to maximum interference, and achieves only a maximum 20% gain in aggregate goodput over the single-channel case (Table 3.3.). If the cards are placed horizontally next to each other, as in Orinoco AP-1000 access points, the interference is minimum leading to almost 100% gain in aggregate goodput. In addition, the degradation due to interchannel interference was found independent of the guard band, i.e., the degradation was almost the same when channel 1 and 6 were used as compared to the case when channel 1 and 11 were used. We suspect this interference arises because of the imperfect frequency filter present in the commodity cards.

Table 3.3   Interference between Two Internal Antenna-Equipped 802.11b Cards Placed on the Same Machine and Operating on Channel 1 and 11. The Last Column Indicates the Total Goodput Achieved As a % of the Sum of Individual Goodputs Without Interference. The Link-Layer Data Rate for All These Experiments was Clamped to 11 Mbps

NIC-1 Action	NIC-2 Action	NIC-1 Goodput	NIC-2 Goodput	% of Max Goodput
Send	Silent	5.52	—	—
Recv	Silent	5.23	—	—
Silent	Send	—	5.46	—
Silent	Recv	—	5.37	—
Send	Send	2.44	2.77	47.6
Recv	Send	2.21	4.02	58.3
Send	Recv	4.22	2.42	61.0
Recv	Recv	4.02	1.89	55.8

This result has implication over the placement of multiple cards on the same machine. The electromagnetic leakage from the cards needs to be taken into account, and one card should not be placed in the zone where the strength of the leakage radiations by the other card is high. One possible way to achieve this is to use USB cards instead of PCI/PCMCIA cards and place them side-by-side in similar configuration as in Orinoco AP-1000 access points.

Another possibility is to equip cards with external antennas and place the external antennas slightly away from each other. Using external antennas alone may not suffice: it is also necessary that the internal antenna of the card is disabled. We used Orinoco Gold PCI adapters, that come with external antennas, that enabled us to build a multichannel WMN using standard PCs.

Yet another option is to use the upcoming Engim chipsets [52] which solve the interference problem at RF-level itself. Engim chipsets receive the complete spectrum, digitize it and process it to compensate for interchannel interference. This wideband spectral processing capability can help build single NIC with multichannel communication capability while introducing minimal interchannel interference.

3.5   FLOW FAIRNESS

The techniques discussed in Section 3.4.4 make the raw bandwidth of multiple IEEE 802.11 channels available to the network layer. The next step is to fairly and efficiently allocate this bandwidth among multiple competing flows. As discussed in Section 3.2, IEEE 802.11 MAC layer demonstrates substantial unfairness in several common scenarios. Without special mechanisms in place, MAC unfairness can not only lead to unfair application flow-level bandwidth allocation but also lead to starvation of some flows. In this section, we study a fairness model suitable for WMNs, and present several of the approaches used to provide flow fairness despite MAC layer unfairness.

3.5.1 Reference Model for Fairness

The notion of fairness itself is dependent on the model one chooses. Gambiroza et al. [38] propose a reference model for fairness in WMN that captures the following four objectives:

1. The granularity of fairness is an ingress-aggregated flow, independent of the number of TCP microflows or mobile devices supported by the ingress node. An ingress node would typically correspond to a single household or a single hotspot, and thus would lead to per-customer service granularity.
2. The spatial reuse of channel must be maximized by reclaiming the unused network bandwidth due to lack of demand on some nodes or other bottlenecks for the flows.
3. Spatial bias must be eliminated to ensure that nodes different hops away from the wired gateways still get similar end-to-end bandwidth shares.
4. Channel time rather than throughput should be considered as the basic network resource to be fairly shared. This objective tackles the bad fish problem discussed in Section 3.2.2.

3.5.2 Implicit Rate-Based Congestion Control

WTCP modifies TCP to utilize a rate-based congestion control [22]. It measures the ratio of inter-packet spacing on the receiver and that on the sender, to determine whether to increase or decrease the sending rate [22]. The instantaneous service rate of the bottleneck link along the path is reflected in the interpacket delay at the receiver. If the sending rate is lower than the available bandwidth, the received packets would maintain their interpacket spacing. Otherwise, the probe packets would queue up behind each other and their spacing would increase [22]. This approach assumes that all flows in the network are serviced in a strict round-robin fashion at the bottleneck links. This assumption does not hold in general on 802.11-based WMNs, because packet transmissions on wireless links tend to be bursty, and traffic bursts arriving at a bottleneck link may be serviced without any interleaving.

3.5.3 Explicit Rate-Based Congestion Control

Ad hoc transport protocol (ATP) [23] follows a cross-layer approach in its core design. In ATP scheme, every intermediate node measures queuing and transmission delays for each packet passing through it. The sum of exponentially averaged queuing and transmission delays yields the average packet service time experienced by all the flows going through the intermediate node. Assuming that each flow has just one packet in the bottleneck link queue, the service time reflects the ideal dispatch interval for all the flows competing over the bottleneck link. Every packet bears the maximum service time encountered on any of the intermediate hops. The bottleneck service time is communicated to the sender which adjusts its packet dispatch interval to match this service time. Contrary to the simulation results, our experiments suggest that an ATP sender fails to utilize the network optimally [44]. This is because ATP includes packet queuing time into the service time measurements that makes it unable to maintain a steady flow rate [45].

EXACT [24] is another explicit rate-based flow control scheme. EXACT routers measure the available bandwidth as the inverse of per-packet MAC contention and transmission time. Each router then runs a proportional max–min fair bandwidth sharing algorithm to divide this measured bandwidth among the flows passing through it. The bandwidth stamping mechanism is similar to ATP. Unlike ATP, EXACT decouples queuing time from the service time measurements and thus achieves much better network utilization.

3.5.4 Ingress Flow Throttling

None of the above congestion control mechanisms addresses the hidden terminal and channel sharing problem discussed in Section 3.2. A promising approach to achieve flow fairness is to throttle flows at their ingress nodes to their network-wide fair shares. Gambiroza et al. [38] provide a rough sketch of such an algorithm. Here, each WMN router measures the average offered load arriving from/to its own mobile users, as well as the average capacity of the links to each of its adjacent routers. The offered loads and link capacities are periodically communicated to other WMN routers. For each link, a router then computes the aggregate time shares in each of the link's contention neighborhoods and chooses the minimum value. The minimum time-share is then converted into rate and transmitted to other routers. Finally, each ingress router determines its end-to-end flow rate, and throttles its flows to this system-wide fair rate. The authors propose to apply this algorithm at layer 2 of each router, as it keeps TCP intact and works even in the presence of uncontrolled transport flows such as UDP.

3.5.5 Neighborhood RED

Xu et al. [29] improve TCP fairness by extending the idea of random early detection (RED) to multihop wireless networks. Simply applying RED to individual node's packet queue does not work because of the shared channel space. Therefore, the authors view the queues on a node and its interfering neighbors as a single distributed queue and apply RED to this distributed queue. The node infers the neighborhood queue size by monitoring the channel utilization. When the utilization goes above a certain threshold, neighborhood congestion is detected. The node then computes its own drop probability, and broadcasts this information to its neighbors. Based on this information, the nodes with highest contribution to the neighborhood channel congestion drop packets from their local queues.

3.5.6 Overlay MAC Layer Approach

The unfairness problems of IEEE 802.11 MAC could be addressed by making it aware of application's priorities. Rao and Stoica [25] propose another approach of forming an overlay MAC layer on top of 802.11 MAC. Based on loose time synchronization and application assigned weights, the overlay MAC runs a distributed time division multiple access (TDMA) algorithm and allocates time slots to competing nodes. Weighted fair queuing is used to assign more time slots to nodes having greater number of flows. Every node in the network decides a set of time slots for all other nodes in the network using a random number generator with the same seed. This allows for a distributed algorithm. The packets are then trickled from the overlay MAC to 802.11 MAC based on this time slot allocation. The advantage of this approach is that it can perform fair bandwidth allocation on top of standard 802.11 MAC. However this scheme faces a common disadvantage faced by most TDMA schemes. It incurs overhead when the owner of a slot has nothing to send. They attempt to reduce the overhead using a timer after which the slot is taken over by the next node.

3.6 OTHER ISSUES

3.6.1 Quality of Service

Many popular applications such as voice-over-IP and video-over-IP have specific end-to-end delay and bandwidth requirements that can not be satisfied by best-effort service. Unlike wired networks, WMN links in a radio neighborhood share the same channel and hence it is not sufficient to just differentiate and prioritize traffic at per-router level. A mechanism is needed to prioritize traffic across nodes. Kanodia et al. [40] devise distributed priority scheduling, where the priority tag of a node's head-of-line packet is piggybacked onto handshake and data packets. Each node also monitors all the transmitted packets and maintains the priority of all nodes in its neighborhood in a scheduling table, and uses the IEEE 802.11 priority back-off scheme to approximate this idealized schedule. To deal with congestion and link errors, the authors further devise multihop coordination in which downstream nodes can increase a packet's relative priority to make up for excessive delays incurred upstream. The authors [40] however suggest modifications to IEEE 802.11 protocol to incorporate these techniques. We believe that both these schemes can indeed be implemented on top of unmodified 802.11 using a mechanism similar to overlay MAC layer [25]. Such a combined solution that would be able to provide differentiated service to multimedia flows traversing an 802.11-based WMN is not yet developed.

3.6.2 Topology Planning

While the position of nodes in a WMN could be chosen randomly and the network would adapt to that given topology, a more planned approach could use the flexibility in initial positioning of nodes to optimize the network performance. Specifically, there are two variables here—the position of regular nodes and the placement of the gateway nodes. The gateway placement problem has been addressed [41,42]. Nodes are grouped into clusters, and a set of cluster heads are selected as the gateways such that the delay and bandwidth requirements of all nodes in the network are met [42]. This problem is addressed by first finding the minimal number of disjoint clusters connecting all the nodes, and then further subdividing the clusters that violate quality of service (QoS). Both the phases are shown to be

NP-hard, and constant-factor approximation algorithms are proposed for each of them. Unlike [42], [41] starts with a set of possible gateways' locations. The goal is to place a minimum number of gateways in these locations such that the bandwidth requirement of every ingress is met. The authors propose greedy approach which ensures that the choice of gateways satisfies the demands of all nodes by bounding the shortest path between any node and the gateway serving it.

The node placement problem is addressed by Basu et al. [43], where network performance is optimized by modifying the existing topology incrementally. The authors model the mean end-to-end delay of a network based on the queue length on individual nodes and the overall network topology, and come up with algorithms that move existing nodes and choose the best possible placement of a new node such that the mean end-to-end delay is minimized.

Each of these solutions addresses a part of the general problem which is to decide the optimal placement of a fraction of regular and gateway nodes, while the placement of remaining ones is given as input.

3.6.3 Advanced Topology Discovery

The most fundamental element of any network management is the ability to visualize the location and interconnectivity of nodes as well as the utilization of network resources (in this case wireless channels) at each location. This is especially important for larger scale WMN as they need to coexist with, rather than compete with, other residential and office wireless LANs. The three research questions here are the following. First, how to utilize multiple radios (and potentially smart antennas) equipped on each mesh node to perform accurate multihop locationing of the entire network? Second, how to determine the accurate radio topology of a network including each node's interference range neighbors? Third, how to produce a comprehensive RF usage map of the network using the inherent RF-sensing capability of the individual mesh node? Apart from network visualization, this information is also useful for network optimizations such as deciding which channels to use on which router.

Localization of network nodes has been a widely researched area. Geographical positioning systems based on satellite information can be used outdoors to give accurate location information. However, presence of obstacles and interference make indoor positioning a challenging task. Haeberlen et al. [46] provide locationing support for mobile clients as they move through a large-scale 802.11 AP deployment. Here, a mobile client sends out the probe requests to the nearby APs and the signal intensities of probe responses received are compared with prior measurements of coarse signal strength distributions in the building. This information is then used to localize the node. Several researchers have investigated the problem of localization in a multihop setting [47,48]. Here the positions of some of the nodes are given as input, while the positions of remaining nodes are determined based on relative signal strengths between nodes. A detailed survey of these techniques is beyond the scope of this chapter.

Finding interference relations between different links is another problem unique to wireless networks. Several interference models have been proposed in literature, which try to model real world interference characteristics. The protocol model assumes that interference is an all or none phenomenon and is some function of distance, while in the physical model interference is measured based on its impact on signal-to-noise ratio [1]. In real world, the time varying nature of the wireless medium makes it harder to compute interfering links. One empirical technique is to determine the interference relationships at startup [49] by measuring the throughput at a receiver for pairs of communicating senders and comparing it with the throughput for each node separately. If throughput is lower in the first case one can conclude that the links interfere. This technique fails to capture the dynamic nature of interference and can only be performed at startup when no other data is traversing the links. De et al. [44] have proposed another scheme to find interfering links. The idea is that a node may not be able to hear transmission from interfering neighbors but can sense them. Empirically, for 802.11a cards if the transmission rate is set to the lowest, and the transmit power is increased by 2 dBm, sense-range neighbors become hearing-range neighbors. Hence sending a broadcast ping at modified power/rate settings and identifying the nodes which respond gives a good approximation of interfering nodes at original power/rate settings.

3.6.4   Long-Distance Wireless Mesh Networks

In recent times long distance mesh networking has been explored for low-cost connectivity in rural areas and in inhospitable terrains. These types of networks are characterized by links that are long distance typically covering several miles, high gain antennas which are used to increase coverage, and multiple radios per node. This model affects different layers of the protocol stack. The MAC protocol has to account for long delays, and higher layers need to be modified if applications such as VoIP are to be deployed on these networks.

One such project is the Digital Gangetic Plains [28] where a long distance mesh network has been set up covering 12 rural villages in India. The longest link extends around 39 km and is achieved using directional antennas and specialized protocols. Long distance links require certain modifications to the 802.11 MAC settings such as increasing the period for an ACK timeout and modification of contention window slot times to accomodate larger link delays. Unfortunately, a carrier sense multiple access (CSMA)-based MAC is not a good fit for this application, since contention resolution is not a major factor and a CSMA MAC would not allow simultaneous operation of multiple directional antennas as it would perceive it as interference. To overcome this limitation, they propose use of a TDMA MAC and come up with a synchronization protocol called SynOP to allow simultaneous operation of multiple antennas per node.

3.7 OPEN ISSUES

The techniques described in this chapter can be used to build a high-capacity WMN using commodity IEEE 802.11 hardware. However, before such networks can become part of mainstream deployments, several other issues need to be addressed. These issues can be broadly categorized into fairness, security, QoS and network management. Many of these issues are generic to any WMN and not just IEEE 802.11-based WMNs. In this section, we focus on those that are specific to IEEE 802.11 wireless mesh networking.

3.7.1 Max–Min Flow Allocation

As discussed earlier, IEEE 802.11 MAC layer exhibits unfair distribution of bandwidth in several common scenarios. Further the bandwidth allocation is not easily controlled by higher layer protocols. The only control that higher layer has is when to release packets. Overlay MAC layer [25] utilizes this fact to alleviate some of the fairness issues with IEEE 802.11 WMNs. However, a complete solution that can achieve max–min fairness on top of unfair 802.11 MAC is yet to be found. Such a solution needs to deal with all the scenarios discussed in Section 3.2.

3.7.2 Interference-Aware Multipath Routing

Most of the multipath routing work has been done with the goal of improving network fault tolerance. Multipath routing can however also improve end-to-end throughput between two nodes if the paths are noninterfering. Absence of interference is essential to ensure that multiple paths can simultaneously transport packets. Given a network, there is no distributed mechanism available that can be used by a source node to discover multiple noninterfering paths to a given destination.

3.7.3 Directional Antenna-Based Mesh Networks

The use of IEEE 802.11 has even been extended to build long-distance WMNs that can be used to provide Internet connectivity to rural areas [28]. Digital Gangetic Plains [28] already deal with many of the IEEE 802.11 assumptions that break down in such settings. The problem of using directional antennas in a general IEEE 802.11 WMN is not fully addressed [26,27]. Specifically, given a WMN where each node is equipped with multiple interfaces each attached to a steerable directional antenna, a topology determination algorithm that determines the direction and channel of each interface is not yet known.

3.7.4 Secure Routing Protocols

Security of IEEE 802.11-based WMN remains an open problem. Although use of cryptography can address most of the outsider attacks, it cannot prevent attacks from compromised mesh nodes. Specifically, a single compromised mesh node can redirect all the traffic of a mesh network toward itself forming a black hole. Research on byzantine fault tolerance of routing protocols is a topic pursued by many researchers [36,37]. However, most of these protocols utilize cryptographic techniques. Another approach to this problem is to introduce redundant computation in the routing protocol. How to utilize redundancy to secure the routing protocol, detect compromised nodes, and isolate them is a research question yet to be answered.

3.7.5 Fault Diagnosis

Automated fault detection and diagnosis has been the holy grail of network management. The problem is complicated in WMN as it is nontrivial to distinguish whether the poor performance is because of bad channel conditions, interference, software bugs, faulty hardware, or compromised nodes. The only work we are aware of in this direction is by Qiu et al. [39]. Here, the authors feed the observed network states such as channel conditions, traffic loads, and network topology to a network simulator. Next, a set of potential network faults are simulated, and the simulated performance is matched with the real workload to determine the actual fault that might have occurred.

3.8 CONCLUSION

Economies of scale make IEEE 802.11 a promising technology for building WMNs. The focus of this chapter had been on understanding the capacity and fairness issues in such networks. Specifically, we surveyed several state-of-the-art routing and transport layer techniques proposed to address these issues.

Substantial research and commercial activities has prompted the formation of a new IEEE 802.11s committee. The goal of the committee is to develop an IEEE 802.11 extended service set (ESS) mesh that would be built on top of the current 802.11a/b/g standards using the IEEE 802.11 WDS. The nodes will be able to automatically discover each other and form mesh networks that support both broadcast/multicast and unicast delivery using radio-aware metrics. For security, all of the APs will be controlled by a single logical administrative entity, and an IEEE 802.11i-based mechanism will be used. QoS standards might also be built into the standards to enable the network to prioritize among different classes of traffic.

REFERENCES

1. P. Gupta, P.R. Kumar. "The Capacity of Wireless Networks," *IEEE Transactions on Information Theory*, Vol. 46, No. 2, 2000, pp. 388–404.
2. J. Jun, M.L. Sichitiu. "The Nominal Capacity of Wireless Mesh Networks," Wireless Communications, *IEEE*, Vol. 10, No. 5, 2003, pp. 8–14.
3. Saad Biaz, Nitin H. Vaidya. "Discriminating Congestion Losses from Wireless Losses Using Inter-Arrival Times at the Receiver," *IEEE Symposium ASSET*, 1999.
4. S. Douglas, J. De Couto, Daniel Aguayo, Benjamin A. Chambers, Robert Morris. "Performance of Multihop Wireless Networks: Shortest Path Is Not Enough", *Proceedings of HotNets*, 2002.
5. S. Xu, Saadawi T. "Does the IEEE 802.11 MAC Protocol Work Well in Multihop Wireless Ad Hoc Networks?" *Communications Magazine, IEEE*, Vol. 39, No. 6, 2001, pp. 130–137.
6. S. Douglas, J. De Couto, Daniel Aguayo, John Bicket, Robert Morris. "A high-Throughput Path Metric for Multi-Hop Wireless Routing," *Proceedings of ACM Mobicom*, 2003.
7. R. Draves, J. Padhye, B. Zill. "Routing in Multi-radio, Multi-hop Wireless Mesh Networks," *Proceedings of ACM MobiCom*, 2004.
8. S. Lee, M. Gerla. "Split Multipath Routing with Maximally Disjoint Paths in Ad Hoc Networks," *Proceedings of the IEEE ICC*, 2001, pp. 3201–3205.

9. H. Hassanein, A. Zhou. "Routing With Load Balancing in Wireless Ad Hoc Networks," *Proceedings of ACM International Workshop on Modeling, Analysis, and Simulation of Wireless and Mobile Systems*, 2001.

10. S.J. Lee, M. Gerla. "Dynamic Load-Aware Routing in Ad Hoc Networks," *Proceedings of Third IEEE Symposium on Application-Specific Systems and Software Engineering Technology (ASSET)*, 2000.

11. Sanjit Biswas, Robert Morris. "Opportunistic Routing in Multi-Hop Wireless Networks," *Proceedings of ACM SIGCOMM*, 2005.

12. Siuli Roy, Somprakash Bandyopadhyay, Tetsuro Ueda, Kazuo Hasuike. "Multipath Routing in Ad Hoc Wireless Networks with Omni Directional and Directional Antenna: A Comparative Study," *Proceedings of International Workshop on Distributed Computing*, 2002.

13. Marc Mosko, J.J. Garcia-Luna-Aceves. "Multipath Routing in Wireless Mesh Networks," *Proceedings of IEEE Workshop on Wireless Mesh Networks (WiMesh)*, 2005.

14. Yaling Yang, Jun Wang, Robin Kravets. "Designing Routing Metrics for Mesh Networks," *Proceedings of IEEE Workshop on Wireless Mesh Networks (WiMesh)*, 2005.

15. Yuan Yuan, Hao Yang, Starsky H.Y. Wong, Songwu Lu, William Arbaugh. "ROMER: Resilient Opportunistic Mesh Routing for Wireless Mesh Networks," *Proceedings of IEEE Workshop on Wireless Mesh Networks (WiMesh)*, 2005.

16. Y. Liu, E. Knightly. "Opportunistic Fair Scheduling over Multiple Wireless Channels," *Proceedings of IEEE INFOCOM*, 2003.

17. J. So, N. Vaidya. "Multi-Channel MAC for Ad Hoc Networks: Handling Multi-Channel Hidden Terminals Using A Single Transceiver," *Proceedings of MobiHOC*, 2004.

18. A. Tzamaloukas, J.J. Garcia-Luna-Aceves. "A Receiver-Initiated Collision-Avoidance Protocol for Multi-channel Networks," *Proceedings of Infocom*, 2001.

19. A. Nasipuri, S. Das. "A Multichannel CSMA MAC Protocol for Mobile Multihop Networks," *Proceedings of IEEE Wireless Communications and Networking Conference (WCNC)*, 1999.

20. Ashish Raniwala, Kartik Gopalan, Tzi-cker Chiueh. "Centralized Algorithms for Multi-Channel Wireless Mesh Networks," *ACM Mobile Computing and Communications Review (MC2R)*, April 2004.

21. Ashish Raniwala, Tzi-cker Chiueh. "Architecture and Algorithms for an IEEE 802.11-based Wireless Mesh Network," *Proceedings of IEEE Infocom*, 2005.

22. Prasun Sinha, Narayanan Venkitaraman, Raghupathy Sivakumar, Vaduvur Bharghavan. "WTCP: A Reliable Transport Protocol for Wireless Wide-Area Networks," *Proceedings of ACM Mobicom*, August 1999.

23. Karthikeyan Sundaresan, Vaidyanathan Anantharaman, Hung-Yun Hsieh, Raghupathy Sivakumar. "ATP: A Reliable Transport Protocol for Ad Hoc Networks," *Proceedings of ACM Mobihoc*, June 2003.

24. K. Chen, K. Nahrstedt, N. Vaidya. "The Utility of Explicit Rate-Based Flow Control in Mobile Ad Hoc Networks," *Proceedings of IEEE Wireless Communications and Networking Conference (WCNC)*, 2004.

25. Ananth Rao, Ion Stoica. "An Overlay MAC Layer for 802.11 Networks," *Proceedings of USENIX Mobisys*, 2005.

26. Gupqing Li, Lily Yang, W. Steven Conner, Bahar Sadeghi. "Opportunities and challenges in Mesh Networks Using Directional Anntenas," *Proceedings of IEEE Workshop on Wireless Mesh Netowrks (WiMesh)*, 2005.

27. Robert Vilzmann, Christian Bettstetter, Christian Hartmann. "On the Impact of Beam-Forming on Interference in Wireless Mesh Networks," *Proceedings of IEEE Workshop on Wireless Mesh Netowrks (WiMesh)*, 2005.

28. P. Bhagwaty, B. Ramanz, D. Sanghi. "Turning 802.11 Inside-Out," *Proceedings of HotNets*, 2003.

29. K. Xu, M. Gerla, L. Qi, Y. Shu. "Enhancing TCP Fairness in Ad Hoc Wireless Networks Using Neighborhood RED," *Proceedings of ACM MOBICOM*, 2003.

30. Mahesh Marina, Samir Das. "A Topology Control Approach to Channel Assignment in Multi-Radio Wireless Mesh Networks," *Proceedings of IEEE Broadnets*, 2005.

31. Anand Subramanian, Rupa Krishnan, Samir Das, Himanshu Gupta. "Minimum Interference Channel Assignment in Multi-Radio Wireless Mesh Networks," Poster Session at ICNP, International Conference on Network Protocols, 2005.

32. J. Tang, G. Xue, W. Zhang. "Interference Aware Topology Control and Qos Routing in Multi-Channel Wireless Mesh Networks," *Proceedings of ACM Mobihoc*, 2005.

33. Arindam K. Das, Hamed M.K. Alazemi, Rajiv Vijayakumar, Sumit Roy. "Optimization Models for Fixed Channel Assignment in Wireless Mesh Networks with Multiple Radios," *Proceedings of Second Annual IEEE Communications Society Conference on Sensor and Ad Hoc Communications and Networks*, 2005.

34. Mansoor Alicherry, Randeep Bhatia, Li (Erran) Li. "Joint Channel Assignment and Routing for Throughput Optimization in Multi-radio Wireless Mesh Networks," *Proceedings of ACM MOBICOM*, 2005.

35. Pradeep Kyasanur, Nitin H. Vaidya. "Routing and Interface Assignment in Multi-Channel Multi-Interface Wireless Networks," *Proceedings of IEEE Wireless Communications and Networking Conference (WCNC)*, 2005.

36. Yih-Chun Hu, David B. Johnson, Adrian Perrig. "SEAD: Secure Efficient Distance Vector Routing for Mobile Wireless Ad Hoc Networks," *Proceedings of IEEE Workshop on Mobile Computing Systems and Applications*, 2002.

37. Yih-Chun Hu, Adrian Perrig, David B. Johnson. "Ariadne: A Secure On-Demand Routing Protocol for Ad Hoc Networks," *Proceedings of ACM MobiCom*, 2002.

38. Violeta Gambiroza, Bahareh Sadeghi, Edward Knightly. "End-to-End Performance and Fairness in Multihop Wireless Backhaul Networks," *Proceedings of ACM MobiCom*, 2004.

39. Lili Qiu, Paramvir Bahl, Ananth Rao, Lidong Zhou. "Troubleshooting Multihop Wireless Networks," *Proceedings of ACM Sigmetrics*, 2005.

40. Vikram Kanodia, Chengzhi Li, Ashutosh Sabharwal, Bahareh Sadegh, Edward Knightly. "Distributed Multi-Hop Scheduling and Medium Access with Delay and Throughput Constraints," *Proceedings of ACM MOBICOM*, 2001.

41. Ranveer Chandra, Lili Qiu, Kamal Jain, Mohammad Mahdian. "Optimizing the Placement of Integration Points in Multi-Hop Wireless Networks," *Proceedings of ICNP, International Conference on Network Protocols*, 2004.

42. Yigal Bejerano. "Efficient Integration of Multi-Hop Wireless and Wired Networks with QoS Constraints," *Proceedings of MobiCom*, 2002.

43. Anindya Basu, Brian Boshes, Sayandev Mukherjee, Sharad Ramanathan. "Network Deformation: Traffic-Aware Algorithms for Dynamically Reducing End-to-End Delay in Multihop Wireless Networks," *Proceedings of MobiCom*, 2004.

44. Pradipta De, Rupa Krishnan, Ashish Raniwala, Krishna Tatavarthi, Nadeem Syed, Srikant Sharma, Tzi-cker Chiueh. "MiNT-m: An Autonomous Mobile Wireless Experimentation Platform," *Proceedings of Fourth International Conference on Mobile Systems, Applications, and Services (Mobisys)*, 2006.

45. Raj Jain, Shiv Kalyanaraman, Ram Viswanatha. "Rate Based Schemes: Mistakes to Avoid," ATM Forum/94-0882, September 1994.

46. Andreas Haeberlen, Eliot Flannery, Andrew Ladd, Algis Rudys, Dan S. Wallach, Lydia E. Kavraki. "Practical Robust Localization over Large-Scale 802.11 Wireless Networks," *Proceedings of Mobicom*, 2004.

47. D. Niculescu, B. Nath. "Error Characteristics of Ad Hoc Positioning Systems (APS)," *Proceedings of ACM MobiHoc*, 2004.

48. R. Bischoff, R. Wattenhofer. "Analyzing Connectivity-Based MultiHop Ad-Hoc Positioning," *Proceedings of the Second Annual IEEE International Conference on Pervasive Computing and Communications (PerCom)*, 2004.

49. Jitu Padhye, Sharad Agarwal, Venkat Padmanabhan, Lili Qiu, Ananth Rao. Brian Zill. "Estimation of Link Interference in Static Multi-Hop Wireless Networks," Short paper in IMC, 2005.

50. Paramvir Bahl, Ranveer Chandra, John Dunagan. "SSCH: Slotted Seeded Channel Hopping for Capacity Improvement in IEEE 802.11 Ad-Hoc Wireless Networks," *Proceedings of Mobicom*, 2004.

51. M.K. Marina, S.R. Das. "Ad Hoc On-Demand Multipath Distance Vector Routing," *Proceedings of IEEE ICNP*, 2001.

52. "Engim Corp"; http://www.engim.com.

PART II

PROTOCOLS

4

ROUTING IN WIRELESS MESH NETWORKS

Michael Bahr, Jianping Wang, and Xiaohua Jia

CONTENTS

4.1 INTRODUCTION

Wireless mesh networks (WMNs) are a new trend in wireless communication promising greater flexibility, reliability, and performance over conventional wireless local area networks (WLANs). Wireless mesh network players are growing in number and there are several "wireless mesh companies" that already sell solutions for WMNs. Moreover, there is broad support for mesh networks in standardization groups such as IEEE 802.11, IEEE 802.15, and IEEE 802.16.

Wireless mesh networking and mobile ad hoc networking use the same key concept—communication between nodes over multiple wireless hops on a meshed network graph. However, they stress different aspects. Mobile ad hoc networks (MANETs) have an academic background and focus on end user devices, mobility, and ad hoc capabilities. WMNs have a business background and mainly focus on static (often infrastructure) devices, reliability, network capacity, and practical deployment. Nevertheless, one can often find both terms or their variations together in many descriptions or articles on this topic.

The core functionality of wireless multihop ad hoc networking as well as Wireless mesh networks is the routing capability. Routing protocols provide the necessary paths through a WMN, so that the nodes can communicate on good or optimal paths over multiple wireless hops. The routing protocols have to take into account the difficult radio environment with its frequently changing conditions and should support a reliable and efficient communication over the mesh network.

Since WMNs share common features with wireless ad hoc networks, the routing protocols developed for MANETs can be applied to WMNs. For example, Microsoft Mesh Networks [1] are built based on Dynamic Source Routing (DSR) [2], and many other companies, e.g., [3] are using Ad hoc On-demand Distance Vector (AODV) routing [4]. Sometimes, the core concepts of existing routing protocols are extended to meet the special requirements of wireless mesh networks, for instance, with radio-aware routing metrics as in the IEEE 802.11s WLAN mesh networking standardization.

Despite the availability of several routing protocols for ad hoc networks, the design of routing protocols for WMNs is still an active research area for several reasons [5]:

- In most WMNs, many of the nodes are either stationary or have minimum mobility and do not rely on batteries. Hence, the focus of routing algorithms is on improving the network throughput or the performance of individual transfers, instead of coping with mobility or minimizing power usage.
- The distance between nodes might be shortened in a WMN, which increases the link quality and the transmission rate. However, short distances also increase the interference among hops, which decreases the available bandwidth on each link. Therefore, new routing metrics need to be discovered and utilized to improve the performance of routing protocols in a multiradio multihop WMN.
- In a multiradio/multichannel WMN, the routing protocol not only needs to select a path among different nodes, but also needs to select the most appropriate channel or radio on the path for each mesh node. Therefore, routing metrics need to be discovered and utilized to take advantage of multiple radios in a wireless mesh network.
- In a WMN, cross-layer design becomes a necessity because the change of a routing path involves the channel or radio switching in a multiradio multichannel mesh node.

This chapter focuses on routing in wireless mesh networks which considers reliability in the wireless environment as well as performance and capacity improvements, without losing the ties to more general routing concepts. It is not possible to give a comprehensive overview of all routing schemes due to limited space. The special properties of WMNs are discussed in Section 4.2. Section 4.3 introduces the general concepts of routing. Section 4.4 discusses different routing metrics in wireless mesh networks. Section 4.5 describes several routing protocols for wireless mesh networks. Section 4.6 presents the routing protocol HWMP that is currently standardized in IEEE 802.11s. And Section 4.7 introduces joint routing and channel assignment for WMNs.

4.2 SPECIAL PROPERTIES OF WIRELESS MESH NETWORKS

Wireless mesh networks forward data packets over multiple wireless hops. Each mesh node acts as relay point/router for other mesh nodes. WMNs can often be found in commercial usage scenarios, i.e., someone sells the mesh network. The most prominent ones are public access networks and municipal wireless networks, where access points are the nodes of the wireless mesh networks.

Reliability and network performance are the important goals for WMNs, especially in the challenging wireless environment. Mobility of mesh nodes is usually not considered ($v = 0$). Static nodes can be mounted on lamp poles, attached to houses, etc., where there is sufficient power supply. With these assumptions and the usage scenario in mind, the wireless ad hoc routing protocols can be optimized with respect to reliability and network performance. They may be extended to use special routing metrics and might even be located on layer 2 to have better access to the information of the MAC and the physical layers.

Mesh nodes can have multiple wireless interfaces in order to increase the capacity of the mesh network. Multiple interfaces reduce the throughput degradation due to the sequential receiving and forwarding of packets in mesh nodes with only a single wireless interface. It might also be possible to use multiple channels. The ad hoc capabilities of WMNs are limited but the still simple installation and the flexibility are an advantage.

Recently, it has become more and more natural that (powerful) client devices are mesh nodes as well. This extends WMNs toward and into the area of classical mobile ad hoc networks. This is not problematic, because the general concepts are the same between MANETs and WMNs. They only use different "values" for network parameters: nodes with mobility from "static" to "moving with v" use wireless communication through one or more interfaces over multiple wireless hops, where paths are determined with self-organizing routing protocols working with different routing metrics.

Three types of wireless mesh networks can be distinguished. *Infrastructure mesh networks* consist of dedicated devices of the network infrastructure, such as access points or relays. Client devices do not participate in the mesh routing. Instead, they connect to the access points in the mesh network by traditional wireless access technologies. *Client mesh networks* consist of client devices such as laptops. The client devices participate in the mesh routing. Furthermore, they might perform functionalities of infrastructure devices. There can be mesh networks that consist of both infrastructure devices and client devices. These could be called *hybrid mesh networks*. Akyildiz et al. [5] describe a similar but different classification of WMNs.

4.3 GENERAL CONCEPTS OF ROUTING PROTOCOLS

4.3.1 Classification of Routing Protocols

The main task of routing protocols is the path selection between the source node and the destination node. This has to be done reliably, fast, and with minimal overhead. Especially, there has to be a path computed if there exists one.

In general, routing protocols can be classified into topology-based and position-based routing protocols (cf. Figure 4.1). Topology-based routing protocols select paths based on topological information, such as links between nodes. Position-based routing protocols select paths based on geographical information with geometrical algorithms. There are routing protocols that combine those two concepts.

Topology-based routing protocols are further distinguished among reactive, proactive, and hybrid routing protocols. Reactive protocols

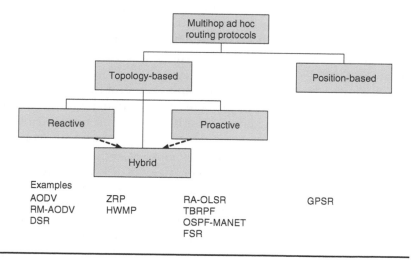

Figure 4.1 Classification of routing protocols.

compute a route only when it is needed. This reduces the control overhead but introduces a latency for the first packet to be sent due to the time needed for the on-demand route setup. In proactive routing protocols, every node knows a route to every other node all the time. There is no latency, but permanent maintenance of unused routes increases the control overhead. Hybrid routing protocols try to combine the advantages of both the philosophies: proactive is used for near nodes or often used paths, while reactive routing is used for more distant nodes or less often used paths.

Other possibilities for the classification of routing protocols are flat vs. hierarchical, distance vector vs. link state, source routing vs. hop-by-hop routing, single-path vs. multipath, or based on the usage scenario.

In principle, mesh networks can deploy any routing protocol from any of the classes described earlier. However, not every protocol will work well. The selection of a suitable routing protocol depends on the anticipated application scenario and the performance requirements.

4.3.2 Routing on Layer 2

According to the OSI layer model and the TCP/IP model, the routing functionality is located at layer 3, the networking layer, that usually uses the Internet protocol (IP). Lately, there are efforts to develop routing protocols for ad hoc mesh networks on layer 2. Although this "violates" the current network layer concept, the following benefits are expected: faster access to more status information of layer 2 and physical layer, faster forwarding, improvements of media access with respect to wireless multihop communication, and synergies between mechanisms (e.g., periodic broadcasts do not have to be done on both layers, just once).

Of course, the benefits do not come without a price. Routing on layer 2 is more difficult to implement, the additional information on the network structure as known from IP addresses is not available in MAC addresses, and it is more difficult to do "Internet" working between heterogeneous networks. Nevertheless, especially the advantages of the better access to the lower layers, which will increase the reliability of wireless ad hoc mesh networks due to faster and more appropriate reactions on changes in the radio environment and of the wireless links, are a strong motivation for layer 2 routing protocols.

The concepts for the path selection are the same, whether on layer 3 or layer 2. The latter only uses MAC addresses. It also means that some mechanisms, so far unknown on layer 2, have to be introduced:

time to live (TTL) for loop prevention, source address and destination address as end points of a wireless multihop path, the extension of mechanisms originally designed for a single wireless hop to multiple hops.

4.3.3 Requirements on Routing in Wireless Mesh Networks

Based on the performance of the existing routing protocols for ad hoc networks and the specific requirements of WMNs, an optimal routing protocol for WMNs must capture the following features [5]:

- *Fault Tolerance*: One important issue in networks is survivability. Survivability is the capability of the network to function in the event of node or link failures. WMNs can ensure robustness against link failures by nature. Correspondingly, routing protocols should also support path reselection subject to link failures.
- *Load Balancing*: Mesh-enabled wireless routers are good at load balancing because they can choose the most efficient path for data.
- *Reduction of Routing Overhead*: The conservation of bandwidth is imperative to the success of any wireless network. It is important to reduce the routing overhead, especially the one caused by rebroadcasts.
- *Scalability*: A mesh network is scalable and can handle hundreds or thousands of nodes. Because the network's operation does not depend on a central control point, adding multiple data collection points, or gateways is convenient. Given thousands of nodes in a WMN, scalability support in the routing protocols is important.
- *QoS Support*: Due to the limited channel capacity, the influence of interference, the large number of users and the emergence of real-time multimedia applications, supporting quality of service (QoS) has become a critical requirement in such networks.

4.3.4 Multipath Routing for Load Balancing and Fault Tolerance

Resiliency is a key attribute of WMNs. A mesh network can support multiple paths among network nodes naturally, hence it is more robust against failures. Mesh nodes can be added to increase redundancy.

Multiple paths are selected between the source node and the destination node. When a link is broken on a path due to bad channel quality or mobility, another path in the set of existing paths can be chosen. Thus, without waiting to set up a new routing path, the end-to-end delay, throughput, and fault tolerance can be improved. However, given a performance metric, the improvement depends on the availability of node disjoint routes between the source node and the destination node.

Another objective of using multipath routing is to perform better load balancing in order to prevent congestion and to reroute traffic around congested nodes.

4.3.5 QoS Routing

QoS routing in multihop wireless networks needs to provide guaranteed bandwidth for a connection request. A connection request would be blocked if we cannot find a route with guaranteed bandwidth. The problem is complicated when the bandwidth for a route is affected by the interference of other routes in the network.

There are two different types of interference in a wireless multihop network: interflow interference and intraflow interference. For a route P, the interflow interference occurs when a link of P uses the same channel with another link that is not of P within their interference range; and the intraflow interference occurs when two links of P within their interference range use the same channel. Intraflow interference is more difficult to deal with because the interference depends on the routing itself, which is not known before the routing is determined.

4.4 ROUTING METRICS

Routing protocols compute or discover *minimum cost* or *minimum weight paths* between the source node and the destination node. The cost/weight is defined through the routing metric. Each path has a *path metric* that is usually the sum of all *link metrics* on the path. Other concatenations of link metrics are possible, too.

Routing metrics for WMNs have to fulfill four requirements [6]:

1. Ensuring route stability, i.e., no frequent route changes
2. Determined minimum cost/weight paths have good performance
3. Efficient algorithms for calculation of minimum cost/weight paths available
4. Ensuring loop free forwarding

Different routing metrics are possible. The exploitation of a certain property of a mesh network might require a special routing metric, for instance, for the use of multiple interfaces or multiple channels. And some metrics might not work with all kinds of routing protocols. The following is a list of some existing routing metrics for WMNs:

Hopcount is the classical routing metric, which is easy to determine. However, it does not give any information about the wireless environment, except that two nodes have a direct link.

Expected Transmission Count (ETX) metric [7] predicts the number of required transmissions for sending a data packet over the link, which includes retransmissions. ETX is calculated from the forward and reverse delivery ratio of a link. The delivery ratios d_f and d_r are measured with dedicated link probe packets of fixed size. They are broadcast with interval τ. The number of received probes for the last w seconds is stored in c. A node can compute all d_r's. They are included in the probe packets to provide the d_f's. The ETX path metric is the sum of all ETX link metrics on the path.

$$\text{ETX} = \frac{1}{d_f \times d_r} \quad \text{with} \quad d_r(t) = \frac{c_{[t-w,t]}}{w/\tau}$$

Expected Transmission Time (ETT) [8] can be considered as a bandwidth-adjusted ETX metric

$$\text{ETT} = \text{ETX} \times \frac{S}{B}$$

where S is the size of a packet. The data rate B can be estimated with the technique of packet pairs. Each node sends two back-to-back probe packets to each of its neighbors periodically, say every minute. The first probe packet is very small, the second is rather large. The time difference between the receipt of the two probe packets is measured and sent to the sender. The size of the second probe packet is divided by the minimum of the last 10 consecutive samples in order to estimate the data rate. The path metric is the sum of all ETT values of the links on the path.

Weighted Cumulative Expected Transmission Time (WCETT) [8] is an extension of the ETT metric. It tries to minimize intraflow interference by penalizing paths that have more transmissions on the same channel or radio. WCETT has been developed for the multi-radio link-quality source routing (MR-LQSR) routing protocol and is described in more detail in Section 4.5.6.

Metric of Interference and Channel-Switching (MIC) [6] improves WCETT and also captures interflow interference. The MIC metric of a path p is defined as

$$MIC(p) = \frac{1}{N \times \min(\text{ETT})} \sum_{\text{link } l \in p} \text{IRU}_l + \sum_{\text{node } i \in p} \text{CSC}_i$$

with the two components interference-aware resource usage (IRU) and channel switching cost (CSC).

Airtime Link Metric [9] is a measure for the consumed channel resources when transmitting a frame over a certain link. This metric is proposed for the upcoming IEEE 802.11s WLAN mesh networking standard and is described in Section 4.6.1.

4.5 ROUTING PROTOCOLS

This section will describe selected routing protocols for wireless multi-hop networks as an illustration of the general concepts of routing protocols as well as some special routing protocols for wireless mesh networks. A comprehensive overview of all routing protocols cannot be done due to limited space.

It is not mentioned for every routing protocol whether it can work with multiple interfaces and with other routing metrics than hop-count. In principle, every routing protocol can be extended to work with multiple wireless interfaces at a mesh node. The same is true for routing metrics, but it is more complicated. Depending on the actual routing metric, more substantial extensions might be necessary, as can be seen in the on-demand part of the Hybrid Wireless Mesh Protocol (HWMP, cf. Section 4.6.2).

4.5.1 Ad hoc On-demand Distance Vector Routing Protocol (AODV)

AODV is a very popular routing protocol for MANETs. It is a reactive routing protocol. Routes are set up on demand, and only active routes are maintained. This reduces the routing overhead, but introduces some initial latency due to the on-demand route setup. AODV has been standardized in the IETF as experimental RFC 3561 [4]. There are several implementations available, for instance, AODV-UU of Uppsala University [10]. Further information on AODV can be found in [11,12]. Recently, an adaptation of AODV has been proposed for WLAN mesh networking (cf. Section 4.6.2).

AODV uses a simple request–reply mechanism for the discovery of routes. It can use hello messages for connectivity information and signals link breaks on active routes with error messages. Every routing information has a timeout associated with it as well as a sequence number. The use of sequence numbers allows to detect outdated data, so that only the most current, available routing information is used. This ensures freedom of routing loops and avoids problems known from classical distance vector protocols, such as "counting to infinity."

When a source node S wants to send data packets to a destination node D but does not have a route to D in its routing table, then a *route discovery* has to be done by S. The data packets are buffered during the route discovery. See Figure 4.2 for an illustration of the route discovery process.

The source node S broadcasts a route request (RREQ) throughout the network. In addition to several flags, a RREQ packet contains the hopcount, a RREQ identifier, the destination address and destination sequence number, and the originator address and originator sequence number. The hopcount field contains the distance to the originator of the RREQ, the source node S. It is the number of hops that the RREQ has traveled so far. The RREQ ID combined with the originator address uniquely identifies a route request. This is used to ensure that a node rebroadcasts an route request only once in order to avoid broadcast storms, even if a node receives the RREQ several times from its neighbors.

When a node receives a RREQ packet, it processes as follows:

■ The route to the previous hop from which the RREQ packet has been received is created or updated.
■ The RREQ ID and the originator address are checked to see whether this RREQ has been already received. If yes, the packet is discarded.
■ The hopcount is incremented by 1.
■ The reverse route to the originator, node S, is created or updated.

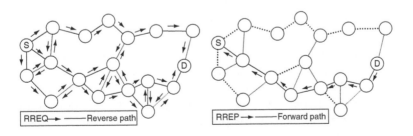

Figure 4.2   AODV route discovery: route request (*left*) and route reply (*right*).

- If the node is the requested destination, it generates a route reply (RREP) and sends the RREP packet back to the originator along the created reverse path to the source node *S*.
- If the node is not the destination but has a valid path to *D*, it issues a RREP to the source depending on the *destination only flag*.

  If intermediate nodes reply to RREQs, it might be the case that the destination will not hear any RREQ, so that it does not have a back route to the source. If the *gratuitous* RREP flag is set in the RREQ, the replying intermediate node will send a gratuitous RREP to the destination. This sets the path to the originator of the RREQ in the destination.
- If the node does not generate a RREP, the RREQ is updated and rebroadcast if TTL is ≥ 1.

On receipt of a RREP message, a node will create or update its route to the destination *D*. The hopcount is incremented by one, and the updated RREP will be forwarded to the originator of the corresponding RREQ. Eventually, the source node *S* will receive a RREP if there exists a path to the destination. The buffered data packets can now be sent to the destination *D* on the newly discovered path.

Connectivity information is provided and maintained by periodically broadcasting routing protocol messages. If a node has not sent a broadcast message, e.g., a RREQ message, within the last hello interval, the node may broadcast a *hello message*. A hello is actually a RREP with TTL = 1 and the node itself as the destination. If a node does not receive any packets from a neighboring node for a defined time, the node considers the link to that neighbor broken.

When a link failure has happened, the node before the broken link checks first whether any active route had used this link. If this was not the case, nothing has to be done. On the other hand, if there have been active paths, the node may attempt local repair. It sends out a RREQ to establish a new second half of the path to the destination. The node performing the local repair buffers the data packets while waiting for any route replies.

If local repair fails or has not been attempted, the node generates a route error (RERR) message. It contains the addresses and corresponding destination sequence numbers of all active destinations that have become unreachable because of the link failure. The RERR message is sent to all neighbors that are precursors of the unreachable destinations on this node. A node receiving a RERR invalidates the corresponding entries in its routing table. It removes all destinations that do not have the transmitter of the RERR as next hop from the list of unreachable destinations. If there are precursors to the destinations in this pruned list, the updated RERR message is forwarded to them.

4.5.2 Dynamic Source Routing Protocol (DSR)

DSR is one of the pioneering routing protocols for MANETs. DSR is being standardized in the IETF MANET working group [2].

DSR is a well-known, reactive routing protocol. It computes a route only if one is needed. The route discovery consists of route request and route reply. The route request is broadcast into the wireless network. However, instead of setting the (reverse) paths in the routing tables of the nodes, the route request collects the addresses of the traversed nodes on its way to the destination. Route reply sends this path back to the source where all paths are stored in a route cache. The path, i.e., the list of addresses from the source to the destination, is included in the header of each packet by the source node. Each node forwards a received packet to the next hop based on the list of addresses in the header (source routing). DSR uses RERR messages for the notification of route breaks [12].

4.5.3 Optimized Link State Routing Protocol (OLSR)

OLSR is a popular proactive routing protocol for wireless ad hoc networks. It has been developed at INRIA and has been standardized at IETF as Experimental RFC 3626 [13]. Further information on OLSR can be found in [14,15].

OLSR uses the classical shortest path algorithm based on the hop-count metric for the computation of the routes in the network. However, the key concept of OLSR is an optimized broadcast mechanism for the network-wide distribution of the necessary link-state information. Each node selects the so-called multipoint relays (MPRs) among its neighbors in such a way that all 2-hop neighbors receive broadcast messages even if only the MPRs rebroadcast the messages. The forwarding of broadcast messages by MPRs only can significantly reduce the number of broadcast messages. Figure 4.3 shows an example where the number of broadcast messages is reduced by half. This optimized forwarding mechanism is used for all broadcasts in an OLSR network. Moreover, the amount of link-state information to be distributed within the network can be reduced with OLSR, because only the link state information to all MPR selectors is necessary for the computation of shortest paths.

Each node periodically broadcasts hello messages for local topology detection. Hello messages are not forwarded (TTL = 1) and

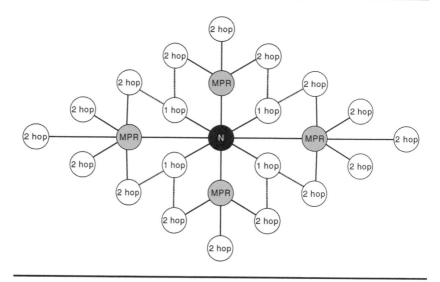

Figure 4.3   Multipoint relay selection in OLSR.

contain a list of the neighbors of the sending node. Each node in the wireless mesh network will know its 2-hop neighborhood through this hello mechanism. It is also possible to verify bidirectionality of links. OLSR attaches the status (asymmetric, symmetric) to each link. Furthermore, each node announces its willingness to forward packets in the hello messages. The information of the hello messages is stored in several information repositories: the link set, the neighbor set and the 2-hop neighbor set.

With this knowledge, each node can now compute its set of multipoint relays. Each node computes its MPR set independently from all other nodes solely based on the received local topology. The only requirement is that the complete 2-hop neighborhood will receive broadcast messages if only MPRs forward them and only symmetric links are considered. It is not necessary that the MPR set is minimal, but a smaller MPR set keeps the protocol overhead lower. OLSR proposes a simple heuristic for the MPR selection in [13] which is described below, but other algorithms are possible.

■ *N*: Neighbors of the node.
■ *N2*: The set of 2-hop neighbors of the node excluding (i) nodes only reachable by members of *N* with willingness WILL_NEVER, (ii) the node performing the computation, and (iii) all the symmetric neighbors: the nodes for which there exists a symmetric link to this node.

- $D(Y)$: Degree of 1-hop neighbor $Y \in N$, which is the number of symmetric neighbors of Y excluding all members of N and excluding the node performing the computation
- Step 1: Start with an MPR set consisting of all members of N with willingness = WILL_ALWAYS
- Step 2: Calculate $D(Y)$, for all $Y \in N$
- Step 3: Add to the MPR set those nodes in N, which are the only nodes to provide reachability to a node in $N2$
- Step 4: Remove the nodes from $N2$ which are now covered by a node in the MPR set
- Step 5: While there still exist nodes in $N2$ which are not covered by at least one node in the MPR set:

    For each node in N, calculate the reachability, i.e., the number of nodes in $N2$ that are not yet covered by at least one node in the MPR set, and which are reachable through this 1-hop neighbor

    Select as an MPR the node with highest willingness among the nodes in N with nonzero reachability. In case of a tie, select the node that provides reachability to the maximum number of nodes in $N2$. In case of multiple nodes providing the same amount of reachability, select the node as MPR whose $D(Y)$ is the greatest

    Remove the nodes from $N2$ that are now covered by a node in the MPR set
- Step 6: As an optimization, each node Y in the MPR set can be checked for omission in increasing order of its willingness. If all nodes in $N2$ are still covered by at least one node in the MPR set excluding node Y, and if the willingness of node Y is smaller than WILL_ALWAYS, then node Y may be removed from the MPR set

The selected multipoint relay nodes are stored in the MPR set. The neighbors that have been selected as MPRs, will have a link status indicating the MPR selection in the hello messages. A node receiving a hello message can derive from this information those nodes that selected it as an MPR. These MPR selectors are stored in the MPR selector set.

Every node periodically broadcasts its link state information through the whole OLSR network by topology control (TC) messages. A TC message contains a list of neighbors of the originating node. This neighbor list must at least contain all MPR selectors of this node to guarantee shortest paths with respect to hopcount. Each TC message has an advertised neighbor sequence number associated with the neighbor list that allows to discard outdated topology information. The information of the TC messages is stored in the topology set.

The OLSR routing table that contains entries for all reachable destinations in the mesh network (proactive routing protocol) is computed from the link set, neighbor set, 2-hop neighbor set, and topology set with a classical shortest path algorithm (e.g., Dijkstra algorithm [16]). If any of the above sets has changed, the routing table has to be recalculated. Furthermore, it might be useful to send a hello or TC message to propagate the change of the topology immediately.

All entries of the information repositories, e.g., the neighbor set, have an expiration time associated with them. This soft state mechanism provides some robustness against the loss of OLSR control packets.

OLSR can also deal with multiple (OLSR) interfaces at a node. Such a node selects the address of any one of its interfaces as the main address and periodically broadcasts multiple interface declaration (MID) messages. MID messages distribute the relationship between the main address and other interface addresses. Obviously, a node with only a single OLSR interface does not have to send MID messages.

4.5.4 Cross-Layer Routing Approach

The interference in wireless networks dramatically degrades the network performance. The interference is directly related to the transmission power. Larger transmission power means more reliable links with higher capacity. On the other hand, larger transmission power also means more interference, thus, less network throughput. Therefore, to provide the routing layer with the information of the lower layers can help to find more reliable and higher capacity paths.

A cross-layer routing algorithm, called mesh routing strategy (MRS), is introduced in [17] to find high throughput paths with reduced interference and increased reliability by optimally controlling transmission power. It is observed [17] that the more (less) the power used, the lower (higher) the packet error rate (*PER*), but the higher (lower) is the interference. MRS searches the optimal trade-off by setting an optimal transmission power level that minimizes the distance from the ideal optimum. MRS processes the local power optimization and routing discovery separately. Such a two-step strategy works as follows [17]:

■ Initially, through neighbor discovery protocol, each node explores its neighborhood, calculates the metrics, such as transmission rate, interference and PER, and determines the local transmission. After that, local links are advertised.

■ Whenever an event triggers a change in the routing metrics of one or more links, the power optimization is performed on the concerned link and the route update process is started.

■ Once the best metric is identified (which is a relatively stable condition), the link is advertised. The MRS routing protocol selects optimal paths to reach any other wireless mesh router of the network by a distance vector approach.

4.5.5 Bandwidth Aware Routing

In a paper on QoS routing for wireless mesh networks [18], the authors discuss interference-aware topology control and QoS routing in multichannel wireless mesh networks. They present a concept of *co-channel interference* and develop a heuristic algorithm to set up a WMN that has the minimum interference among all *K*-connected topologies. Then, they introduce the bandwidth-aware routing problem for QoS routing with bandwidth requirements. If the traffic for a connection request is splittable, i.e., using multiple routes to satisfy the bandwidth request, they show that the problem can be solved by a linear programming (LP) formulation. For the case where the bandwidth for a request can only be satisfied by a single a route, a heuristic algorithm is developed by identifying the maximum bottleneck capacity for a route. Simulation results are also provided to show the effectiveness of the proposed design.

4.5.6 Multi-Radio Link-Quality Source Routing (MR-LQSR) Protocol

In a wireless mesh network, some degradation in throughput might be expected over five or six hops. Channel interference could result in lower throughput if the nodes are too close to each other or if the power is too high for the area. WMN routing protocols should select paths based on observed latency and wireless environment as well as other performance factors, resulting in the best possible throughput across the network.

To increase the capacity of the wireless mesh networks, nodes might have multiple radios, preferably working on different channels or different bands. However, there are some issues with channel diversity through multiple radios, which routing protocols have to take into account. Different transmission rates and bands have different transmission ranges. Shortest path routing protocols will prefer the links with larger transmission range, which are usually the links with the lower transmission rate. Furthermore, there will be no increase in capacity if multiple radios of a mesh node use the same channel due to interference. Therefore, paths with channel (radio) diversity have to be preferred.

The WCETT metric [8] takes into account the link quality, channel diversity, and the minimum hopcount. It can achieve a good trade-off between delay and throughput because it considers channels with good quality and channel diversity at the same time. MR-LQSR [8] has been developed for multiradio multihop WMNs based on the WCETT metric.

MR-LQSR assigns a weight to each link, which is the expected amount of time it would take to successfully transmit a packet of some fixed size S on that link. This time depends on the transmission rate of the link and the loss rate. Given a link i from node x to node y, the expected transmission time ETT of the packet on this link is measured. We denote this value by ETT_i.

The path metric in MR-LQSR tries to balance the trade-off between throughput and delay. However, the sum of the ETT of all links of a path reflects only the end-to-end delay. The impact of channel diversity is not considered, and hence needs to be extended.

It is observed that transmission time on a link is determined by the available bandwidth, which is further determined by the interference of a channel. More specifically, the transmission time is inversely proportional to the available bandwidth in a link. Conservatively, consider an n-hop path, assume that any two hops among those n hops interfere with each other if they share one channel. Define X_j as

$$X_j = \sum_{\text{Hop } i \text{ is on channel } j} ETT_i, \quad 1 \le j \le k$$

Thus, X_j is the sum of transmission times of hops on channel j. The higher X_j indicates lower available bandwidth in each link using channel j. The total path throughput will be dominated by the bottleneck channel, which has the largest X_j.

In MR-LQSR, WCETT is defined as:

$$WCETT = (1 - \beta) \times \sum_{i=1}^{n} ETT_i + \beta \times \max_{1 \le j \le k} X_j$$

The first term reflects the latency of this path. The second term represents the path throughput. The weighted average tries to balance the two.

4.5.7 Other Topology-based Routing Protocols for Wireless Mesh Networks

Many routing protocols have been proposed for wireless multihop networks. This section lists a few other routing protocols. It is by no means exhaustive. For extensive information, the reader is referred to the World Wide Web.

Topology Dissemination Based on Reverse-Path Forwarding (TBRPF) has been standardized in IETF as experimental RFC 3684 [19]. It is a proactive routing protocol. TBRPF is used in a few installations and products of WMNs.

Dynamic on-demand MANET routing protocol (DYMO) is currently developed at the IETF MANET working group [20]. It is a reactive routing protocol and contains the basic route discovery and maintenance features similar to AODV and a mechanism for future enhancements.

OSPF-MANET is an ongoing effort at the IETF to adapt Open Shortest Path First (OSPF) to wireless multihop networks. The advantage of OSPF-MANET is the easy integration of WMNs and MANETs into existing (wired) OSPF networks. Similar to OLSR, the flooding of link state advertisements will be reduced. However, OSPF-MANET concentrates on connected dominating sets (CDS) for the reduction of rebroadcasts. Obviously, OSPF-MANET is a proactive routing protocol.

Fisheye State Routing (FSR) [21] is a proactive routing protocol that uses the "fisheye" concept for a reduction of broadcast messages needed for the distribution of topology information. The nodes closer to a node N receive topology information more frequently than faraway nodes. This is done by incrementing the TTL of the messages for each flood until the maximum value before it continues with the initial, small value.

The hybrid concept of the zone Routing Protocol (ZRP) consists of proactive routing in the close neighborhood of the nodes and of reactive routing for further away nodes, which minimizes the disadvantages of both methods while making use of their advantages.

4.5.8 Position-based Routing Protocols

In this class of routing algorithms, packets are forwarded based on the geographical positions of the forwarding node, its neighbors, and

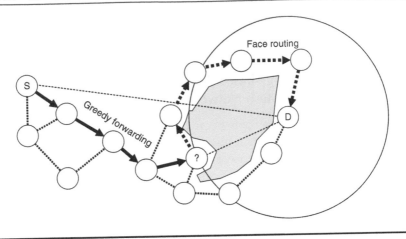

Figure 4.4 Position-based forwarding.

the destination. This requires that every node knows its own geographical position. The position of the destination has to be provided by a *location service*. A simple forwarding algorithm such as *greedy forwarding* can be used with this position information. The packet is sent to the neighbor closest to the destination. However, the simple forwarding algorithm may get stuck in a local minimum and cannot reach the destination although a path to the destination exists, as illustrated in Figure 4.4.

Face routing [22] is usually used as a fallback strategy in such a case. The network graph is logically segmented into so-called faces, where the considered links do not cross each other. This planarization of the network graph can be done locally with distributed algorithms. Packets can proceed out of a local minimum by being forwarded around these faces toward the destination. Face routing is proven to reach the destination if there exists a path [22,23], but it might not be the shortest one due to the "diversion" into the local minimum.

One of the first practical position-based routing protocols for wireless networks is Greedy Perimeter Stateless Routing (GPSR) [24]. It combines greedy forwarding with face routing as fallback. Landmark guided forwarding [25] combines topology-based proactive routing for nearby nodes with position-based forwarding for faraway nodes. Mauve et al. [26] provide a good overview about position-based routing and location services.

4.6 PROPOSED ROUTING FOR IEEE 802.11s WLAN MESH NETWORKING

In July 2004, the WLAN mesh networking study group of the IEEE 802.11 working group had its first meeting as task group "s." Its goal is to develop a standard for IEEE 802.11 mesh networks with up to 50 mesh nodes* [27] that can be deployed in different usage scenarios: residential, office, public access and community networks, and public safety networks [28]. Routing is one of the major functionalities besides MAC enhancements for multihop communication and security.

Note: The standardization of WLAN mesh networks in IEEE 802.11s was still in progress at the time of writing. The following descriptions are based on the draft version 0.01 from the March 2006 IEEE 802.11 Meeting in Denver, CO (USA) [9]. While the general concepts of the routing protocol seem to be quite fixed, changes in the details are possible.

The routing protocol of the upcoming IEEE 802.11s standard is located at layer 2 and uses therefore MAC addresses. The mesh data frames use the four-address frame format of the IEEE standard 802.11–1999 (Reaff 2003) [29] extended by the IEEE 802.11e QoS header field and new mesh extensions. The latter include a time to live field for the ultimate avoidance of loops and a mesh end-to-end sequence number for the control of broadcast flooding and to enable ordered delivery of data frames. The four-address format provides address fields for transmitter and receiver (current link) as well as for source and destination (path). Routing messages are sent as action management frames [29].

Legacy IEEE 802.11 devices can connect to a mesh through mesh access points using the conventional methods of the standard. The mesh access point acts as a proxy mesh node for these devices. Mesh access points are mesh nodes with additional access point functionality.

4.6.1 Airtime Routing Metric

The airtime link metric is proposed as the default radio-aware routing metric for basic interoperability between IEEE 802.11s devices. It reflects the amount of channel resources consumed for transmitting a frame over a particular link. The path with the smallest sum of airtime link metrics is the best path.

---

* The correct term for nodes of IEEE 802.11s mesh networks is *mesh point*. In this chapter, we use the term *mesh node*.

The airtime cost c_a for each link is calculated with the following formula:

$$c_a = \left[O_{ca} + O_p + \frac{B_t}{r} \right] \frac{1}{1 - e_{fr}}$$

The channel access overhead O_{ca}, the MAC protocol overhead O_p, and the number of bits B_t in a test frame are constants whose values depend on the used IEEE 802.11 transmission technology. The transmission bit rate r in Mbit/s is the rate at which the mesh node would transmit a frame of size B_t based on the current conditions with the frame error rate e_{fr}.

4.6.2 Hybrid Wireless Mesh Protocol (HWMP)

HWMP is the default routing protocol for WLAN mesh networking. Every IEEE 802.11s compliant device will be capable of using this routing protocol. The hybrid nature and the configurability of HWMP provide good performance in all anticipated usage scenarios [28].

The foundation of HWMP is an adaptation of the reactive routing protocol AODV [4] to layer 2 and to radio-aware metrics called radio metric AODV (RM-AODV). A mesh node, usually a mesh portal,* can be configured to periodically broadcast announcements, which sets up a tree that allows proactive routing towards this mesh portal.

The reactive part of HWMP follows the general concepts of AODV as described in [4] and Section 4.5.1. It uses the distance vector method and the well-known route discovery process with route request and route reply (cf. Figure 4.2). Destination sequence numbers are used to recognize old routing information. However, there are some significant differences in the details. Figure 4.5 shows the structure of an HWMP route request to illustrate the new features.

HWMP uses MAC addresses as a layer 2 routing protocol instead of IP addresses. Furthermore, HWMP can make use of more sophisticated routing metrics than hopcount such as radio-aware metrics. A new path metric field is included in the RREQ/RREP messages that contains the cumulative value of the link metrics of the path so far. The default routing metric of HWMP is the airtime metric (cf. Section 4.6.1) where the separate link metrics are added up to get the path metric.

Since a radio-aware metric changes more often than the hopcount metric, it is preferable to have only the destination to answer to a

---

* A *mesh portal* is a mesh node that provides a connection to the outside of the mesh (other networks).

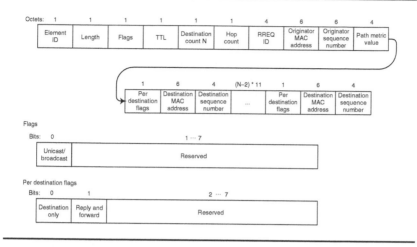

Octets:	1	1	1	1	1	1	4	6	6	4
	Element ID	Length	Flags	TTL	Destination count N	Hop count	RREQ ID	Originator MAC address	Originator sequence number	Path metric value

	1	6	4	(N–2) * 11	1	6	4
	Per destination flags	Destination MAC address	Destination sequence number	...	Per destination flags	Destination MAC address	Destination sequence number

Flags

Bits:	0	1 ··· 7
	Unicast/ broadcast	Reserved

Per destination flags

Bits:	0	1	2 ··· 7
	Destination only	Reply and forward	Reserved

Figure 4.5   HWMP route request.

RREQ so that the path metric is up to date. For this reason, the destination only flag is set (DO = 1) by default in HWMP.

By explicitly setting the destination only flag to DO = 0, it is possible to let intermediate nodes reply. This gives a shorter latency of the route discovery, but the path metric is not up to date. Therefore, the intermediate node that replied with a RREP will forward the RREQ to the destination. This is controlled by the reply and forward flag. It is set by default (RF = 1), but can be unset to get the traditional AODV behavior. The destination only flag in the forwarded RREQ has to be set (DO = 1). This prevents further intermediate nodes from generating route replies which could be many.

Any received routing information (RREQ/RREP) is checked for validity with a sequence number comparison. Routing information is valid if the sequence number is not smaller than the sequence number in the previous information. If the sequence numbers are the same and the routing information, which is the path metric, is better, then the new information will be used and the new message will be processed.

HWMP can use periodic maintenance RREQs to maintain a best metric path between the source and the destination of active paths. This is an optional feature.

HWMP allows multiple destinations in RREQ messages, which reduces the routing overhead when a mesh node has to find routes to several nodes simultaneously. This is the case for repairing broken links and for maintenance RREQs.

Some flags can have different values for each destination. Therefore, per destination flags are associated with each destination and its sequence number. These are the flags specifically related to the generation of route reply messages.

An explicit time to live (TTL) field is necessary since there is none in the header as in traditional AODV.

The use of the proactive extension to RM-AODV is configurable. The proactive extension uses the same distance vector methodology as RM-AODV and makes use of routing messages of RM-AODV.

To use the proactive extension, at least one mesh portal has to be configured to periodically broadcast mesh portal announcements. This triggers a root selection and arbitration process, out of which a single root portal evolves. The root portal sets the *announcement type* flag to 1 (root) in its periodic mesh portal announcements. On receipt of such a root portal announcement, a mesh node will set up a path to the root portal through the mesh node it received the root portal announcement with the best path metric. A path to the announcing mesh portal can also be set up on receipt of portal announcements with announcement type flag set to 0 (portal). The path setup will lead to a tree rooted in the root (mesh) portal.

If the *registration* flag is not set in the announcement message (non-registration mode), the processing of the root announcements stops here. When a mesh node wants to send data frames to the root portal, it can send a gratuitous RREP to the root portal immediately before the first data packet. This will set up the backward path from the root portal to the source node.

If the registration flag is set in the announcement message (registration mode), the mesh node waits a certain time for further root announcement messages to arrive or it might also issue a RREQ with TTL = 1 to explicitly ask its neighboring nodes for routes to the root portal. The mesh node chooses the path with the best path metric to the root portal. It registers with the root portal by sending a gratuitous RREP to the root portal. The registration has to be done every time the node changes its parent node.

An overview of the different configuration options of HWMP is shown in Figure 4.6.

4.6.3 Radio Aware Optimized Link State Routing (RA-OLSR)

RA-OLSR protocol is an optional, proactive routing protocol of the emerging IEEE 802.11s standard. It follows closely the specification

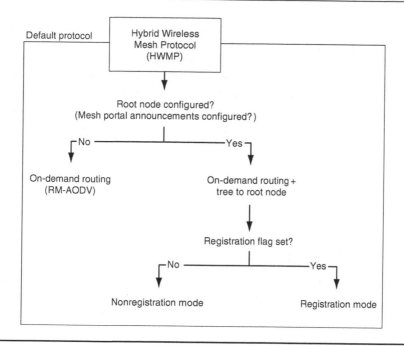

Figure 4.6   Configurability of HWMP.

of the OLSR protocol (RFC 3626 [13]) as described in Section 4.5.3. Instead of IP addresses as in [13] it uses MAC addresses and can work with arbitrary routing metrics such as the air-time metric of Section 4.6.1. Furthermore, it defines a mechanism for the distribution of addresses of nonmesh WLAN clients in the RA-OLSR mesh.

The link state is the value of the link metric and is used in the shortest path computation. Therefore, a link metric field is associated to each reported neighbor in hello messages and TC messages. The value of the link metric is also stored in the corresponding information repositories; the link set and the topology set. The link metric is also used in the heuristic for the selection of the multipoint relays.

Each mesh access point maintains a local association base (LAB) that contains all legacy IEEE 802.11 stations associated with this mesh AP. It broadcasts local association base advertisement (LABA) messages periodically, in order to distribute the association information in the mesh network. The information received from LABA messages is stored in the global association base (GAB) in each node. The information of both LAB and GAB is used in the construction of the routing table and provides routes to legacy stations associated with mesh access points. To save bandwidth, it is possible to advertise only the checksum of the blocks of the LAB. If there is a mismatch between a received checksum and the checksum in the GAB, the node requests an update of the corresponding block of the LAB of the originating node.

RA-OLSR also utilizes the frequency control for link state flooding as known from Fisheye State Routing [21] for the distribution of the topology control messages. The idea is that nearer mesh nodes receive topology information more frequently than faraway nodes. Therefore, the TTL in the topology control messages is set to 2, 4, and maximum TTL sequentially.

4.6.4 Extensibility

One of the conclusions from research in mobile ad hoc networking is that there will not be a single routing protocol that is optimal in every useful scenario. Therefore, an extensibility mechanism is proposed that allows flexibility but still provides interoperability between mesh nodes of different vendors.

Each mesh network announces its used routing protocol and routing metric to new nodes by corresponding IDs. Only those nodes that support the used routing protocol and routing metric are allowed to join the mesh network. All IEEE 802.11s devices will be able to use the default routing protocol and the default routing metric. Other routing protocols or routing metrics, which are better suited for some scenarios, can be used in addition to the default ones (cf. Figure 4.7.)

4.7 JOINT ROUTING AND CHANNEL ASSIGNMENT

In a multiradio, multichannel WMN, besides routing, another main design problem is to assign a radio channel to each network interface to achieve efficient utilization of available channels, the so-called channel assignment. In the literature, there are numerous studies that consider routing and channel assignment. Some only consider routing to maximize the throughput [7,8,30]. Some only consider channel assignment to minimize the interference [31]. Some study both routing and channel assignment [18,32–35]. Raniwala, and Chiueh [32] consider routing first, followed by channel assignment. Since channel assignment is performed after routing, the load in each link is given. Therefore, such channel assignment is also called load-aware channel assignment. Tang et al. [18] consider channel assignment first, followed

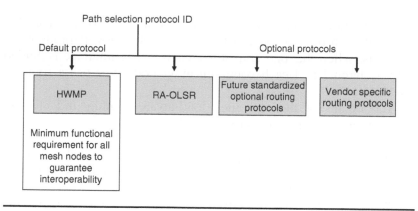

Figure 4.7 Extensibility of IEEE 802.11s routing protocols.

by routing. Given a channel assignment, the interference among links is determined, therefore, routing under such a constraint is also called interference-aware routing.

However, in a multiradio multichannel WMN, routing and channel assignment are interdependent. Routing depends on the capacity of the virtual link which is determined by channel assignment. On the other hand, channel assignment depends on the expected load of the virtual link, which is affected by routing. Due to the circular dependence between routing and channel assignment, a joint routing and channel assignment mechanism to maximize the network throughput is desirable [33–35].

In this section, we first introduce the combined load-aware routing and channel assignment mechanism [33], which iterates over channel assignment and routing algorithms until each link's capacity matches its expected load as closely as it can. We then introduce the joint channel assignment and routing [34,35]. Due to the hardness of the joint routing and channel assignment, both papers solve the linear programming (LP) relaxation of the problem optimally, then have some postprocessing to have feasible channel assignment.

4.7.1 Combined Load-aware Routing and Channel Assignment

The combined load-aware routing and channel assignment [33] starts with an initial estimation of the expected load on each virtual link regardless the link capacity, then iterates over channel assignment and routing algorithms until the channel assignment is feasible. The inputs to the combined channel assignment and routing algorithm include an estimated traffic load metrix, a WMN topology, and the available radios at each node as well as the number of nonoverlapping radio channels. When the algorithm stops, each 802.11 interface is binded with a channel and each communicating node pair has a route in the WMNs.

The details of the combined load-aware routing and channel assignment are presented as follows [33]:

- ■ Step 1: The routing algorithm computes the initial routes for each node pair given a set of node pairs and the expected traffic load between each node pair.
- ■ Step 2: Given the input from the routing step, the radio channel assignment algorithm assigns a radio channel to each interface such that the available bandwidth at each virtual link is no less than its expected load.
- ■ Step 3: The new channel assignment is fed back to the routing algorithm to reach more informed routing decisions.
- ■ Step 4: Recalculate the link load based on the routing information. If some of the link loads are more than their capacities, go to Step 2, otherwise, Stop.

The above combined mechanism can be used as a generic approach to combine any channel assignment and routing solutions together to achieve better network throughput. For example, in Step 2, any load-aware channel assignment solutions can be applied. In Step 3, any interference-aware routing algorithms can be applied.

4.7.2 Joint LP-based Routing and Channel Assignment

Kodialam and Nandagopal [34] and Alicherry et al. [35] propose the LP-based joint routing and channel assignment. It is assumed that potentially there is traffic demand for any pair of nodes [34]. (Alicherry et al. [35] assume that communication is only involved to and from wired gateways, rather than involving pairs of end nodes. In a WMN, the wireless users are mostly interested in connecting to the Internet; therefore, an asymmetric traffic pattern is reasonable and it indeed simplifies the LP model. In this section, we only discuss the work of Alicherry et al. [35].

4.7.2.1 Models and Assumptions

A WMN can be described by a graph $G = (N, E)$ where N represents a set of nodes each being equipped with multiple wireless radios, and E represents direct communication links between a pair of nodes. There is one gateway node $u_0 \in N$ that is connected to the Internet, and the traffic from any other node to the Internet is directed through u_0.

In practice there may be multiple gateway nodes in the system, and our results can easily be generalized to such a case. For the aim of simplicity, it is assumed that there is only a single gateway node.

A node can communicate with some other nodes through wireless communication links. Specifically, there is a communication link $e = (u, v) \in E$ if and only if nodes u, v are within the valid transmission range of a radio. Suppose that there are K orthogonal wireless channels, and the bandwidth of each channel is c. It is assumed that the system works in a periodical synchronous time-slotted mode where each cycle contains T time slots. Obviously, the model we introduce here provides an upper bound for a system in asynchronous operation.

For the routing and channel assignment problem, there are three decisions to make: assign a set of wireless channels to each link $e \in E$, determine whether each (link, channel) pair is active for each time slot $\tau = 1, \ldots, T$, and assign the communication traffic to the active (link, channel) pairs over different time slots.

The objective of routing and channel assignment is to maximize the throughput of the system by the above decisions. In particular, we assume that each node $u \in N$ has an average traffic demand d_u that needs to be routed to the gateway node u_0, and we want to accommodate these traffic as much as possible. As we are to discuss later, there are different criteria to measure the maximum throughput.

The approach contains two steps: interference-aware routing followed by postprocessing of adjusting channel assignment.

First, we solve an LP that determines a routing solution by considering the traffic load and the impact of the interference. Such a solution also provides a channel assignment but may not be feasible due to the relaxation in the LP.

Second, in postprocessing, we adjust the channel assignment to obtain a feasible solution. In doing so, we keep the routing solution so that the traffic throughput will not be reduced.

4.7.2.2 Integer Programming Formulation and Linear Programming Relaxation

We need the following notations to define our problem.

N_u: Set of nodes that are within the valid transmission range of node u.

N'_u: Set of nodes that are within the interference range of node u.

For notational convenience, we assume $u \in N'_u$.

R_u: The number of radios that node u has.

We first use integer linear programming (ILP) to formulate the problem.

y_{uv}^{kt}: Binary variable, $y_{uv}^{kt} = 1$ if and only if link $(u,v) \in E$ is active on channel k at time slot τ.

Any channel assignment must satisfy the following two constraints:

1. *Radio Constraint*: At any time, a node can use at most R_u different channels to send packages.

$$\sum_{v \in N_u} \sum_{k=1}^{K} y_{uv}^{kt} < R_u, \quad \forall \, u \in N, \quad t = 1, \ldots, T \qquad (4.1)$$

2. *Interference Constraint*: At any time, two interference links cannot be active at the same channel.

$$\sum_{u' \in N'_u \cup N'_u} \sum_{v' \in N_{u'}} y_{u'v'}^{kt} \le 1, \quad \forall (u, v) \in E, \quad k = 1, \ldots, K, \quad t = 1, \ldots, T \quad (4.2)$$

Equation 4.1 and Equation 4.2 characterize the necessary and sufficient conditions for a feasible channel assignment. It defines a time-varying network over which routing can be conducted for sending communication packages. Although it helps us to clearly define and understand the problem, the time-indexed binary variables make the size of the model very large and thus difficult to solve. In the following, the ILP formulation is relaxed to a LP so that a solution can be easily obtained. The LP solution serves as a necessary condition for a feasible channel assignment.

We can see that $\sum_{t=1}^{T} y_{uv}^{kt}$ is the aggregation of active time slots for link (u, v) on channel k within a cycle, and thus $\frac{1}{T} \sum_{t=1}^{T} y_{uv}^{kt}$ is the percentage usage of link (u, v) on channel k. Recall that the bandwidth of each channel is c, then $\frac{c}{T} \sum_{t=1}^{T} y_{uv}^{kt}$ is the corresponding available bandwidth.

We define a new variable $x_{uv}^{k} = \frac{c}{T} \sum_{t=1}^{T} y_{uv}^{kt}$. Then a necessary condition for Equation 4.1 to hold is

$$\sum_{v \in N_u} \sum_{k=1}^{K} x_{uv}^{k} \le cN_u, \quad \forall \, u \in N \qquad (4.3)$$

and a necessary condition for Equation 4.2 to hold is

$$\sum_{u' \in N'_v \cup N'_v} \sum_{v' \in N_{u'}} x_{u'v'}^{k} \le c, \quad \forall (u, v) \in E, \quad k = 1, \ldots, K \qquad (4.4)$$

Now, we can add the routing decision for the traffic demand d_u over the communication network defined by Equation 4.3 and Equation 4.4. To this end, we need to enforce the flow conservation constraint for each node.

$$\sum_{v \in N_u} \sum_{k=1}^{K} x_{uv}^{k} + d_u' = \sum_{v \in N_u} \sum_{k=1}^{K} x_{uv}^{k}, \quad \forall u \in N, \quad u \neq u_0 \qquad (4.5)$$

$$0 \leq d_u' \leq d_u, \quad \forall u \in N, \quad u \neq n_0 \qquad (4.6)$$

where d_u' is the traffic sent out from node u.

Equation 4.3 through Equation 4.6 define the basic structure for a routing and channel assignment. There may be different criteria and considerations for the throughput maximization, which can be handled by defining objective functions and introducing some other constraints.

1. *Maximizing the Total Throughput*: We can define the objective function as

$$\max \sum_{u \in N} d_u'$$

2. *Ensuring Fairness*: We can define a parameter λ so that each node can send at least λ portion of its traffic demand.

$$\max \lambda$$

subject to $0 \leq \lambda \leq 1$ and $d_u' \geq \lambda d_u$ for all $u \in N$.

3. *Maximizing the Worst Case*: We can maximize the lowest traffic to be sent.

$$\max d$$

subject to $d \leq d_u'$ for all $u \in N$.

4.8 OUTLOOK AND OPEN ISSUES

Wireless mesh networking is a topic that now attracts great attention from industrial companies and universities. Big efforts are under way to develop working mesh standards. There are many companies selling wireless mesh devices. And there are already many working installations of wireless mesh networks around the world.

The core idea of wireless mesh networks—forwarding of packets over multiple wireless hops—is a new quality in wireless communications. It can make devices "truly wireless." The necessary routing protocols have a sound basis, which is a good foundation for the continuous growth of the number of deployments of WMNs. Due to their great flexibility and robustness, wireless mesh networks will be an important part of future (wireless) network architectures.

Although the area of wireless mesh networking can build on the huge amount of results from a decade of research in mobile ad hoc networking, there are still many open research issues. The special properties of WMNs require and allow optimizations in order to meet the performance goals for the use of wireless mesh networks. Each application scenario may require different optimizations. High network throughput and network capacity are the important requirements in practical deployments. New routing metrics have to be developed and utilized in order to support necessary improvements. Mobility comes into play when client devices are integrated into wireless mesh networks. Better and more powerful devices can have multiple radio interfaces and can make use of channel diversity. This has to be supported by routing protocols and routing metrics. Last but not least, cross-layer design is important in order to get better access to the layers that have a high influence on the routing—MAC and physical layer.

REFERENCES

1. Microsoft Mesh Networks. Available at: http://research.microsoft.com/mesh/.
2. David B. Johnson, David A. Maltz, and Yih-Chun Hu, "The Dynamic Source Routing Protocol for Mobile Ad hoc Networks (DSR)," *IETF Internet Draft*, draft-ietf-manet-dsr-10.txt, July 2004, work in progress.
3. Kiyon Autonomous Networks. Available at: http://www.kiyon.com.
4. Charles E. Perkins, Elizabeth M. Belding-Royer, and Samir R. Das, "Ad hoc On-Demand Distance Vector (AODV) Routing," IETF Experimental RFC 3561, July 2003.
5. Ian F. Akyildiz, Xudong Wang, and Weilin Wang, "Wireless Mesh Networks: A Survey," *Computer Networks*, vol. 47, no. 4, March 2005.
6. Yaling Yang, Jun Wang, and Robin Kravets, "Designing Routing Metrics for Mesh Networks," *Proceedings of IEEE WiMesh 2005*, Santa Clara, CA, September 2005.
7. Douglas S.J. Couto, Daniel Aguayo, John Bicket, and Robert Morris, "A High-Throughput Path Metric for Multi-Hop Wireless Routing," *Proceedings of ACM MobiCom 2003*, San Diego, CA, September 2003.

8. R. Draves, J. Padhye, and B. Zill, "Routing in Multi-Radio Multi-Hop Wireless Mesh Networks," *Proceedings of ACM MobiCom 2004*, Philadelphia, PA, September 2004.

9. IEEE P802.11s^{TM}/D0.01, draft amendment to standard IEEE 802.11^{TM}: ESS Mesh Networking, March 2006.

10. Uppsala University Ad hoc Implementation Portal. Available at: http://core. it.uu.se/AdHoc/ImplementationPortal.

11. Ad hoc On-Demand Distance Vector Routing. Available at: http://moment. cs.ucsb.edu/AODV/aodv.html.

12. Charles E. Perkins, Elizabeth M. Royer, Samir R. Das, and Mahesh K. Marina, "Performance Comparison of Two On-Demand Routing Protocols for Ad hoc Networks," *IEEE Personal Communications*, February 2001.

13. Thomas Heide Clausen and Philippe Jacquet (eds.), "Optimized Link State Routing Protocol (OLSR)," IETF Experimental RFC 3626, October 2003.

14. P. Jacquet, P. Mühlethaler, T. Clausen, A. Laouiti, A. Qayyum, and L. Viennot, "Optimized Link State Routing Protocol for Ad hoc Networks," *Proceedings of INMIC 2001*, Pakistan 2001.

15. OLSR Routing Protocol (RFC 3626). Available at: http://hipercom.inria.fr/olsr/.

16. R.K. Ahuja, T.L. Magnanti, and J.B. Orlin, *Network Flows: Theory, Algorithms, and Applications*. Prentice-Hall, Upper Saddle River, NJ, 1993.

17. L. Lannone and S. Fdida, "MRS: A Simple Cross-Layer Heuristic to Improve Throughput Capacity in Wireless Mesh Networks," *Proceedings of ACM CoN-EXT'05*, 2005, pp. 21–30.

18. J. Tang, G. Xue, and W. Zhang, "Interference-aware Topology Control and QoS Routing in Multi-Channel Wireless Mesh Networks," *Proceedings of ACM MobiHoc'05*, 2005.

19. Richard G. Ogier, Fred L. Templin, and Mark G. Lewis, "Topology Dissemination Based on Reverse-Path Forwarding (TBRPF)," IETF Experimental RFC 3684, February 2004.

20. Ian Chakeres, Elizabeth Belding-Royer, and Charlie Perkins, "Dynamic MANET On-Demand (DYMO) Routing," *IETF Internet Draft*, draft-ietf-manet-dymo-03.txt, October 2005, work in progress.

21. Atsushi Iwata, Ching-Chuan Chiang, Guangyu Pei, Mario Gerla, and Tsu-Wei Chen, "Scalable Routing Strategies for Ad hoc Wireless Networks," *IEEE Journal on Selected Areas in Communications*, vol. 17, no. 8, August 1999.

22. P. Bose, P. Morin, I. Stojmenović, and J. Urrutia, "Routing with Guaranteed Delivery in Ad hoc Wireless Networks," *ACM Wireless Networks*, vol. 7, no. 6, November 2001.

23. Fabian Kuhn, Roger Wattenhofer, and Aaron Zollinger, "Asymptotically Optimal Geometric Mobile Ad-Hoc Routing," *ACM Dial-M*, 2002.

24. Brad Karp and H.T. Kung, "GPSR: Greedy Perimeter Stateless Routing for Wireless Networks," *Mobicom 2000*, Boston, MA, August 2000.

25. Menghow Lim, Adam Greenhalgh, Julian Chesterfield, and Jon Crowcroft, "Landmark Guided Forwarding," *ICNP 2005*, Boston, MA, November 2005.

26. Martin Mauve, Jörg Widmer, and Hannes Hartenstein, "A Survey on Position-Based Routing in Mobile Ad hoc Networks," *IEEE Network*, November/ December 2001.

27. Jim Hauser, Dennis Baker, and W. Steven Conner, "Draft PAR for IEEE 802.11 ESS Mesh," IEEE P802.11 Wireless LANs, document 11-04/0052r2, January 2004.

28. W. Steven Conner et al., "IEEE 802.11 TGs Usage Models," IEEE P802.11 Wireless LANs, document 11-04/0662r16, January 2005.

29. "IEEE Wireless LAN Edition," a compilation based on IEEE Std 802.11™-1999 (R2003) and its amendments, IEEE, Standards Information Network, IEEE Press.

30. A. Adya, P. Bahl, J. Padhye, A. Wolman, and L. Zhou, "A Multi-Radio Unification Protocol for IEEE 802.11 Wireless Networks," *IEEE Broadnets*, 2004.

31. M. Marina and S. Das, "A Topology Control Approach for Utilizing Multiple Channels in Multi-Radio Wireless Mesh Networks," *IEEE Broadnets*, 2005.

32. A. Raniwala and T. Chiueh, "Architecture and Algorithms for an IEEE 802.11-based Multi-Channel Wireless Mesh Network," *Proceedings of IEEE Info-Com'05*, 2005.

33. A. Raniwala, K. Gopalan, and T. Chiueh, "Centralized Channel Assignment and Routing Algorithms for Multi-Channel Wireless Mesh Networks," *ACM Mobile Computing and Communication Review MC2R*, vol. 8, pp. 50–65, 2004.

34. M. Kodialam and T. Nandagopal, "Characterizing the Capacity Region in Multi-Radio Multi-Channel Wireless Mesh Networks," *ACM MobiCom*, 2005.

35. M. Alicherry, R. Bhatia, and L. Li, "Joint Channel Assignment and Routing for Throughput Optimization in Multi-Radio Wireless Mesh Networks," *ACM MobiCom*, 2005.

5

MEDIUM ACCESS CONTROL IN WIRELESS MESH NETWORKS

Michelle X. Gong, Shiwen Mao,
Scott F. Midkiff, and Brian Hart

CONTENTS

Wireless mesh networking is the ideal technology for providing quick-and-easy network access where network infrastructure is hard to install or has been destroyed. Wireless mesh network (WMN) is equally suited to the low-cost extension of network access to a wide area. Typical deployment scenarios include public safety networks, home networks, office networks, and public access networks. For example, in public access networks, WMN deploys wireless routers, called wireless mesh points (MPs) on light poles along streets to form a multihop wireless backbone with acceptable connectivity and performance.

In wireless mesh networking, the medium access control (MAC) protocol plays an important role in coordinating channel access among mesh nodes. Most traditional medium access protocols are designed for nodes with omnidirectional antennas and for sharing a single channel. Examples include Aloha, Slotted Aloha, carrier sense multiple access (CSMA), and CSMA with collision avoidance (CSMA/CA). The two MAC protocols defined in the IEEE 802.11 standards, i.e., the IEEE 802.11 MAC protocol and the IEEE 802.11e quality of service (QoS) enhancement MAC protocol, are single-channel MAC protocols designed for nodes with omnidirectional antennas. Even though single-channeled MAC protocols are robust and easy to implement, WMNs based on such rudimentary MACs may suffer low throughput due to collisions and interference caused by multihop routing. For instance, the maximum throughput for a daisy-chained network could be only one-seventh of the nominal link bandwidth when the IEEE 802.11 MAC is used [27]. As a result, congestion in such networks would be more frequent and persistent, making it a great challenge to support bandwidth-intensive applications (e.g., video communications).

To address the low throughput problem in multihop mesh networks, MAC protocols that explore alternative physical layer technologies, such as directional antennas (including smart antennas), have been proposed. The basic idea is to reduce the transmitter's interference range and to improve channel spatial reuse by using directional antennas. Another effective solution to the low throughput problem is to use multiple channels at mesh nodes, allowing concurrent transmissions on these channels. In fact, many current physical layer standards do provide multiple channels at the physical layer. For example, the IEEE 802.11b PHY standard for wireless local area networks (WLANs) provides three orthogonal channels (channel 1, 6, and 11) for use in the United States, while IEEE 802.11a provides 12 nonoverlapping channels. Such orthogonal channels could be used simultaneously in a neighborhood without interfering with each other. Consequently, there has been substantial effort on developing such multichannel MAC protocols that can efficiently assign channels to mesh nodes and coordinate the sharing of these channels.

This chapter presents a survey of existing MAC schemes to provide a high-level picture for both researchers and practitioners, and to facilitate the research and standardization efforts in this active area. The MAC protocols reviewed in this chapter include both protocols proposed by the academic community and protocols currently being studied and evaluated by the wireless industry for inclusion in the IEEE 802.11s WMN standard.

We first give an introduction to WMNs and wireless MAC protocols in Section 5.1. We discuss the design objectives and a number of technical challenges in Section 5.2. In Section 5.3, we classify existing MAC protocols into two categories: conventional MAC protocols and advanced MAC protocols, including protocols designed for WMNs with directional antennas, multichannel MAC protocols, and contention-free MAC protocols. In addition to classifying existing schemes, we also present the operation, pros and cons of representative schemes in each category. Section 5.3 also describes three advanced MAC features defined in the current 802.11s standard draft. Design trade-offs are discussed in Section 5.4. Section 5.5 concludes by discussing future directions and standardization trends in this active area.

5.1 DESIGN OBJECTIVE AND CHALLENGES

While WMNs can extend network coverage and potentially increase network capacity, they can also impose unique challenges on MAC protocol design. The first and foremost challenge stems from the ad hoc nature of WMNs. In the absence of fixed infrastructure that characterizes traditional wireless networks, control and management of WMNs have to be *distributed* across all nodes. Distributed MAC is a much more challenging problem than centralized MAC. For multichannel MAC protocols, distributed channel selection and/or channel assignment adds another level of difficulty.

The second challenge is due to the multihop transmissions in WMNs. As nodes may not necessarily be within each other's radio range, packets have to be relayed from one node to another before they can reach the destination. In wireless networks, radio signals attenuate over distance. Therefore, simultaneous transmissions may lead to collisions at the receiver even though both senders sense the channel to be idle. Due to the inefficiency of carrier sensing in multihop wireless networks, the "hidden node" problem and "exposed node" problem may occur [9].

A hidden node (node B in Figure 5.1) is a node that is out of range of a transmitter node (node A in Figure 5.1), but within the range of a receiver node (node C in Figure 5.1). As nodes A and B are out of each other's sensing range, they may transmit at the same time, which causes a collision at the receiver as illustrated in Figure 5.1. Such hidden nodes can lead to high-collision probability and can cause substantial interference. The exposed node problem occurs when a node that overhears any data transmission has to refrain from transmitting or receiving, even though its transmission may not interfere with the ongoing data transmission at all. An exposed node (node C in Figure 5.2) is a node that is out of the range of a receiver node (node A in Figure 5.2), but within the range of a transmitter node (node B in Figure 5.2). The dotted circles illustrate the radio range of nodes in the center of the circle. On detecting a transmission from node B, node C defers its transmission to node D, even though a transmission from

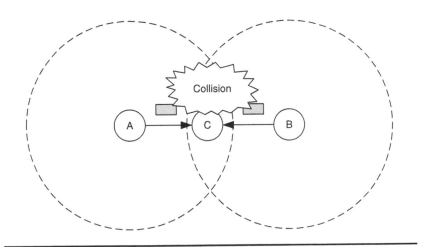

Figure 5.1   An illustration of the hidden node problem.

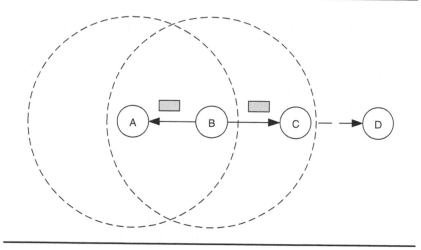

Figure 5.2 An illustration of the exposed node problem.

node C does not interfere with the reception at node A. Due to the exposed node problem, the link utilization can be significantly impaired, which leads to low end-to-end throughput and high packet-delivery latency.

In addition to the hidden node problem and the exposed node problem, nodes that utilize directional antennas or a multichannel MAC protocol may suffer from a "deafness" problem. In a network where nodes utilize directional or smart antennas, a transmitter may fail to communicate to its intended receiver because the receiving antenna is pointed in a direction away from the transmitter. In a multichannel network, the deafness problem occurs when two neighboring nodes choose different channels for transmitting and receiving. Therefore, even though two nodes are within each other's radio range, they cannot detect each other. Without a proper design of the MAC protocol, the deafness problem may cause significant throughput reduction and unfairness in the network.

The third challenge is introduced by the dynamical nature of the WMNs, such as variations in link quality, changing congestion levels, and user mobility. When the network environment changes, a MAC protocol should quickly adapt.

In addition to the above challenges that are unique to WMNs, MAC design faces challenges due to the error-prone nature of wireless channels. Link-layer retransmission schemes are often used in wireless MAC protocols to make the wireless link more reliable.

5.2   CONVENTIONAL WIRELESS MAC PROTOCOLS

The main responsibility of a MAC protocol is to ensure fair and efficient resource sharing. There are two major categories of MAC schemes: (1) contention-based protocols and (2) collision-free channel partition protocols. Contention-based protocols assume there is no central entity to allocate channel resources in the network. To transmit, each node must contend for the medium. Collisions result when more than one node tries to transmit at the same time. The well-known contention-based wireless MAC protocols include Aloha, Slotted Aloha, CSMA, and CSMA/CA. In contrast, collision-free protocols assign dedicated channel resources to each node that wishes to communicate. Collision-free protocols can effectively eliminate collisions at the cost of possibly low channel utilization for bursty data traffic. Examples of such protocols include time division multiple access (TDMA), frequency division multiple access (FDMA), and code division multiple access (CDMA). As collision-free protocols often require a centralized algorithm, which is hard to achieve in a multihop environment, most existing MAC protocols for WMN belong to the contention-based protocol category.

5.2.1   Aloha and Slotted Aloha

The first contention-based MAC protocol is Aloha, developed in the early 1970s by Abramson at the University of Hawaii [6]. The basic operation of Aloha is as follows: nodes can transmit whenever they have a packet to send. The receiver needs to acknowledge successful receipt of the data packet. If a collision occurs and the packet is corrupted, the sender does not receive an acknowledgment (ACK) within the time-out period. Then, the sender waits for a random amount of time and retransmits the packet. Aloha is easy to implement but suffers low throughput problems. For instance, the maximum achievable throughput is only 18.4% of the total available bandwidth.

Slotted Aloha improves on the performance of pure Aloha by synchronizing the transmission time slots of all nodes. In Slotted Aloha, nodes can only transmit at the beginning of a time slot. By restricting the starting time of frame transmissions, collisions can occur only when two frames are transmitted in the same time slot. Therefore, the vulnerable period for slotted Aloha is only one time slot vs. two time slots as in Aloha. The introduction of time slots doubles the throughput by reducing the probability of collisions by one-half compared to Aloha. However, both Aloha and Slotted Aloha still suffer from a stability problem that can occur when a large number of nodes have backlogged frames to transmit. This performance degradation occurs mainly because all nodes transmit at will without considering transmissions at other nodes.

As nodes in a network need to collaborate to transmit, knowledge of the behavior of other nodes can help a node to make a better decision. For instance, if a node can detect whether other nodes are currently transmitting or not, it can adapt its behavior accordingly. CSMA that is introduced next is based on this idea.

5.2.2 CSMA and CSMA/CA

A family of CSMA protocols was proposed in the 1970s by Kleinrock and Tobagi [21]. In CSMA, a node first senses the channel to make sure that it is idle before transmitting. If the channel is busy, the node defers its transmission. The exact behavior of a node that senses a busy channel leads to different versions of CSMA, described by terms such as 0-persistent CSMA, 1-persistent CSMA, and p-persistent CSMA. Using carrier sensing, a node can successfully avoid collisions with transmitting stations within its carrier sense range. Therefore, CSMA performs better than Aloha and Slotted Aloha.

CSMA/CA leverages the performance benefits of CSMA and extends CSMA to further reduce the likelihood of collisions. In a wireless network, as radio signals attenuate over distance, simultaneous transmissions may lead to collisions at the receiver even though both senders have sensed an idle channel. This is called the hidden node problem. By utilizing two small control packets, i.e., request-to-send (RTS) and clear-to-send (CTS), CSMA/CA can effectively mitigate the hidden node problem [20]. To avoid a collision, a sender that has a data packet to send first senses whether the channel is idle and then sends out an RTS message. On overhearing the RTS message, all nodes within the sender's radio vicinity refrain from transmitting for a period of time specified in the RTS. The designated receiver replies back with a CTS message, which silences all the nodes within the receiver's radio vicinity. After the RTS/CTS exchange, the sender's data packet is transmitted normally. With this approach, although there still could be collisions on small control packets (i.e., RTS), collisions on larger data packets can be largely eliminated. Thus, CSMA/CA often achieves improved performance over CSMA, especially when the data packets are large.

5.2.3 IEEE 802.11 DCF Protocol

The IEEE 802.11 standard specifies two medium access methods: (1) distributed coordination function (DCF) that builds on CSMA/CA and (2) point coordination function (PCF) providing contention-free access [1]. Because PCF requires a central control entity, i.e., a point coordinator, it is rarely used in WMNs. Instead, the DCF protocol is widely adopted in WLANs and WMNs. Additionally, because of its robustness and flexibility, many advanced MAC protocols are also based on the IEEE 802.11 DCF protocol.

The IEEE 802.11 DCF protocol is based on the CSMA/CA principle and it operates in a similar way. A node wishing to transmit first senses the channel. If the medium is sensed busy, it defers its transmission. If the medium is free for a specified period of time called distributed inter frame space (DIFS), the node is allowed to transmit. Upon correctly receiving the data packet, the receiver returns an ACK after a fixed period of time called short inter frame space (SIFS). Receipt of the ACK indicates the correct reception of the data packet. If no ACK is received, the sender assumes a collision has occurred and doubles the size of its contention window. Then, the sender chooses a random back-off number between 0 and its contention window size. The sender is allowed to retransmit the packet when the channel is free for a DIFS period of time augmented by the random back-off time. The packet is dropped after a given number of failed retransmissions.

To further reduce the collisions, the standard defines a virtual carrier sense mechanism as shown in Figure 5.3. A node wishing to transmit first transmits a short control packet called RTS, which includes source, destination, and duration of the packet transmissions that will follow (i.e., the CTS, the data packet, and the corresponding ACK). If the medium is free, the receiver responds with a CTS message that includes the duration of the data packet and its ACK. Any node receiving the RTS and/or CTS messages sets its network allocation vector (NAV) to the given duration. Once set, the NAV counts down to zero unless the NAV is set to another value by a new RTS or CTS. A node cannot transmit until its NAV equals zero. This virtual carrier sensing mechanism can effectively reduce the probability of a collision at the receiver by a station that is hidden from the transmitter; this is because nodes within the receiver's radio vicinity can hear the CTS and they mark the medium as busy until the end of the data transmission. The duration information carried by the RTS protects the transmitter from collisions when receiving the ACK. Because the RTS and CTS

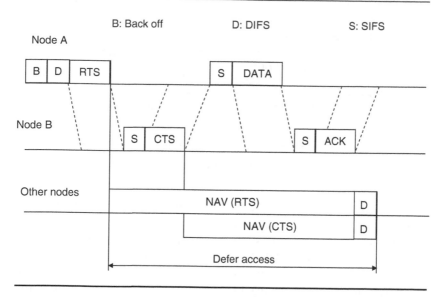

Figure 5.3 The IEEE 802.11 MAC handshake and channel reservation mechanism.

are short frames, the overhead of collisions on RTS/CTS is relatively smaller compared to the bigger data packets.

5.2.4 IEEE 802.11e MAC Protocol

The IEEE 802.11e standard draft defines a number of QoS enhancements to IEEE 802.11 [4]. Two main functional blocks are defined in IEEE 802.11e: (1) the channel access functions and (2) the traffic specification (TSPEC) management. The channel access function defines a new coordination function called the hybrid coordination function (HCF). HCF has two modes of operation: a contention-based protocol called enhanced distributed channel access (EDCA) and a polling mechanism called HCF controlled channel access (HCCA). Both access functions enhance or extend the functionalities of the original 802.11 access method, i.e., DCF, and operate on top of DCF (Figure 5.4). Because the operation of HCCA and PCF requires a central control entity and synchronization among nodes, they are hard, if not impossible, to realize in a WMN. Hence, we focus mainly on the EDCA access method.

EDCA enhances the original DCF by providing prioritized medium access based on different traffic classes, also called access categories (ACs). The IEEE 802.11e defines four ACs, each of which has its own

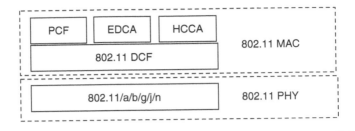

Figure 5.4 The IEEE 802.11e MAC architecture.

queue and its own set of EDCA parameters. The differentiation in priority between ACs is realized by setting different values for the EDCA parameters. The EDCA parameters include: (1) arbitration inter-frame space number (AIFSN), (2) minimum contention window (CW_{min}), (3) maximum contention window (CW_{max}), and (4) transmission opportunity (TXOP) limit. AIFS is the period of time the wireless medium is sensed idle before the start of a frame transmission. For instance, an AIFSN of 2 corresponds to DIFS and an AIFSN of 1 corresponds to PIFS. The relationships between different IFS are illustrated in Figure 5.5. A TXOP is a bounded time interval in which a node is allowed to transmit a series of frames.

As shown in Table 5.1, real-time traffic such as video and voice has more aggressive EDCA parameters, which is to ensure QoS traffic has a better chance to acquire the medium than the best-effort or background traffic. This basic idea of supporting prioritized traffic is similar to DiffServ [3]. As illustrated in Figure 5.6, when data arrives, the IEEE 802.11e MAC classifies the data with the appropriate AC and puts it in

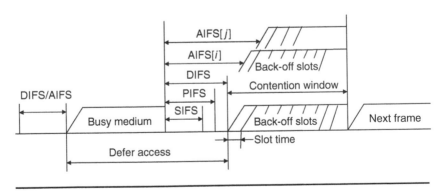

Figure 5.5 Relationships of different IFSs.

Table 5.1 Default EDCA Parameter Set

AC	CW_{min}	CW_{max}	AIFSN	TXOP Limit For 11b PHY	For 11a PHY
Background (AC_BK)	CW_{min}	CW_{max}	7	0	0
Best effort (AC_BE)	CW_{min}	CW_{max}	3	0	0
Video (AC_VI)	$(CW_{min} + 1)/2 - 1$	CW_{min}	2	6.016 ms	3.008 ms
Voice (AC_VO)	$(CW_{min} + 1)/4 - 1$	$(CW_{min} + 1)/2 - 1$	2	3.264 ms	1.504 ms

the corresponding AC transmit queue. Data packets from different AC queues first contend internally among themselves based on each queue's AIFSN, the contention window, and the random back-off time. The AC with the smallest back-off wins the internal contention. The winning AC then contends externally for the wireless medium. The external contention is similar to that in DCF. With the tuning of AC parameters provided by IEEE 802.11e, traffic performance from different ACs is optimized and prioritization of traffic achieved.

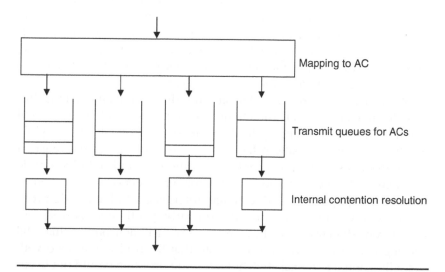

Figure 5.6 An EDCA reference implementation model.

5.3 ADVANCED MAC PROTOCOLS FOR WMNs

Conventional MAC protocols may suffer from low throughput in WMNs, due to contention introduced by the multihop operation. Therefore, MAC protocols assuming enhanced physical layers and/or multichannel MAC protocols have attracted considerable interest in recent years. In addition, under the assumption that mesh nodes are synchronized with each other, contention-free MAC protocols, such as TDMA-based schemes, have been proposed.

5.3.1 Protocols for Mesh Nodes Equipped with Directional Antennas

MAC protocols designed for WMNs typically assume omnidirectional antennas that transmit radio signals to and receive them from all directions. When two nodes are communicating, all other nodes in the vicinity have to remain silent, which has a negative impact on the network capacity. Therefore, the network capacity diminishes with an increase in the number of nodes [15]. With directional antennas (including smart antennas), two pairs of nodes located in each other's radio vicinity may potentially communicate simultaneously, depending on the directions of transmission. Due to this spatial reuse, directional antennas offer tremendous potential to improving the performance of WMNs without a significant increase in hardware cost. Furthermore, due to high antenna gains, directional antennas are expected to provide increased coverage and connectivity among mesh nodes, and diminished sensitivity to nonmesh 802.11 devices. However, harnessing these potentials requires new mechanisms at the MAC layer for intelligently and adaptively exploiting the antenna system.

As shown in Figure 5.7, unlike an omnidirectional antenna that has a uniform gain over all directions, the main lobe of a directional antenna has much higher gain than the side and back lobes. While this type of antenna pattern facilitates more efficient spatial reuse, it also introduces three new problems: a different kind of hidden node problem, a deafness problem, and a higher directional interference. For nodes equipped with directional antennas, a hidden node problem occurs when two transmitters are nearby yet their antennas point in different directions, so they are invisible to each other while still causing collisions at the receiver. Note that for nodes equipped with omnidirectional antennas, this hidden node problem would not occur because as long as two transmitters are within each other's radio range, they can detect each other. The deafness problem occurs

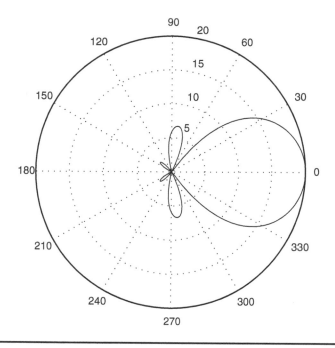

Figure 5.7   Antenna pattern of a directional antenna.

when the transmitter fails to communicate with the receiver because the receiver is listening in another direction. Higher directional interference is caused by the higher antenna gains. Existing MAC protocols try to address these three new problems in different ways.

Most MAC protocols in this class are based on the IEEE 802.11 DCF protocol, which typically comprises the RTS–CTS–DATA–ACK procedure. However, unlike the DCF protocol that transmits control and data messages omnidirectionally, MAC protocols utilizing directional antennas use different combinations of directional and omnidirectional messages.

The Directional MAC (D-MAC) is one of the first MAC protocols designed for ad hoc networks with directional antennas [22]. D-MAC assumes that each node is equipped with several directional antennas, but only one is allowed to transmit at any given time, depending on the location of the intended receiver. RTS, DATA, and ACK are sent directionally to reduce the number of exposed nodes and to improve spatial channel reuse, whereas CTS is transmitted omnidirectionally to reduce the number of hidden nodes. Alternatively, RTS packets are sent omnidirectionally if none of the directional antennas at the transmitter are blocked. This is to reduce the collision probability of control packets.

The tone-based directional MAC (Tone DMAC) protocol proposes to use omnidirectional out-of-band tones to indicate deafness to blocked transmitters [11]. The RTS, CTS, DATA, and ACK are all transmitted directionally. Nodes that overhear any of the directional transmissions update their records to avoid interfering with communication in those directions. After a DATA/ACK exchange, both the transmitter and the receiver switch back to the omnidirectional mode and send out omnidirectional tones to indicate that they were recently engaged in communication. Thus, a neighboring node can realize deafness and reduce its contention window to its minimum value. Without receiving the tones, a blocked node may keep on increasing its contention window size and lose its fair chance to access the medium. Transmitters may be identified by the different tones they use.

The directional virtual carrier sensing (DVCS) approach adds three primary capabilities to the original IEEE 802.11 MAC protocol: caching the angle of arrival (AOA), beam locking and unlocking, and use of directional NAV (DNAV) [34]. Each node caches AOAs from neighboring nodes when it overhears a signal. Based on the cached location information, a sender sends an RTS directionally to the receiver for up to four times. After four times, the RTS is sent omnidirectionally. A node can adapt its beam pattern during the RTS/CTS exchange and lock its beam pattern for data transmission and reception. The beam patterns at both the sender and the receiver are unlocked after the ACK frame transmission is completed. In DVCS, each DNAV is associated with a direction, a beamwidth and a duration. Multiple DNAVs can be set for a node.

Korakis et al. [23] propose to use circular directional RTS to inform the neighbors about the intended transmission. The transmitter transmits RTS on each of its antennas sequentially until the transmission of RTS covers all the area around the transmitter. The RTS contains the duration of the intended four-way handshake (as in 802.11). As the information is spread around by the circular RTS, the neighbors are informed about the intended transmission. The neighbors can then decide whether they should defer their transmissions in the direction of the transmitter or receiver. Because the neighbors are aware of the intended handshake, the number of hidden nodes can be significantly reduced.

In addition to the above protocols, many other MAC protocols that try to take advantage of directional beam-forming have been

Table 5.2   Comparison of Existing Algorithms

Protocols	Carrier Sensing	RTS	CTS	DATA	ACK	Tones
Directional MAC (D-MAC) [22]	O[a]	D/O[b]	O/O	D/O	D/O	No
Tone-based directional MAC (Tone DMAC) [11]	O	D/O	D/O	D/O	D/O	Yes
Directional virtual carrier sensing (DVCS) [34]	D[c]	D/D	D/D	D/D	D/D	No
Circular directional RTS [23]	O	Circular-D/O	D/O	D/D	D/D	No

[a]O: Omnidirectional.

[b]D/O: Directional transmission and omnidirectional receiving.

[c]D: Directional.

proposed [17,26,29,30]. Table 5.2 compares some of the existing protocols based on whether they use directional transmissions or out-of-band tones, where "D" represents directional, "O" represents omnidirectional, and "D/O" means directional transmission and omni-directional receiving [37]. Even though most of the proposed protocols improve performance compared to the standard 802.11 MAC, support-ing directional antennas in multihop WMNs is a difficult task and future work in this area is needed to fully exploit the benefits of beam-forming.

5.3.2   Multichannel MAC Protocols

Traditionally, most MAC protocols are designed to operate on only one channel, which could greatly limit the transport capacity of a wireless network. Under the current IEEE 802.11 standards, multiple nonoverlapping channels are available and can be utilized simultan-eously, which offers a great potential for improving the capacity of WMNs. In fact, in any multi-access point WLAN deployment, for improved capacity, adjacent WLAN cells use orthogonal channels to enable simultaneous transmissions.

Some of the early work in multichannel MACs, such as Sousa and Silvester [33] and Joa-Ng and Lu [19], assume that no channel assignment or selection is needed because every node can have its own unique channel. However, in reality, the number of available channels is limited and channels have to be assigned to each node dynamically to avoid contention and collisions and to enable optimal spatial reuse of available channels.

We classify the existing multichannel MAC protocols according to the channel selection techniques. Specifically, based on the way the channel is selected, existing approaches can be classified into three categories: (1) handshake-based channel selection, (2) channel hopping, and (3) cross-layer channel assignment. Existing schemes can also be classified based on other criteria. For instance, some protocols use a common control channel for all nodes, while others do not. The purpose of using a common control channel is to transmit control packets that assign data channels to mobile nodes. Some protocols require multiple transceivers, while others require only one transmitter and multiple receivers, or one transceiver. Later, this section discuss the representative multichannel MAC schemes in each category according to the channel selection techniques.

5.3.2.1 Handshake-Based Channel Selection

Many multichannel MAC protocols utilize a handshake between the transmitter and the receiver for channel selection. Examples include the dynamic channel assignment (DCA) protocol [39], the multichannel CSMA MAC [18], and multichannel MAC (MMAC) [31]. Just like the IEEE 802.11 standard, the handshake mechanism is realized by exchanging control messages between senders and receivers.

In DCA, the total bandwidth is divided into one control channel and n equivalent data channels [39]. Furthermore, DCA assumes that each node is equipped with two half-duplex receivers, one is used for the control channel and the other can switch between different data channels. A five-way handshake procedure is employed to select channel and transmit data packets (Figure 5.8). In Figure 5.8, D indicates a period with duration given by a DIFS time during which carrier sensing is performed. B is the back-off period. S indicates when performing carrier sensing is performed for a period given by the SIFS. All nodes maintain two data structures, a channel usage list (CUL) and a free channel list (FCL), to keep track of the data channels that are being used and channels that are free. A node builds its CUL by listening to neighboring nodes' control messages that carry channel usage information. After performing carrier sensing

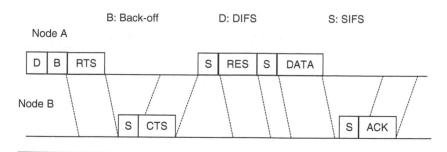

Figure 5.8   Five-way handshake procedure of DCA. (From Wu, S.-L., Lin, C.-Y., Tseng, Y.-C., and Sheu, J.-P., in *Proc. International Symposium on Parallel Architectures, Algorithms and Networks (ISPAN) 2000,* **Dallas/Richardson, Texas, December 2000, pp. 232–237. With permission.)**

and back-off, the transmitter sends an RTS message with a list of free channels. Upon receiving the RTS message, the receiver checks to see which channel in the transmitter's FCL can be used in its own radio vicinity. Then, the receiver sends back a CTS message indicating the channel to use. Before transmitting the data packet, the transmitter sends out a reservation (RES) packet to inform its neighbors of the reserved data channel. The RTS, CTS, and RES messages are transmitted to the control channel, while data and ACK packets are transmitted to the data channel. Collisions on data channels can be mitigated through the use of multiple channels. In addition, separating control and data channels can alleviate the hidden node problem.

The multichannel CSMA MAC [18] has a channel selection procedure similar to that of DCA. Before sending an RTS, a sender first senses the carrier on all data channels and builds a list of data channels that are available for transmission. If none of the data channels are free, the sender should enter back-off. Otherwise, the sender sends out an RTS with an FCL. On successful reception of the RTS message, a receiver creates its own FCL by sensing all the data channels. It then compares its own FCL with the one contained in the RTS message. If there are free channels in common, the receiver selects the best free channel based on the channel condition. The receiver then sends back a CTS message to inform the sender the channel to be used. Because a node needs to sense all data channels at the same time, this protocol requires that each node be equipped with one transmitter and multiple receivers.

In multichannel MAC (MMAC), time is divided into beacon intervals [31]. Each beacon interval is further divided into two smaller

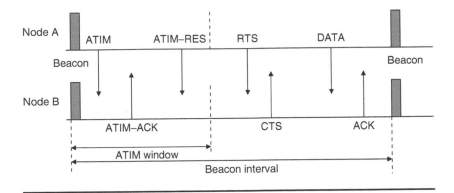

Figure 5.9 Process of channel negotiation and data exchange in MMAC. (From So, J., and Vaidya, N., in *Proc. ACM International Symposium on Mobile Ad Hoc Networking and Computing (Mobi-Hoc) 2004*, Tokyo, Japan, May 2004, pp. 222–233. With permission.)

intervals (Figure 5.9). The first interval is used for channel negotiation and the second interval for transmitting data packets. A small window, called an ad hoc traffic indication message (ATIM) window, is placed at the beginning of each beacon interval. Here, all nodes transmit and receive on a common control channel. Each node maintains a preferable channel list (PCL) that records the usage of channels inside the transmission range of the node. Channels are categorized into three different preferences based on predetermined criteria. If a node A has buffered data for node B, it sends an ATIM message that contains its PCL during the window. Upon receiving the ATIM message, the destination node B selects a channel based on the sender's PCL and its own PCL. Node B then returns an ATIM–ACK message with the selected channel information. If node A selects the channel specified in the ATIM–ACK message, it sends out an ATIM–RES message to notify its neighboring nodes which channel will be used. In MMAC, all channel negotiations occur in ATIM windows over the common control channel. The RTS, CTS, ACK, and data are all transmitted on the negotiated data channel. MMAC solves the hidden node problem by synchronizing all nodes in the network and by allowing the nodes to negotiate channels at the same time. However, MMAC has stringent synchronization requirements that cannot be easily satisfied in wireless ad hoc networks. In addition, MMAC does not support broadcasting, which is required by most routing protocols.

Handshake-based channel selection is widely used in MMAC protocols and it works even when nodes are not synchronized. However, the channel negotiation is usually on a per-packet basis, which may incur high control overhead.

5.3.2.2 Channel Hopping

Some MMAC protocols use channel hopping to achieve data exchange between two nodes. Two such examples are receiver-initiated channel-hop with dual polling (RICH-DP) [36] and slotted seeded channel hopping (SSCH) [7].

RICH-DP requires all nodes in a network to follow a common channel-hopping sequence and each hop lasts just long enough for the nodes to receive a collision-avoidance control packet from a neighbor [36]. A node ready to poll any of its neighbors sends out a ready to receive (RTR) message over the current channel hop. Upon successful reception of the RTR message, the polled node starts sending data to the polling node immediately and over the same channel hop, while other nodes hop to the next channel hop. When the data transmission from the polled node is completed, the polling node may start transmitting its own data to the polled node over the same hop. After the data transmissions between the two nodes is over, both nodes resynchronize to the common channel hop. If the polled node has no data to send, the polling node rejoins the rest of the network at the current channel hop.

SSCH assumes that each node can calculate and update its channel hopping sequence based on an initial channel index and a seed. Nodes then switch channels from one time slot to the other based on their channel hopping sequences [7]. Nodes may have different channel hopping sequences. However, the channel hopping sequences are designed in a way that there will always be at least one overlapping channel between any two nodes at some time instances. Then, nodes can learn each other's hopping schedules by broadcasting their channel schedules. If a node has a packet queued for a destination node, the sender attempts to change part of its own channel hopping schedule to match that of the destination node. When the sender and the receiver start to share overlapping channels, they can start to transmit to each other. SSCH requires no dedicated control channel, but needs clock synchronization among nodes.

5.3.2.3 Cross-Layer Channel Assignment

Distributed channel assignment problems have been proven to be *NP*-complete and are thus computationally intractable [8,16]. There exist only a few heuristic solutions, all with considerable complexity, especially for the mobile environment [8,13,16]. However, one way to achieve effective channel selection with little control overhead is by combining channel assignment with the routing protocol [14]. Because the channel assignment is performed by routing, the MAC protocol only needs to manage MAC. As a result, the design of MAC protocol is significantly simplified. A second advantage of this approach, i.e., the separation of channel assignment and MAC, is that it enables optimization of different modules separately. For instance, channel assignment can be combined with different reactive or proactive routing protocols. The MAC protocol can also be designed independently without the knowledge of how channels are assigned to individual nodes. In addition, the separation of functions makes it possible to design backward compatible and practical MMAC protocols. MMAC protocol is based on the idea of cross-layer channel assignment [14].

Like many other MMAC protocols, MMAC assumes one common control channel and multiple data channels. All nodes in the network share the same common control channel and each node is equipped with two half-duplex transceivers. One transceiver listens on the common control channel all the time, whereas the other transceiver can switch from one data channel to another. Nodes are assigned data channels by the routing protocol. When a node is ready to transmit, it

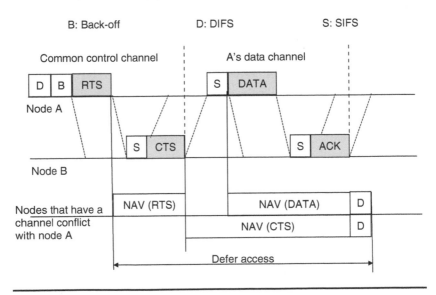

Figure 5.10 Four-way handshake procedure of MMAC.

first informs the destination node of its assigned data channel. As in Figure 5.10, when the sender node A intends to transmit, it first uses the control channel to broadcast an RTS message, carrying its own data channel index, c_A. Upon receiving the RTS message, the destination node B uses the control channel to return a CTS message carrying c_A and switches its receiving channel to c_A. After node A receives the CTS message, it switches to the confirmed data channel c_A and starts data transmission. Neighboring nodes that overhear the RTS/CTS exchange but do not share the same data channel with node A should defer only for the duration of the control message transmission. The sender listens to the data channel until an ACK is received or a timeout occurs. Then, it switches to the common control channel. Similarly, after sending the ACK message on the data channel, the receiving node also switches back to the control channel.

In MMAC, because channel assignment information is piggy-backed in routing control messages, it can propagate to nodes more than one-hop away, whereas most channel selection schemes pro-posed in MMAC protocols only consider nodes within each other's radio range. Because the channel selection is performed not on a per-packet basis, but on a per-route setup basis, MMAC has less computational overhead. In addition, MMAC has lower communica-tion overhead than existing distributed channel assignment protocols since all channel assignments are piggybacked onto routing control messages. If applied properly, cross-layer design can yield very prac-tical and efficient schemes.

In addition to the above described MAC protocols, there has been a growing trend in designing routing protocols that are suitable for multihop multichannel WMNs [25,28,32]. Traditional routing protocols do not account for channel diversity and, thus, they cannot take full advantage of the multiple channels available in a network. Due to the multiple available channels, different radio links may experience vari-ous interference and contention levels. A common approach is to allow routing algorithms to take both hop count and channel diversity into consideration, such that channel utilization is maximized and the system performance is improved. This body of work still utilizes the IEEE 802.11 MAC protocol, i.e., a single-channel MAC, instead of a multichannel MAC protocol.

Raniwala and Chiueh [28] propose to utilize local traffic load infor-mation to dynamically assign channels to different interfaces and to route packets according to routing metrics such as path capacity. By combining routing with intelligent channel assignment, the routing algorithms can achieve a factor of 6 to 7 throughput improvement

Table 5.3 Important Design Feature of Existing Algorithms

Protocols	Medium Access	Channel Selection	Hardware Requirement
Dynamic channel assignment (DCA) [39]	CSMA/CA	Per packet	2 Transceivers
Multichannel MAC (MMAC) [14]	CSMA/CA	Per-route setup	1 or 2 Transceivers
Multichannel MAC (MMAC) [31]	CSMA/CA	Per beacon interval	1 Transceiver synchronization required
Multichannel CSMA [18]	CSMA/CA	Per packet	1 Transmitter multiple receivers
Receiver-initiated channel-hop with dual polling (RICH-DP) [36]	Channel hopping	Hopping sequence	1 Transceiver synchronization required
Slotted seeded channel hopping (SSCH) [7]	Channel hopping	Hopping sequence	1 Transceiver synchronization required

compared to a conventional single-channel scheme. Similar problems have also been investigated by So and Vaidya [32] as well as Kyasanur and Vaidya [25]. While So and Vaidya [32] study routing and channel assignment with single-Network Interface Card (NIC) devices, Kyasanur and Vaidya [25] propose a routing protocol that can work in general multiradio WMNs.

In addition to utilizing different channel selection techniques, existing MMAC protocols have different MAC techniques and hardware requirements. Table 5.3 summarizes some important features of existing MMAC protocols. There is no general rule as to which scheme is better than another. Simpler schemes with reduced hardware requirements are easy to implement, whereas complex schemes with greater hardware requirements often yield better performance.

5.3.3 Contention-Free MAC Protocols for Synchronized Mesh Networks

Currently, most of the current WMNs are based on 802.11 MAC or its derivatives due to the asynchronous operation of mesh nodes. However, if all mesh nodes are synchronized with each other, contention-free MAC protocols can be used to improve the network performance. One of the most popular contention-free MAC protocols is defined in IEEE 802.16 [2], also called WiMax.

The 802.16 MAC protocol is designed to support point to multi-point (PMP) and mesh network models. Utilizing TDMA and originally proposed for licensed bands, the 802.16 MAC protocol tries to achieve efficient use of channel bandwidth. The 802.16 MAC protocol is essentially connection oriented. Upon entering the network, each subscriber station (SS) creates one or more connections, identified by their unique connection IDs (CIDs), over which their data are transmitted to and from the base station (BS). The MAC layer schedules the usage of the airlink resources and provides QoS differentiation. The 802.16 PHY allows the use of two different duplexing schemes: frequency division duplex (FDD) and time division duplex (TDD). Time division multiplexing (TDM) and TDMA are used for downlink data bursts, while the uplink is shared among SSs in a TDMA fashion. Because each downlink burst may contain data for several stations, each station is required to recognize data packets with known CIDs. As illustrated in Figure 5.11, the uplink period is divided into three different periods: the initial maintenance opportunity, the request contention opportunity, and the data grants period. A SS sends access bursts in the initial maintenance opportunity to determine network delay and to request power or profile changes. In the request contention opportunity, a SS requests bandwidth in response to polling from BS. Because SSs need to contend to transmit in both the initial maintenance opportunity and the request contention opportunity, collisions may occur. After a SS is granted bandwidth by the BS, it can transmit data bursts in the granted

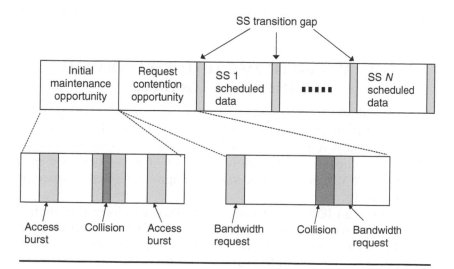

Figure 5.11   The 802.16 uplink super frame structure.

interval, during which collisions do not occur. Note that SSs can request bandwidth in three different ways: (1) use the request contention opportunity upon being polled by the BS, (2) send a standalone MAC message called the "bandwidth (BW) request" in an already-granted time slot, or (3) piggyback a BW request message on a data packet. Upon receiving the BW request, the BS can grant bandwidth in one of two ways: (1) grant per SS (GPSS) or (2) grant per connection (GPC).

Another major function of the 802.16 MAC layer is to perform link adaptation and automatic repeat request (ARQ) functions to maintain desired bit error rates (BERs) while maximizing the data throughput. Transmission parameters, such as modulation and forward error correction (FEC) settings, can be modified on a frame-by-frame basis for each SS. In addition, the 802.16 MAC layer performs both fragmentation of MAC service data units (SDUs) and packing of MAC SDUs to utilize bandwidth more efficiently. Small SDUs are packed to fill up airlink allocations and large SDUs are fragmented when they do not fit into an airlink allocation.

Even though IEEE 802.11 is still the dominant technology in WMNs, a few startup companies, such as Strix Systems, SkyPilot Networks, and Kiyon, have announced plans to build WiMax or WiMax-like mesh backhauls to achieve longer range and better network performance. Additional research has also been conducted to improve the performance of WiMax in WMNs. Examples include applying interference-aware routing and scheduling in WiMax mesh networks [38] and using concurrent transmissions for throughput enhancement [35].

5.4 ADVANCED MAC FEATURES PROPOSED BY THE 802.11 TGs Group

The original 802.11 MAC and recent MAC enhancements (e.g., 11e, 11i, and 11k) are designed primarily for one-hop wireless networks. However, WMNs require the coordination and collaboration of mesh APs over multiple hops. Therefore, new MAC features designed specifically for WMNs become necessary to improve the performance of such networks.

Recently, an IEEE 802.11 task group (TGs) was formed to draft a standard for wireless mesh networking. The first baseline draft of 802.11s supports the 802.11 DCF protocol and the 802.11e EDCA protocol with several additional MAC features [5]. The three most important MAC features include an intramesh congestion control scheme that seeks to relieve the congestion situations among wireless mesh nodes; an optional multichannel MAC protocol, i.e., common channel framework (CCF), which bears some resemblances to the MMAC protocol; and an optional mesh deterministic access (MDA) scheme that offers better QoS. The TGs group is currently working vigorously to refine these initial proposals and to include more advanced MAC features in the future.

5.4.1 Intramesh Congestion Control

With 802.11 MAC, each MP contends for the channel independently, without any regard for what is happening in the upstream or downstream nodes. One of the consequences is that a sender with backlogged traffic may rapidly inject many packets into the network, which would result in local congestion for downstream nodes. In wired networks and WLANs, one of the effective tools to combat congestion has been end-to-end flow control implemented at the higher layers of the network stack. For instance, the TCP sliding window is the primary example for end-to-end flow control at the transport layer.

However, a multihop wireless network, such as a WMN, cannot simply rely on higher layer end-to-end flow control to solve the congestion problem. First, most multimedia applications (video and voice) utilize UDP that does not have any form of congestion control. Figure 5.12 shows that data flows originating from source node 1, 2, and 7 all share a common link, i.e., link 5–6. Without a proper congestion control mechanism, this bottleneck link can easily be overwhelmed. Congestion control for UDP may not be as critical in a wired network because the wired network has much higher bandwidth per link and each individual hop in the wired network is

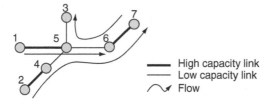

Figure 5.12 An illustration of the congestion problem in wireless mesh networks.

isolated from other hops. However, the bandwidth of wireless networks is very limited and the neighboring hops in the WMN share the same medium. Thus, how to schedule across these neighboring links to maximize the network throughput becomes much more important for WMNs. Secondly, recent studies have shown that TCP congestion control does not work well across a multihop wireless network [12]. To mitigate the congestion problem, a hop-by-hop congestion control mechanism that operates at the data link layer is described in the 802.11s standard draft.

To effectively control or avoid congestion in the network, each mesh node monitors its local/neighborhood congestion condition so that, when necessary, it can notify the neighborhood/upstream nodes of congestion by transmitting a broadcast "neighborhood congestion announcement" and/or a unicast "congestion control request." The standard does not mandate how to monitor and detect the congestion situation and it is up to the implementors to decide what scheme should be used. Two different monitoring and congestion detection mechanisms are provided as example implementations, but implementors are free to design their own congestion detection schemes as well. One way for detecting congestion is for each mesh node to keep track of its own effective MAC transmission and receiving rate for the packets to be forwarded (excluding the received packets destined for this MP), and to monitor the backpressure, which is the difference between the aggregate receive and transmit rates. The other suggested method for congestion detection is based on queue size. If the queue size is above a predefined upper threshold, the node informs its previous hop neighbors by sending unicast signaling messages "congestion control request messages" to each of its upstream nodes, so that the upstream nodes can decrease their transmission rate to it according to a local rate control mechanism. If the queue size is between the predefined lower and upper thresholds, the node again declares congestion, but this time, sends the unicast signaling message congestion control request to each of the upstream nodes with a certain probability.

The goal of rate control is to maintain near zero backpressure at the local node. When the backpressure builds up significantly at the local node, the node informs its previous hop nodes or neighbors so that the recipient nodes can decrease their transmission rate according to a local rate control mechanism. Upon receiving either congestion control request or neighborhood congestion announcement message, the receiving node needs to reduce its effective MAC transmission rate, accordingly, by locally rate limiting its traffic. Again, the standard only provides two example local rate control mechanisms and gives the implementors freedom to design their own local rate control mechanisms. The first example of the local rate control mechanism is based on dynamically adjusting EDCA parameters depending on the congestion condition in the node and/or the neighborhood. The EDCA parameters to be adjusted can be AIFSN, CW_{min}, or both. Use of such a mechanism would be especially effective when the source of a congestion-causing flow is a station (STA) associated with a MP. A MP implementation may adjust the EDCA parameters for the basic service set (BSS) to alleviate congestion due to traffic generated by associated STAs. Thus, STAs do not require explicit knowledge of the congestion control scheme. Another method of rate control, in response to a congestion control request, is to utilize a shaper that includes a traffic meter. The meter measures the temporal properties of a rate-controlled AC traffic against the mean data rate of the congestion control request message. If the packet is out-of-profile, the back-off timer's timeout does not activate a TXOP transmission. Otherwise the back-off timer's timeout activate a TXOP transmission.

5.4.2 Common Channel Framework

An optional common channel framework (CCF) is proposed to enable the operation of single-radio devices in a multichannel environment. The CCF assumes that each node is equipped with a single half-duplex transceiver and nodes in the network or in the same cluster share a common control channel. To legacy devices (STAs and AP) and MPs that do not support the CCF, the common channel appears as any other 802.11 channel and their operation remains unaffected. Using the CCF, node pairs or clusters, select a different channel and switch to that channel for a short period of time, after which they return to the

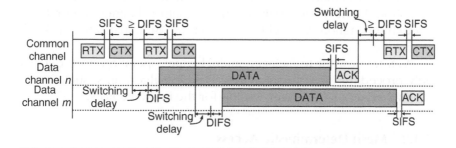

Figure 5.13   Dynamic channel selection on the common control channel.

common channel. During this time, nodes exchange one or more DATA frames. The channel coordination itself is carried out on the common channel by exchanging control frames or management frames that carry information about the destination channel. In this way, simultaneous transmission on multiple channels is achieved which in turn results in increased aggregate throughput.

Within the CCF, multichannel operation of single-radio MPs is facilitated by defining a channel coordination window (CCW) that is repeated periodically. As shown in Figure 5.13, MPs are synchronized to each other and utilize the common control channel to select an available data channel. At the start of a CCW, CCF-enabled MPs tune to the common channel and establish communication with each other. In addition, at the start of CCW, the channel occupancy status is reset and MPs can renegotiate channels. The CCW is repeated with a period P, and the duration of CCW is usually a fraction of P. Once on the common channel, an arbitrary MP can initiate transmission by sending an request-to-switch (RTX) frame carrying information of the destination data channel on which the communication can take place. The destination MP accepts this request by transmitting a clear-to-switch (CTX) frame carrying the same destination data channel. The destination MP can also decline this request by sending a CTX message with the destination channel index set to the common channel. If the receiving MP accepts the RTX request, the MP pair switches to the destination channel together. Once on the destination channel, the transmitting MP performs a clear channel assessment (CCA) for the duration of DIFS. If there are activities on the destination channel (e.g., channel is assessed as BUSY during CCA), the transmitting MP should mark the channel as unavailable and should not return to this channel for a predefined period. If the channel is found clear, the transmitting MP can send a DATA frame. If the receiving MP does not receive the data frame or the transmitting MP does not receive an ACK, they should switch back to the common control channel. After returning to the common channel the transmitting MP initiates the back-off procedure.

As a framework to facilitate multichannel operation, CCF defines only procedures and the interfaces. Detailed algorithms, such as the channel selection algorithm, are left for the implementors.

5.4.3 Mesh Deterministic Access

The MDA is an optional access method that allows MPs to access the channel with lower contention than otherwise in selected periods.

The goals are to utilize channel bandwidth more efficiently and to provide better QoS for periodic traffic. The method sets up contention-free time periods called MDA opportunity (MDAOP) in mesh neighborhoods where a number of MDA-supporting MPs may potentially interfere with each other. For each MDAOP period, supporting MPs are allowed to access the channel in a contention-free manner using MDA access parameters (CW_{min}, CW_{max}, and AIFSN). To use the MDA method for access, supporting MPs must synchronize to each other.

MDAOP is a period of time within every Mesh DTIM interval that is set up between a transmitter and a receiver. The setup of an MDAOP set is initiated by the intended transmitter, and is accepted or rejected by the intended receiver. Once accepted, the transmitter is referred to as the owner of the MDAOP. The setup procedure for an MDAOP set is as follows:

1. The MP that intends to be the transmitter in a new MDAOP set builds a map of neighborhood MDAOP times in the Mesh DTIM interval after hearing advertisements from all of its neighbors that have MDA active. If no advertisement was heard from a neighbor in the last dot11MDAdvertPeriodMax, the MP may request the neighbor for MDAOP advertisement.

2. Based on its traffic characteristic, the transmitter then chooses the MDAOP starting point and duration in the Mesh DTIM interval that do not overlap with either its neighborhood MDAOP times or the neighbor MDAOP interfering times of the intended receiver. It also avoids using times that are known to it as being used by itself or one of its neighbors for other activities such as beacon transmissions.

3. Before utilizing the new MDAOP set, the transmitter verifies that the new MDAOP set will not cause the mesh access fraction (MAF) limit to be crossed for any of its neighbors. If the MAF limit would be crossed for any of its neighbors, due to the new MDAOP set, it suspends the setup process. If the MAF limits at all neighbors are respected by the new MDAOP set, the transmitter transmits an MDAOP setup request Information Element (IE) to the intended receiver with the chosen MDAOP location and duration.

4. The receiver of the MDAOP setup request IE checks to see if the proposed MDAOP times have any overlap with its neighborhood MDAOP times. The receiver also checks if the new MDAOP set will cause the MAF limit to be crossed for any of its neighbors. The MDAOP setup reply IE is used to reply to a

setup request. The receiver rejects the setup request if there are any overlaps of the requested MDAOP set with its neighborhood MDAOP times, or other times that it knows are set to be used by itself or its neighbors for activities such as beacon transmissions. It may suggest alternate times by including the optional field alternate suggested request IE in the MDAOP setup reply element. The receiver also rejects the setup request if the MAF limit of itself or any of its neighbors will be exceeded due to the new setup.

5. If suitable, the receiver accepts the setup. After successful setup, both the MDAOP owner (the transmitter) and the receiver advertise the MDAOP set times in the transmit-receive (TX-RX) times report field of the MDAOP advertisement IE.

Once the setup of an MDAOP is advertised, all MPs that hear these advertisements except the transmitter that set up the MDAOP are required to not initiate any new transmission during the TXOP initiated in the MDAOP. This can be done by setting their NAVs for the duration of the MDAOP at the beginning of the MDAOP, or by using enhanced carrier sensing (ECS) that achieves the same result. The transmitter that sets up the MDAOP uses CSMA/CA and back-off to obtain a TXOP as described in 802.11e and using parameters $MDACW_{min}$, $MDACW_{max}$, and MDAIFSN. The ranges of values allowed for $MDACW_{min}$, $MDACW_{max}$, and MDAIFSN parameters are identical to that allowed for EDCA in 802.11e. If the MP successfully captures an MDA TXOP before the end of its MDAOP, it may transmit until the end of the MDAOP or until a time less than MDA TXOP limit from the beginning of the MDA TXOP, whichever is earlier. The retransmission rules for access in an MDAOP are the same as that of EDCA in 802.11e. Specifically, if there is loss inferred during the MDA TXOP, retransmissions require that a new TXOP be obtained using the MDA access rules in the MDAOP. No MDA TXOPs may cross MDAOP boundaries. If the MP intends to end the TXOP with enough time before the end of the MDAOP, it is responsible for relinquishing the remaining MDAOP time by using any of the methods that reduce NAV as defined in 802.11e and 802.11-1999.

MDA provides an effective way to utilize channel and to provide better QoS in a wireless mesh environment. However, further investigations are needed to see how devices supporting MDA and 802.11 can coexist in the same radio vicinity.

5.5 TRADE-OFFs AND CONSTRAINTS

Generally, different network environments may have different design objectives. Even for the same design objective, there could be different approaches, especially when it comes to the design of complex protocols such as wireless MAC protocols.

For instance, for protocols supporting directional antennas, there is a fundamental trade-off between spatial reuse and packet collisions. Directional transmissions can reduce interference and increase the spatial reuse. However, directional transmission and/or reception reduce the reserved region, which introduces extra hidden nodes and the deafness problem. Due to the increase in the number of hidden nodes, the number of collisions increases, which reduces the throughput. The negative impact of the deafness problem is even more profound, leading to excessive packet drops, large delay variance, and unfairness in the network, which can significantly reduce the network throughput. The existing MAC protocols designed to support directional antennas seek to find a balance between spatial reuse and an increased number of hidden nodes. However, this is not a trivial task and future work is needed to further exploit the benefits of directionality.

For multichannel MAC protocols, there is a trade-off between hardware complexity and network capacity. Providing one node with multiple NICs may achieve better performance and higher channel utilization. For instance, Raniwala and Chineh [28] report a factor of 6 to 7 improvement, while in some cases Kyasanur and Vaidya [25] observe a factor of 8 improvement, compared to the single-channel schemes. When the number of available transceivers per node is less than the number of available channels, network capacity degrades in many scenarios [24]. However, small mobile devices may not have multiple NICs due to high-cost and high-power consumption. Furthermore, multiple NICs that can transmit and receive simultaneously on the same device can generate nonnegligible interference to each other, to the extent that transmission on one channel typically precludes simultaneous reception on any nearby channel. So far, little work has been done on modeling interchannel interference generated by multiple transceivers that are on the same device. Additionally, equipping a device with multiple transceivers can dramatically increase the hardware cost.

Another design trade-off is between system performance and control overhead, which includes communication, computation, and storage overhead. Protocols that are more complex can sometimes achieve better performance. However, if the communication overhead becomes too high, it can degrade performance. For example, if too many control messages are sent in the network or the size of control messages becomes too large, these control messages may consume a significant portion of network capacity, which may result in reduced data throughput and increased delay. Furthermore, high-computation overhead may drain more power and take more processor time, while high-storage overhead requires more allocated memory. Thus, protocol performance and complexity is another important trade-off that needs to be considered in the design process. Other design trade-offs may be involved with compatibility with legacy systems such as IEEE 802.11 networks, scalability, and so on. For instance, due to the widespread popularity of IEEE 802.11 networks, a MAC protocol that is compatible with IEEE 802.11 might be adopted more easily.

By and large, even though some general design guidelines should be carefully followed, there is no "one-size-fits-all" solution that can meet all design requirements. Furthermore, different design objectives may need to be addressed in different scenarios. For example, in sensor networks, energy saving may be even more important than improving throughput.

5.6 CONCLUSIONS AND FUTURE DIRECTIONS

Even though wireless MAC protocols have been extensively studied since the 1970s, advanced MAC protocols suitable for WMNs did not gain much attention until recently. Such MAC design, however, is a challenging task due to the ad hoc nature of WMNs.

In this chapter, we presented the design objectives and technical challenges for effective MAC protocols in WMNs. Moreover, we classified and reviewed existing MAC protocols, and discussed several trade-offs and constraints in MAC design. There is no general solution that can meet all design requirements, so different design objectives must be emphasized in different scenarios.

Due to the multihop and ad hoc nature of WMNs, supporting directional antennas is not an easy task and most existing designs tend to be complex. In the future, simple and more efficient MAC protocols may be built assuming a simplified mesh architecture or multiradio support at each node. Another important future direction is the capacity of wireless networks that utilize directional antennas or multiple channels. There is some initial work on the capacity of multichannel wireless networks, assuming nodes are equipped with multiple NICs [24]. The performance of the IEEE 802.11 MAC has been a subject of research for years [10]. However, little work has been done in terms of theoretical modeling and analysis of MAC protocols utilizing directional antennas or MMAC protocols. Such analysis is important for studying the performance limits and for providing useful insights that can guide the design of practical and effective MAC protocols. The IEEE 802.16 MAC protocol utilizes TDMA and is designed to achieve more efficient bandwidth utilization than contention-based protocols. However, TDMA also has its inherent weaknesses. Because data traffic is highly bursty, MPs may have to invoke the scheduling algorithm frequently, which has a high computational burden. Furthermore, when the 802.16 MAC operates in the connection-oriented mode, setting up and tearing down connections incur overhead. In addition, further research is still needed to better understand the performance of WiMax in a multihop mesh networks.

This chapter also covers the three most important MAC features included in the first 802.11s standard draft. Within the 802.11 TG standards group, advanced MAC features that improve network performance yet are backward compatible with 802.11 devices are still under discussion. It is expected that, in the future, these MAC features will be further refined and more advanced MAC features will be included in the standard to improve the performance and to enhance the functionality of WMNs.

REFERENCES

1. IEEE Std. 802.11, Wireless LAN media access control (MAC) and physical layer (PHY) specifications, 1997.
2. IEEE Std. 802.16-2004, IEEE standard for local and metropolitan area networks—Part 16: Air interface for fixed broadband wireless access systems, 2004.
3. S. Blake, D. Black, M. Carlson, E. Davies, Z. Wang, and W. Weiss, *An architecture for differentiated services*, IETF RFC 2475, 1998.
4. IEEE 802.11e/D13.0, Medium access control (MAC) quality of service (QoS) enhancements, 2005.
5. IEEE 802 11-06/328r0, Joint SEE-Mesh/Wi-Mesh Proposal to 802.11 TGs, 2006.
6. N. Abramson, "The ALOHA system—another alternative for computer communications," in *Proc. AFIPS Fall Joint Computer Conf.*, Montvale, NJ, pp. 281–285, 1970.
7. P. Bahl, R. Chandra, and J. Dunagan, "SSCH: Slotted seeded channel hopping for capacity improvement in IEEE 802.11 ad-hoc wireless networks," in *Proc. IEEE/ACM the 10th Annual International Conference on Mobile Computing and Networking MobiCom*, Tokyo, Japan, pp. 216–230, May 2004.

8. A. Bertossi and M. Bonuccelli, "Code assignments for hidden terminal interference avoidance in multihop packet radio networks," *IEEE/ACM Trans. on Networking*, vol. 3, no. 4, pp. 441–449, August 1995.

9. V. Bharghavan, A. Demers, S. Shenker, and L. Zhang, "MACAW: A media access protocol for wireless LANs," in *Proc. ACM SIGCOMM*, London, UK, pp. 212–225, August 1994.

10. G. Bianchi, "Performance analysis of the IEEE 802.11 distributed coordination function," *IEEE Journal on Selected Areas in Communications*, vol. 18, no. 3, pp. 535–547, March 2000.

11. P.R. Choudhury and N.H. Vaidya, "Deafness: A MAC problem in ad hoc networks when using directional antennas", University of Illinois at Urbana Champaign, Tech. Rep., 2003.

12. Z. Fu, P. Zerfos, H. Luo, S. Lu, L. Zhang, and M. Gerla, "The impact of multihop wireless channel on TCP throughput and loss," in *Proc. IEEE INFO-COM'03*, San Francisco, CA, pp. 209–221, March/April 2003.

13. J.J. Garcia-Luna-Aceves and J. Raju, "Distributed assignment of codes for multihop packet-radio networks," in *Proc. IEEE MILCOM*, Monterey, CA, pp. 450–454, November 1997.

14. M. Gong, S. Mao, and S. Midkiff, "A cross-layer approach to channel assignment in wireless ad hoc networks," *ACM Mobile Networks and Applications (MONET)*, in press.

15. P. Gupta and P.R. Kumar, "The capacity of wireless networks," *IEEE Trans. on Information Theory*, vol. 46, no. 2, pp. 388–404, March 2000.

16. L. Hu, "Distributed code assignments for CDMA packet radio networks," *IEEE/ACM Trans. on Networking*, vol. 1, no. 6, pp. 668–677, December 1993.

17. Z. Huang, C.C. Shen, C. Srisathapornphat, and C. Jaikaeo, "A busy-tone based directional MAC protocol for ad hoc networks," in *Proc. IEEE MILCOM*, Anaheim, CA, pp. 1233–1238, October 2002.

18. N. Jain, S.R. Das, and A. Nasipuri, "A multichannel CSMA MAC protocol with receiver-based channel selection for multihop wireless networks," in *Proc. IEEE International Conference on Computer Communications and Networks (IC3N)*, Scottsdale, AZ, pp. 432–439, October 2001.

19. M. Joa-Ng and I.-T. Lu, "Spread spectrum medium access protocol with collision avoidance in mobile ad-hoc wireless network," in *Proc. IEEE INFOCOM*, New York, NY, pp. 21–25, March 1999.

20. P. Karn, "MACA: A new channel access protocol for packet radios," in *Proc. ARRL/CRRL Amateur Radio 9th Comp. Net. Conf.*, pp. 134–140, 1990.

21. L. Kleinrock and F. Tobagi, "Packet switching in radio channels: Part I—carrier sense multiple-access modes and their throughput-delay characteristics," *IEEE Trans. on Communications*, vol. COM-23, no. 12, pp. 1400–1416, December 1975.

22. Y.B. Ko, V. Shankarkumar, and N.H. Vaidya, "Medium access control protocols using directional antennas in ad hoc networks," in *Proc. IEEE INFOCOM 2000*, Tel Aviv, Israel, pp. 13–21, March 2000.

23. T. Korakis, G. Jakllari, and L. Tassiulas, "A MAC protocol for full exploitation of directional antennas in ad-hoc wireless networks," in *Proc. ACM MobiHoc 2003*, Annapolis, MD, pp. 98–107, June 2003.

24. P. Kyasanur and N. Vaidya, "Capacity of multi-channel wireless networks: Impact of number of channels and interfaces," University of Illinois at Urbana-Champaign, Tech. Rep., 2005.
25. P. Kyasanur and N. Vaidya, "Routing and interface assignment in multi-channel multi-interface wireless networks," in *Proc. IEEE Wireless Communications and Networking Conference (WCNC) 2005*, New Orleans, LA, pp. 2051–2056, March 2005.
26. D. Lal, R. Thoshniwal, R. Radhakrishnan, D. Agrawal, and J. Caffery, "A novel MAC layer protocol for space division multiple access in wireless ad hoc networks," in *Proc. IC3N 2002*, Miami, FL, pp. 614–619, October 2002.
27. J. Li, C. Blake, D.S.J. De Coute, H.I. Lee and R. Morris, "Capacity of ad hoc wireless networks," in *Proc. IEEE/ACM International Conference on Mobile Computing and Networking (MobiCom)*, Rome, Italy, pp. 61–69, July 2001.
28. A. Raniwala and T. Chiueh, "Architecture and algorithms for an IEEE 802.11-based multi-channel wireless mesh network," in *Proc. IEEE INFOCOM 2005*, Miami, FL, pp. 2223–2234, March 2005.
29. S. Roy, D. Saha, S. Bandyopadhyay, T. Ueda, and S. Tanaka, "A network-aware MAC and routing protocol for effective load balancing in ad hoc wireless networks with directional antennas," in *Proc. the 4th ACM international symposium on Mobile ad hoc networking and computing (MobiHoc) 2003*, Annapolis, MD, pp. 88–97, June 2003.
30. M. Sanchez, T. Giles, and J. Zander, "CSMA/CA with beam forming antennas in multihop packet radio," in *Proc. Swedish workshop on wireless ad hoc networks*, 2001.
31. J. So and N. Vaidya, "Multi-channel MAC for ad hoc networks: Handling multi-channel hidden terminals using a single transceiver," in *Proc. ACM International Symposium on Mobile Ad Hoc Networking and Computing (MobiHoc) 2004*, Tokyo, Japan, pp. 222–233, May 2004.
32. J. So and N. Vaidya, "Routing and channel assignment in multi-channel multi-hop wireless networks with single-NIC devices," University of Illinois at Urbana-Champaign, Tech. Rep., 2005.
33. E.S. Sousa and J.A. Silvester, "Spreading code protocols for distributed spread-spectrum packet radio networks," *IEEE Trans. on Communications*, vol. 36, no. 3, pp. 272–281, March 1988.
34. M. Takai, J. Martin, A. Ren, and R. Bagrodia, "Directional virtual carrier sensing for directional antennas in mobile ad hoc networks," in *Proc. the 3rd ACM international symposium on Mobile ad hoc networking and computing (MobiHoc) 2002*, Lausanne, Switzerland, pp. 183–193, June 2002.
35. J. Tao, F. Liu, Z. Zeng, and Z. Lin, "Throughput enhancement in WiMax mesh networks using concurrent transmission," in *Proc. International Conference on Wireless Communications, Networking and Mobile Computing*, vol. 2, pp. 871–874, 2005.
36. A. Tzamaloukas and J.J. Garcia-Luna-Aceves, "A receiver-initiated collision-avoidance protocol for multi-channel networks," in *Proc. IEEE INFOCOM*, 2001, Anchorage, AK, pp. 189–198, April 2001.
37. R. Vilzmann and C. Bettstetter, "A survey on MAC protocols for ad hoc networks with directional antennas," *EUNICE Open European Summer School*, Colmenarejo, Spain, July 2005.

38. H. Wei, S. Ganguly, R. Izmailov, and Z. Haas, "Interference-aware IEEE 802.16 WiMax mesh networks," in *Proc. IEEE VTC Spring'05*, Stockholm, Sweden, pp. 3102–3106, May 2005.

39. S.-L. Wu, C.-Y. Lin, Y.-C. Tseng, and J.-P. Sheu, "A new multi-channel MAC protocol with on-demand channel assignment for multihop mobile ad hoc networks," in *Proc. International Symposium on Parallel Architectures, Algorithms and Networks (ISPAN) 2000*, Dallas/Richardson, TX, pp. 232–237, December 2000.

6

SECURITY IN WIRELESS MESH NETWORKS

Rainer Falk, Chin-Tser Huang,
Florian Kohlmayer, and Ai-Fen Sui

CONTENTS

The potential of wireless mesh networking cannot be exploited without considering and adequately addressing the involved security issues. This chapter describes the security issues relevant for wireless mesh networking and corresponding solutions. Section 6.1 starts with a brief summary of relevant IT security technology, describing security services and security building blocks. The security objectives of a wireless mesh network (WMN) depend to a large degree on their usage scenario, i.e., how it is employed (administrative domains, and infrastructure integration). Section 6.2 describes relevant usage scenarios and the respective security requirements. Section 6.3 explains the security technology developed specifically for WMNs. Section 6.4 describes concrete commercial and research security proposals and Section 6.5 concludes with a summary of open issues.

6.1 SECURITY TECHNOLOGY OVERVIEW

This section gives a short overview of basic security technology that is relevant for WMNs. After describing security services and cryptographic algorithms, commonly used security technology is summarized.

WMN is exposed to the same basic threats common for both wired and wireless networks: messages can be intercepted, modified, delayed, replayed, or new messages can be inserted. A network and provided resources could be accessed without authorization, and they could be made unavailable by denial of service (DoS) attacks.

Generic security services defeating these threats are:

■ *Confidentiality*: Data is only revealed to the intended audience.
■ *Integrity*: Data cannot be modified without being detected.

- *Authentication*: An entity has in fact the identity it claims to have.
- *Access Control*: Ensures that only authorized actions can be performed.
- *Nonrepudiation*: Protects against entities participating in a communication exchange can later falsely deny that the exchange occurred.
- *Availability*: Ensures that authorized actions can in fact take place.

Further security services are user privacy (anonymity, pseudonymity, user profiling, and tracing), traffic flow confidentiality (which entities are communicating), or even protection of the fact that communication is taking place.

The protection of communication traffic involves confidentiality (encryption), authentication of communication partners, as well as protecting the integrity and authenticity of exchanged messages. Integrity protection refers not only to the integrity of a single message, but also to the correct ordering of related messages (replay, reordering, or deletion of messages). This section describes well-known technology for protecting communication traffic. These technologies can also be used within a mesh network to authenticate mesh nodes (MNs) and to establish session keys that protect the confidentiality and integrity of traffic exchanged between MNs.

Communication traffic can be protected at different layers (link layer, network layer, transport layer, by application): especially wireless systems (GSM, UMTS, DECT, IEEE 802.11 WLAN, Bluetooth, 802.16 WiMax) include means to protect the wireless link. These use different frame encapsulation schemes, different authentication protocols, and different cryptographic algorithms.

Wireless local area network (WLAN) based on IEEE 802.11i (Wi-Fi Protected Access: WPA, WPA2) supports two security modes [19]: either a shared key is configured on the WLAN devices (preshared key [PSK]), which is often used in home networks, or the users can be authenticated against an authentication server (AAA server). For this purpose, the extensible authentication protocol (EAP) is used [1]. The actual authentication takes place between the mobile station (MS) and the AAA server using EAP (see Figure 6.1). The EAP is transported between the MS and the access point (AP) using EAPOL, and between the AP and the AAA server by the RADIUS protocol. If successful, a master session key (MSK) is derived, which is sent from the authentication server (AS) to the WLAN AP. It is used as input to the WLAN

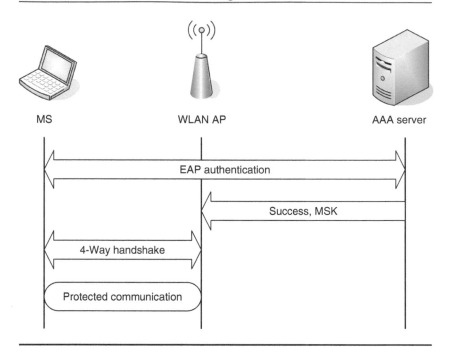

Figure 6.1 EAP-based WLAN access.

4-way handshake protocol that establishes a temporal session key for protecting the wireless link. This key is actually used for protecting the user traffic, using either temporal key integrity protocol ([TKIP], part of WPA) or AES-based CCMP (CTR with CBC-MAC protocol, part of WPA2). Various EAP methods exist for an authentication based on digital certificates (e.g., EAP-TLS), passwords (e.g., EAP-MSChapV2), or reusing mobile network authentication protocols (e.g., EAP-SIM, EAP-AKA). The EAP-based WLAN access is used in particular for enterprise networks and public hot-spots where a user data base is available.

The communication traffic can also be protected above the link layer. IPsec protects IP traffic at the network (IP) layer. The IPsec architecture specifies two security protocols: encapsulation security payload (ESP) and authentication header (AH). In the case of ESP, it is possible to either encapsulate only the payload of an IP packet (transport mode) or the entire IP packet (tunnel mode). An IPsec security association (SA) defines the cryptographic keys and algorithms to use. A SA is identified by a triple consisting of a destination IP address, a protocol (AH or ESP) identifier, and a security parameter index.

This unidirectional SA can be configured explicitly, or it can be established dynamially, e.g., by the Internet key exchange (IKEv2) protocol [22]. A common application of IPsec are virtual private networks (VPN) to securely access a company Intranet. Communication traffic can be protected at the transport layer using the Transport Layer Security (TLS) protocol [9], which is based on and is very similar to the Secure Socket Layer (SSL). Its main application is for protecting HTTP over TLS/SSL (https), but it may also be used as a standalone protocol. The TLS/SSL protocols include authentication and key establishment based on digital certificates. Recently, support for pre-shared keys (PSK-TLS) was also introduced [10]. It is also possible to protect traffic at higher layers. This allows to perform application-specific security operations. For example, e-mails can be encrypted (confidentiality protection) and/or signed (authentication, integrity, and nonrepudiation of origin) using S/MIME or PGP.

More information on applied cryptography can be found, e.g., in Menezes et al. [29] and Schneier [42].

6.2   MESH USAGE SCENARIOS

Security objectives and appropriate security measures largely depend on the environment/scenario in which WMN is used. Analysis of the security issues, relevant threats, and development of a suitable security architecture are possible only with respect to a specific application domain of mesh networks. This can easily be seen when considering different usage scenarios for mesh network technology: e.g., a military mesh network where even traffic flow must be kept confidential, a personal area mesh network where all nodes belong to a single person, an operator-controlled mesh network where the nodes constituting the mesh network are under a single administrative control, a mesh network of a small set of mutually trusting users, or a mesh network with users who do not trust or even know each other. It may also be necessary to distinguish different types of nodes, e.g., nodes under user control, infrastructure nodes provided to improve network coverage, or nodes that act as gateways (GW) to other networks. It may be that only some nodes are allowed to participate in maintaining routing information or in forwarding packets/frames. Perhaps nodes differ in their capabilities with respect to user interface, processing power, memory, and battery capacity.

Figure 6.2 shows relevant entities of WMN that are constituted of MNs. These may be dedicated MNs, i.e., nodes that fulfill only mesh-related tasks of routing and forwarding of user traffic, or user nodes

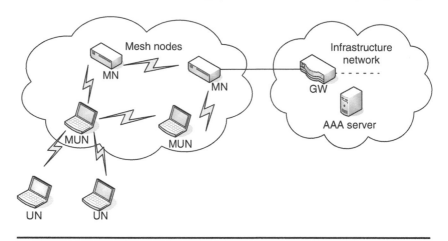

Figure 6.2   Mesh network scenario.

(MUN) that also perform mesh-related tasks. There may also exist pure user nodes (UNs) that make use of the communication service provided by the mesh network, but are not part of the mesh network itself, i.e., they do not participate in the mesh routing. Connectivity with an infrastructure network is provided by a GW. Although shown as separate entity in the figure, it may be integrated with an MN that happens to have also a wired connection. In the general case, more than one GW may be deployed. Shown is also a AAA server (authentication, authorization, and accounting) that holds entries for the different entities (i.e., for users and optionally also for MNs). The AAA server is used during network access to authenticate users/ nodes so that only authorized ones are accepted. The AAA server assigns authorization policies for the different entities defining their role (assigned permissions), and it may be used for accounting.

Ad hoc mesh networks can be classified in managed and open networks [32]: in a managed or closed ad hoc network, a central, commonly trusted authority exists. Only authenticated, known nodes are accepted to join an ad hoc mesh network. In an open ad hoc mesh network, however, any node may participate. Security relies on giving incentives for cooperation, reputation schemes, and robustness to misbehaving behavior. Considering a possible integration of ad hoc mesh networks with an infrastructure network, and the possibility to set up wireless MNs separately, from end-user nodes leads to a large number of relevant usage scenarios with partially different security requirements. These are relevant for mesh networks, as—in contrast to pure MANET-like ad hoc networks—many proposals are built on dedicated meshed nodes that provide UN with access to an infrastructure network.

Factors distinguishing the usage scenarios are

■ *Administrative Domains*: Do the nodes forming the mesh network belong to a single administrative domain or multiple administrative domains? Within a single administrative domain, the MNs can be assumed to be well behaving. With MNs belonging to multiple cooperating administrative domains, some MNs may show malicious behavior (by error or intention).

■ *Node Roles*: Are UNs and MNs distinguished, or are they similar? This means whether user devices participate in the mesh routing, or only dedicated MNs, e.g., meshed APs, participate in the mesh networking, providing their service transparently to users.

■ *User Classes*: Different user classes may have to be handled differently by the mesh network. For example, when the same mesh network is used for public access and public safety application, the traffic within the mesh network has to be kept separated.

■ *Self-Contained or Integrated*: A mesh network may be self-contained, or it may extend an infrastructure network. When it is integrated with an infrastructure network, often a trusted party will exist that can provide also a security infrastructure.

The following sections describe relevant usage scenarios and their respective relevant security requirements, giving an overall view of mesh network system. General issues like governance on setup and management procedures or administration are not considered.

6.2.1 Single Administrative Domain

From a security perspective, the simplest case is a self-contained mesh network where all nodes belong to a single administrative domain. Practical use cases are a single meshed home network, or a closed set of MNs of a public safety unit.

The security objective can be summarized as "keep the outsiders out." This requires some cryptographic credentials that allows an MN to authenticate another node as the one belonging to its own administrative domain. In the simplest case, this may be a shared key that the MNs have configured (like a pre-shared key in WLAN home network).

For larger node sets, a more scalable approach is to rely on node certificates that are verified by neighbor nodes.

If a dedicated AS is available within the mesh network, an alternative approach is to use it for MN authentication and key establishment. Similarly for WLAN user access, it would verify for MNs the node credentials and provide cryptographic keys for communicating with other MNs.

6.2.2 Network Infrastructure Extension

A mesh network may be set up specifically as an extension of an infrastructure network. The MNs are operated by a single entity, i.e., they belong to a single administrative domain. This may be the same as that of the infrastructure network, or a separate one when the mesh multihop extension is operated by a separate mesh operator. In the latter case, the compensation of the mesh operator is an important factor. The mesh network is used by end users to access the infrastructure network, but the end users are not part of the administrative domain of the MNs. The same access restrictions apply as for direct access to the infrastructure network. So the UN may only use the communication services provided by the mesh network for accessing the infrastructure network, but without participating in the mesh network itself. Within the mesh extension, it has to be ensured that only authorized MNs are included in the mesh extension. The communication within the mesh extension (backhaul security) has to be secured.

The security objectives can therefore be distinguished in security of infrastructure network access by end users, and security within the mesh extension (backhaul) itself. Possibly, the traffic of users belonging to different groups may be required to be handled separately. This may be the case when a single mesh extension is used for different purposes, e.g., to access a public safety network and for public Internet access. Depending on the trust relationships between the mesh extension and the end users, the user traffic can be protected hop-by-hop within the mesh extension, or a cryptographic tunneled may be established through the mesh extension toward a GW located in the infrastructure network.

Typical use cases are meshed WLAN APs (wireless distribution system): some of the APs are connected to a fixed infrastructure network. The APs build a mesh between them to provide connectivity with the infrastructure network for all APs. The APs are used by users. For the users, no difference exists from a security perspective between an AP with a direct connection to the infrastructure network and an AP that is connected through other APs to the infrastructure network.

6.2.3 Mesh Federation

In the case of mesh federation, the MNs forming a mesh network belong to different administrative domains (operators). So MNs belonging to different, known domains are configured to form a common mesh network. So federation takes place under control of the involved operators. The degree of federation can vary. It may be a full integration where all network functions including routing are integrated. Or it may rather be a cooperation where each of the two domains operates independently on its own, but cooperates in a controlled way with the other domain, exchanging, e.g., network status information and providing connectivity. The security requirements depend on the choice how close the federation shall be: in a full integration, the situation is similar as described in Section 6.2.2 in a controlled federation, the MNs have to distinguish to which domain an MN belongs to and apply appropriate policies. The security objectives for the end UNs are the same as described in Section 6.2.2.

An example scenario is where MNs were deployed initially independently by a provider for public Internet access and by a public safety organization. To safe costs and increase coverage, these two operators may decide to couple (federate) their originally separated mesh networks for their mutual benefit.

6.2.4 Community Mesh

From a security perspective, the most challenging usage scenario is a community mesh. It consists of end-user devices where "anybody" may be an end user (pure ad hoc case), i.e., nodes in the general case are not knowing or even trusting each other.

In this case, security cannot rely on administratively set restrictions allowing to identify "trusted" entities. Approaches known from pure ad hoc networks are applicable as incentives for cooperation, and measures increasing attack robustness against malicious behavior (detection and isolation of malicious nodes, reputation-based schemes). One issue of these measures is that future behavior may be different from the previous one (a node that has earned a good reputation might show malicious behavior in the future), the behavior may be different with respect to the affected nodes (a node may behave well to one node, but maliciously toward another), and that it may be easy to obtain identities (so that, once a bad reputation has been obtained, a node can start again with a new, neutral identity; a single attacker may obtain many identities and state mutually a good reputation for these identities).

6.3   MESH SECURITY ISSUES

WMNs aim to diversify the capabilities of ad hoc networks. On the one hand, ad hoc networks can actually be considered as a subset of WMNs. They share common features, such as multihop, wireless, dynamic topology, and dynamic membership. On the otherhand, mesh may have wireless infrastructure/backbone and have less mobility. The existing security schemes proposed for ad hoc networks can be adopted for WMNs. However, most of the security solutions for ad hoc networks are still not mature enough to be implemented practically. Moreover, the different network architectures between WMNs and ad hoc networks may render a solution for ad hoc networks ineffective in WMNs [2,12].

In this section, security challenges and attacks for mesh networks are identified. Authentication, secure MAC secure routing, key management, communication security and intrusion detection technologies are discussed in detail.

6.3.1   Security Challenges

The security challenges of WMNs are rooted from their topology features. By analyzing the characteristics of WMNs and comparing them with other networking technologies [41], the authors show that the new security challenges are mainly due to the multihop wireless communications and by the fact that the nodes are not physically protected.

Multihopping is indispensable for WMNs to extend the coverage of current wireless networks and to provide non-line-of-sight (NLOS) connectivity among users [2]. Multihopping delays the detection and treatment of the attacks, makes routing a critical network service and may lead to severe unfairness between the nodes [41]. In addition, nodes rely on each other to communicate, and thus the cooperation of node is indispensable.

While the use of wireless links renders a mesh network susceptible to attacks, the physical exposure of the nodes allows an adversary to capture, clone, or tamper with these devices.

Other specific challenges for WMNs include:

■ WMN may be dynamic because of changes in both its topology and its membership (i.e., nodes frequently join and leave the network). Any security with a static configuration would not suffice.
■ In WMNs, mesh routers and mesh clients hold significantly different characteristics such as mobility and power constraints. As a result, the same security solution may not work for both mesh routers and mesh clients.
■ There are also issues introduced by MN belonging to different authorities, such as selfish and greedy behavior, and trust management.

6.3.2 Overview of Potential Attacks to WMNs

There are two sources of threats to WMNs. First, external attackers not belonging to the mesh network may jam the communication or inject erroneous information. Second, more severe threats come from internal compromised nodes, since internal attacks are not as easy to prevent as external ones.

The attack can be rational, i.e., the adversary misbehaves only if misbehaving is beneficial in terms of price, obtained quality of service or resource saving; otherwise it is malicious [41].

Passive and active attacks can be distinguished. Passive attacks intend to steal information and to eavesdrop on the communication within the network. In active attacks, the attacker modifies and injects packets into the network.

Furthermore, attacks might target various protocol layers. At the physical layer, an attacker may jam the transmissions of wireless antennas or simply destroy the hardware of a certain node. Such attacks may be easily detected and located and are not discussed in detail. At the MAC layer, an attacker may abuse the fairness of medium access by sending mass MAC control and data packets or impersonate a legal node. An attacker could also exploit the protocols of the network layer. One type of attacks is to intimate knowledge of the routing mechanisms. Another type is packet forwarding, i.e., the attacker may not change routing tables, but the packets on the routing path may be led to a different destination that is not consistent with the routing protocol. Moreover, the attacker may sneak into the network, and impersonate a legitimate node and does not follow the required specifications of a routing protocol. At the application layer, an attacker could inject false or fake information, thus undermining the integrity of the application.

Attack types are summarized for ad hoc networks, which are also applicable to WMNs [6]:

- ■ *Impersonation*: Impersonation is an attack in which an adversary attempts to assume the identity of a legitimate node in the mesh network. If the adversary spoofs a legitimate UN, the adversary may gain disallowed access to the network or receive messages intended for the spoofed node. If the adversary spoofs an MN, then legitimate UN or MNs may be attacked and controlled by the adversary, causing worse problems than the former case does. Consider the following scenario in which a compromised AP in an 802.11 mesh network pretends to behave normally and as required by 802.11i obtains the pairwise master keys (PMKs) of connected wireless stations (WSs). Normally a WS and an AP have the option to cache the PMK for a period of time. With this information, the AP can dupe the WSs and get authenticated using the stored PMK. The compromised AP can thus gain control over this WS by connecting it to an adversary network.
- ■ *Sinkhole Attack*: A sinkhole attack is launched when a malicious MN (either a compromised or an adversary impersonating a legitimate node) convinces neighboring nodes that it is the "logical" next hop for forwarding packets. The malicious node then arbitrarily drops the packets forwarded by neighboring nodes. This attack also has the potential to strand large areas of the mesh network that are geographically distant from the malicious node by pulling messages from their intended paths.
- ■ *Wormhole Attack*: A wormhole attack attempts to convince nodes to use a malicious path through legitimate means. An adversary with fast forwarding capabilities can quickly forward a message through a low latency link. Fast forwarding is accomplished through the collusion of multiple units with fast out-of-band communication or a strongly powered device with a larger range of communication positioned between the base station and its target nodes. Once the target nodes are convinced that the adversary node is on a "better" path to the base station, all communications of the target nodes are attracted to the adversary node. An adversary can selectively forward packets in an effort to disrupt the mesh network.

- *Selfish and Greedy Behavior Attack*: A node increases its own share of the common transmission resource by failing to adhere to the network protocols or by tampering with their wireless interface.
- *Sybil Attack*: In a Sybil attack, a malicious node pretends the identity of several nodes, by doing so undermining the effectiveness of fault-tolerance schemes, such as the redundancy of many routing protocols. Sybil attacks also pose a significant threat to geographic routing protocols. Location aware routing often requires nodes to exchange coordinate information with their neighbors to efficiently route geographically addressed packets. By using the Sybil attack, an adversary can act in more than one place at the same time.
- *Sleep Deprivation*: Sleep deprivation attacks are to request services from a certain node, over and over again, so it cannot go into an idle or power preserving state, thus depriving it of its sleep and exhausting its battery.
- *DoS and Flooding*: DoS attacks can be caused by flooding, i.e., overloading nodes. More advanced DoS attacks are based on intelligently tampering protocol messages. For example, sinkholes are one of the major ways to initiate selective forwarding or nonforwarding of messages.

6.3.3 Authentication

Authentication is a significant issue in WMNs. On the one hand, we do not want to allow unauthorized users to get free connection through the network. On the other hand, we do not want to allow an adversary to sneak into the network and disrupt the normal operation of the network. However, authentication in WMNs is hard to provide because of the open nature of wireless communication: an adversary can easily eavesdrop messages exchanged in the network and inject messages (fabricated or replayed) into the channel.

Generally, there are two common approaches used to provide authentication to WMNs. The first approach is PSK *authentication* in which a key previously shared between two or more nodes is used to prove one nodes identity to other nodes. There are two possibilities for the shared key: pair-wise key and group key. A pair-wise key is shared between two nodes, either between a UN and the AS or between two neighboring MNs. A group key is shared among all members belonging to the same group. In both cases, a PSK should be renewed after a certain period of time to thwart statistical attacks by an adversary. The second approach is *certificate authentication* in which a user is required to show an MN the certificate that contains the users public key verified and signed by a trusted certificate authority (CA) to prove its identity. Normally, each legitimate user is issued a certificate from the CA of its domain. However, the user may sometimes roam to other domains. How to enable the authentication across different domains is an issue that needs to be considered.

Moreover, since one of the biggest benefits offered by WMNs is the support of user node mobility, we have to provide authentication to roaming UNs. For a roaming node, continuous discovery is needed to discover neighbors that are currently in communication range. When a new neighbor is found, mutual authentication should be performed between the two nodes. However, if this node moves back to the range of a previous neighbor, it is necessary to perform re-authentication to prevent an adversary from taking advantage of the gap between the last association and the current association with this neighbor to launch an impersonation attack.

Most of the authentication schemes discussed above focus on the authentication of UNs, but address little about the authentication of MNs. Such neglect may render the network vulnerable to the aforementioned impersonation attacks, sinkhole attacks, and wormhole attacks. To counter these attacks, it is necessary to provide authentication for UNs to authenticate MNs or for one MN to authenticate another. On the other hand, it is important to be mindful of the overhead caused by authentication because wireless user or MNs are often constrained by limited battery, computing power, or memory space. Two specific authentication protocols usable within mesh networks, the "Wireless Dual Authentication Protocol" (WDAP) [53] and the "Secure Unicast Messaging Protocol" (SUMP) [21] are described in Section 6.4.2.

6.3.4 Secure MAC Layer

Since IEEE 802.11 MAC, i.e., CSMA/CA with RTS/CTS (request-to-send/clear-to-send), is a widely accepted radio technique for WMNs [2], most of the following discussions are focused on IEEE 802.11 MAC.

For ad hoc mesh network, IEEE 802.11 uses the distributed coordination function (DCF) mode to schedule the wireless resource, which is based on the exchange of RTS–CTS control messages. The RTS–CTS exchange can avoid hidden terminal problem. A source node sends RTS message to apply for transmission, and a destination sends a CTS message if it is able to receive the message. Any node overhearing a CTS cannot transmit for the duration of the transfer. A source node senses the channel and backs off in accordance to the binary exponential back-off scheme if the channel is unavailable. The binary exponential scheme favors the last winner amongst the contending nodes. That means, nodes that are heavily loaded tend to capture the channel by continually transmitting data thereby causing lightly loaded neighbors to back off again and again [14]. This leads to the capture effect that brings the problem of unfairness. The capture effect may be exploited to launch a DoS attack by injecting a large amount of packets in the network. Those selfish nodes can drop packets to save its own energy and those greedy nodes can disobey the protocol specification to obtain a higher throughput than the other honest nodes.

Specific attacks targeting MAC layer may take the following forms:

- *Flooding Attack*: An attacker sends a lot of MAC control messages or data packets to its neighbor nodes. The victim node may suffer from DoS because of the unfair nature of medium access. This attack also exhausts the victim node's battery. And the channel bandwidth resource may be exhausted as well.
- *Jamming Attack*: A MAC layer attacker may prevent a node from accessing the channel by jamming the RTS signal. This is a kind of effective DoS attack to the destination node. The attack may degrade the performance of wireless network much more due to the cascade effect caused by the random back-off algorithm [33].
- *Sleep Deprivation Attack*: A misbehaving node intentionally selects one neighboring node to relay spurious data. The intention of this attack is to drain battery power and computational power of the victim node.
- *Packet Dropping Attack*: A malicious node can refuse to take its own responsibility, e.g., not relaying packets, or selective dropping the received packet. This action may affect the performance of the wireless network, even makes some victim nodes undergo DoS.

All the attacks could be categorized to selfish and greedy behaviors that disobey protocol specifications for selfish gains, and DOS attacks. The main countermeasures are listed.

6.3.4.1 Countermeasures to Selfish Misbehavior

The problem of selfish behavior in multihop wireless networks is studied and a solution called "Catch" is proposed [27]. It assumes that most of the nodes are honest and cooperative. They collectively prevent a minority of selfish nodes from malicious action, e.g., packet dropping. In game theory parlance, Catch assures that cooperation is an evolutionarily stable strategy. The scheme uses anonymous messages, in which the identity of the sender is hidden, to tackle two critical problems. First, Catch allows a cooperative node to determine whether its neighbors are selfish nodes, i.e., dropping packets that should be relayed. Second, it makes the cooperative neighbors of a selfish node to disconnect it from the rest of the network.

6.3.4.2 Countermeasures to Greedy Misbehavior

Kyasanur and Vaidya [25] made a detection and correction proposal that modifies 802.11 for facilitating the detection of misbehaving nodes. The main idea is to let the receiver assign the back-off value to be used by the sender, so the former can detect any misbehavior of the latter and penalize it by increasing the back-off value.

A similar idea is proposed as a scheme "DOMINO" to solve the problem of a greedy sender in IEEE 802.11 wireless local network, with a possible extension to multihop wireless networks, based on a reciprocal monitoring of sender and receiver [39]. The solution entrusts the receiver with the task of calculating the back-off value. Instead of the sender choosing random back-off values to initialize the back-off counter, the receiver selects a random back-off value and sends it to the sender in the CTS and ACK packets. The sender is requested to use this assigned back-off value in the next transmission to the receiver. In this way, a receiver can observe the number of idle slots between consecutive transmissions from the sender and identify the senders waiting for less than the assigned back off.

Konorski [23] gave a misbehavior-resilient back-off algorithm based on game theory, which assumes that all nodes can hear each other. This method also requires modification of the current MAC protocol.

6.3.4.3 Countermeasures to MAC-Layer DoS Attacks

Two kinds of DoS attacks may be launched to mesh networks, as described in [55]. One kind is single adversary attack. A single adversary intrudes into a mesh network and sends enormous flows to legitimate nodes and hence drain the energy of legitimate nodes as well as significantly degrade the performance of network communication. The second kind of attack exploits the unfairness possible with IEEE 802.11. Two colluding adversaries may send enormous data flows directly to each other, and hence exhaust the network bandwidth in their vicinity which is named as a colluding adversaries attack.

There are several ways to resist MAC-layer DoS attacks:

■ *Fair MAC Protocol*: A perfectly Fair MAC (FAIRMAC) [14] is employed to prevent some of the DoS attacks. The simulated results show that the fair protocol can improve the performance of a wireless network on DoS, which can be applied to mesh network. With this fair protocol, the legitimate traffic flow will not be entirely suppressed even though the network through-put would degrade to some extent.

■ *Protecting Traffic Flow*: With protecting traffic flow strategy, some colluding DoS attacks may be mitigated. In fact, it is hard to attack multiple traffic flows at the same time when there are only two colluding nodes. In other words, once multiple paths exist in the network, the legitimate traffic flow can be protected.

■ *Distance Adjustment*: When the distance between a sender and a receiver is beyond a certain threshold, interference attacks could be launched. In this case, attacks can be avoided by moving sender and receiver close to each other.

6.3.5 Secure Routing

Most routing protocols for mesh networks are cooperative and rely on implicit "trust your neighbor" relationships [50]. Malicious nodes can easily paralyze a network by inserting erroneous routing updates, replaying or changing routing updates, or advertising incorrect routing information [50]. Some specific threats relevant for ad hoc mesh routing functionality are [26]:

■ *Eavesdropping*: An attacker tries to discover information by listening to network traffic. Routing data can reveal information about the relation and location of the nodes, and about the network topology in general.

■ *Sinkhole, Wormhole*: In the attack, a malicious node uses the routing protocol to advertise itself as having the shortest path to the node whose packets it wants to intercept. Then the malicious can choose to drop the packets to perform a DoS attack (black hole), or selectively forward the packets (gray hole), or alternatively use its place on the route as the first step in a man-in-the-middle attack.

■ *Routing Table Overflow*: In a routing table overflow attack the attacker attempts to create routes to nonexistent nodes. The goal is to create enough routes to prevent new routes from being created or to overwhelm the protocol implementation.

■ *Rushing Attack*: A rushing attack is an effective Dos attack against on-demand routing protocols [18]. To limit the overhead of Route Request (RREQ) flood, each node typically forwards only the first RREQ originating from any route discovery. Utilizing this property, an attacker that forwards RREQs more quickly than legitimate nodes can do so, can increase the probability that routes including the attacker will be discovered rather than other valid routes.

■ *Sleep Deprivation*: The sleep deprivation attack is briefly introduced in [44]. Usually, this attack is practical only in mesh networks where battery life is a critical parameter. An attacker can attempt to consume batteries by requesting routes, or by sending unnecessary packets to the victim node using, e.g., a black hole attack. Although when a device does not offer services at all or offer services only to some authorized peers, it must nevertheless participate in the routing process and spend battery power for that purpose.

■ *Location Disclosure*: A location disclosure attack reveals information about the location of nodes or about the structure of the network.

Lundberg [26] mentions the following criteria that a secure ad hoc mesh routing protocol should fulfill:

■ *Certain Discovery*: If a route between two points in a network exists, it should always be possible to find it. Also, the node that requested the route should be able to be sure to find a route to the correct node.

■ *Isolation*: The protocol should be able to identify misbehaving nodes and make them unable to interfere with routing. Alternatively, the routing protocol should be designed to be immune to malicious nodes.

■ *Lightweight Computations*: Devices connected to a mesh network may be battery powered with limited computational abilities, restricting the possibility to carry out expensive computations.

■ *Location Privacy*: The information carried in message headers, which are not encrypted, may be as valuable as the message itself. The routing protocol should protect information about the location of nodes in a network and the network structure.

■ *Self-Stabilization*: The self-stabilization property requires that a routing protocol should be able to automatically recover from any problem in a finite amount of time without human intervention. That is, it should not be possible, by injecting a small number of malicious packets, to permanently disable a network. So an attacker has to remain active, sending malicious data continuously.

■ *Byzantine Robustness*: A routing protocol should be able to function correctly even if some of the nodes participating in routing are intentionally disrupting its operation. Byzantine robustness can be seen as a stricter version of the self-stabilization property: the routing protocol must not only automatically recover from an attack, but should also not cease from functioning even during the attack. Clearly, if a routing protocol does not have the self-stabilization property it cannot have Byzantine robustness either.

The design of securing routing solutions focuses on providing countermeasures against specific attacks, or sets of attacks. They can be classified as cryptography-based solutions, reputation-based solutions, add-ons to existing protocols, and countermeasures to specific attacks [3].

6.3.5.1 Cryptography-Based Solutions

Security of mesh routing can be enhanced using cryptographic measures protecting the integrity and authenticity, and potentially also the confidentiality of routing messages. Cryptographic measures prohibit an adversary from manipulating or even intercepting routing messages, but they require some real-world trust relations to distinguish "trustworthy" and "malicious" nodes [3].

Authenticated routing for ad hoc networks (ARAN), proposed in [11], utilizes cryptographic certificates to achieve authentication and nonrepudiation. Every node that forwards an RREQ or an RREP must also sign it. The method introduces heavyweight computations and the size of the routing messages increases at each hop. In addition, it is prone to replay attacks if the nodes do not have time synchronization.

Secure routing protocol (SRP) is a set of security extensions that can be applied to any ad hoc routing protocol that utilizes broadcasting as its route querying method [11]. SRP requires that for each route discovery the source and the destination have a SA between them, which can be used to establish a shared secret key. But it does not mention route error messages, thus any node can forge error messages with other nodes as source.

Secure Efficient Ad hoc Distance vector (SEAD) [16] is a secure ad hoc network routing protocol based on Destination-Sequenced Distance-Vector (DSDV) algorithm. It employs the use of hash chains to authenticate hop counts and sequence numbers.

Secure ad hoc on-demand distance vector routing (SAODV) [51] is a proposal for security extensions to the AODV protocol. The proposed extensions utilize digital signatures and hash chains to secure AODV packets. Signatures are used for authenticating the nonmutable fields of the messages, while a new one-way hash chain is created for every route discovery process to secure the hopcount field, which is the only mutable field of an AODV message.

6.3.5.2 Reputation-Based Solutions

Watchdog and Pathrater [28] consist of two extensions to the DSR routing protocol that attempts to detect and mitigate the effects of nodes that do not forward packets although they have agreed to do so. The watchdog extension is responsible for monitoring that the next node in the path forward data packets by listening in promiscuous mode. It identifies as misbehavior nodes those nodes that fail to do so. The pathrater assesses the results of the watchdog and selects the most reliable path for packet delivery. One of the base assumptions of this scheme is that malicious nodes do not collude to circumvent it and perform sophisticated attacks against the routing protocol.

The CONFIDANT protocol [5] was motivated by "the selfish gene" by Dawkins. Each node determines whether its neighbor is misbehaving and ALARMs its "friends" when a misbehaving node is detected. Each node maintains reputation ratings for other nodes that are reduced on receipt of ALARMs. It is a good concept but cannot circumvent the difficulties in diagnosing misbehavior accurately [46].

6.3.5.3 Add-Ons to Existing Protocols

The add-on mechanisms address specific security problems in ad hoc routing techniques and extensions to existing approaches.

Security-aware ad hoc routing (SAR) [49] utilizes a security metric for the route discovery and maintenance functions. But security properties like time stamp, sequence number, authentication, integrity, etc., have a cost and performance penalty.

Techniques for intrusion resistant ad hoc routing algorithms (TIARA) [38] is a set of design techniques mainly against DoS attacks. Each node has a policy that defines the list of authorized flows that can be forwarded by the node. The design principles include: flow-based route access control (FRAC), multipath routing, source-initiated flow routing, flow monitoring, the use of sequence numbers, and referral-based resource allocation.

Packet leashes [17] adds some extra information to each packet sent to allow a receiving node to determine if a packet has traversed an unrealistic distance. Two kinds of leashes: temporal and geographical are defined.

6.3.5.4 Countermeasures to Specific Attacks

If a node "agrees" to join a route (e.g., by forwarding RREQ in DSR), but fails to actually forward packets correctly, this is called misroute. A node may do so to conserve energy, or to launch a DoS attack, due to failure of some sort, or because of overload. In [28], a watchdog approach is exploited to verify whether a node has forwarded a packet or not. But misbehaving hosts are not punished. In best-effort fault tolerant routing (BFTR) [47], path redundancy is used to tolerate misbehavior by using disjoint routes. The target of a route discovery is required to send multiple route replies, and thus the source can discover multiple routes.

A node may make a route appear too long or too short by tampering with RREQ. In Ariadne [15,17], source–destination (S–D) pairs share secret keys and one-way hash function are used to ensure that RREQ and RREP follow the known route. But it does not ensure that the nodes on the route will deliver packets correctly.

Geographical leashes can help to resist wormhole attack, i.e., each transmission from a host should be allowed to propagate over a limited distance. In the SCAN [48] approach, the neighbors collaboratively authorize a token to the node before it joins the network activities, which can also help to resist wormhole attack.

6.3.6 Key Management and Communication Security

Key management is the set of techniques and procedures supporting the establishment and maintenance of keying relationships between authorized parties [29]. Key management plays a fundamental role as keys are the basis for cryptographic techniques providing confidentiality, entity authentication, data origin authentication, data integrity, and digital signatures. The goal of a good cryptographic design is to reduce more complex problems to the proper management and safekeeping of a small number of cryptographic keys, ultimately secured through trust in hardware or software by physical isolation or procedural controls.

Key management is based on keys that are initially established by noncryptographic, out-of-band techniques (e.g., in person, by a trusted courier or an administrator). For secret keys, confidentiality and authenticity must be ensured, whereas for public keys confidentiality is not required.

Key management usually involves some form of infrastructure to distribute initial keys (e.g., stored on a SIM card) and to provide infrastructure services (e.g., certificates, certificate repositories, or security servers as the authentication center in GSM networks).

Key management solutions for mesh networks have to take their operational environment into account. In the pure ad hoc case, no security infrastructure at all is assumed on which security can be based. Here nodes/users do not even know or trust each other. However, the fact that mesh networks may be built ad hoc does not necessarily imply that no infrastructure exists on which security can be based. A suitable infrastructure that can be used for setting-up security is available in the following cases:

- When all nodes of a mesh network belong to a single stakeholder, he can configure required key material (e.g., sensor networks [36], public safety networks where all nodes of a mesh network belong to the same party, or operator-controlled mesh network).
- It may be possible to access and reuse an existing security infrastructure, as e.g., a public-key infrastructure (PKI). Possibly even online security servers can be accessed, e.g., using GW services provided by some MNs, or when a mesh network is only one way for communication (e.g., mobile phones with additional mesh capabilities).
- A security infrastructure can be built in an ad hoc way that is valid only in a specific environment. For example, in a classroom scenario the teacher could build a mesh security infrastructure. While a base key for the "security master" will have to be distributed manually anyhow, the master node can be used to provide further key material for communication between nodes. Similarly, in the case of an event, the organizer could provide a mesh infrastructure.

In mobile mesh networks, signaling/routing traffic and user traffic can be distinguished. protection of these may be handled in different ways.

For routing traffic, the following basic options exist:

- Provide no security at all, i.e., rely solely on the robustness of the routing protocol with respect to misbehaving nodes.
- Protect integrity of routing messages through a MAC, i.e., using a secret key. Each party who can verify the MAC can also

compute valid MACs. If all nodes of a group of nodes trust each other they can use a shared secret key. From a security perspective, only nodes belonging to the group and nodes not belonging to the group are distinguished. The identity of individual nodes of the group cannot be verified. If nodes of a group do not trust each other, a different key should be used for every link.

■ Protect integrity of routing messages through a digital signature in a hop-by-hop mode. This means that each node signs its sent routing messages, but the origin of routing information received from other nodes cannot be verified by the receiver.

■ Protect integrity of routing messages through a digital signature in an end-to-end mode. This means that each node signs only its own routing data, while propagating routing data received from other nodes unchanged, including its digital signature. This allows a receiver to verify the origin of each piece of routing information.

■ Ensure also confidentiality of routing messages. Even confidentiality of routing messages may be protected to prevent that an eavesdropper can learn the topology of the mesh network by observing routing messages.

If user data is protected only hop-by-hop, then each intermediate node can access the plaintext data. Furthermore, each node has to decrypt and re-encrypt user data. For the protection of user data, the following main options exist:

■ No security at all (actually not a realistic option).
■ Secure communication within a group that shares a secret group key. So protection is achieved with respect to entities not belonging to this group, but not with respect to nodes within this group. A group consisting of only two nodes is a special case, i.e., two communicating nodes share a secret key.
■ Secure end-to-end communication using public-key cryptography. Here, usually the private/public keys are used to establish a secret session key. This session key is then employed to protect the actual data exchange.

6.3.7 Intrusion Detection

To enhance the security of WMNs, two strategies need to be adopted: to embed security mechanism into network protocols such as secure routing and MAC protocols; and to develop security monitoring and response systems to detect attacks, monitor service disruption, and respond quickly to attacks [2]. The intrusion detection system proposed by [52] can detect misbehavior at multilevel protocol layers. It uses "training" data to determine characteristics of normal routing table updates (such as rate of change of routing info), and normal MAC layer (such as access patterns by various hosts). But the efficacy of this approach is not evaluated, and is debatable. Detection technology can also be integrated to prevent or penalize the attacker. In WLAN based on IEEE 802.11 MAC DCF, a commercial example of intrusion detection systems is Siemens HiPath Wireless Security [43] and AirDefense Guard [8], in which distributed sensors, placed near APs, monitor the wireless medium and send reports to a central server. This idea can be applied to the IDS design of mesh network.

6.3.8 Other Security Technologies for WMNs

There are ways to preventing misbehaviors and thus helpful for trust management in mesh. One important way is to use the reputation concept. Reputation is a tool for motivating cooperation between nodes and for dictating a good behavior within the network. To be more precise, a node could be assigned a reputation value determined by its neighbors. Based on how "good" this value is, a node can be used or not in a given service provision. In addition, if a node does not pay attention to its reputation and keep acting maliciously, it will be isolated and discarded. The Packet Purse Model [7] is a cost-based approach for motivating collaboration between mobile nodes. The Packet Purse Model assigns a cost to each packet transfer, and the link-level recipient of a packet pays the link-level sender for the service. A reciprocal principle also helps to prevent misbehavior. A node wanting to send packets must send a certain number of packets for others. This rational exchange ensures that a misbehaving party cannot gain an advantage from misbehavior, and thus it will not have any incentives to misbehave, which have been summarized in [6,54].

6.4 CONCRETE PROPOSALS

Section 6.4.1 summarizes the security approach followed by some concrete commercial mesh networking solutions and of selected research proposals. Section 6.4.2 presents two authentication protocols usable for mesh networks.

6.4.1 System Proposals

This section describes the security approach followed by two commercial mesh network systems and of some research proposals.

6.4.1.1 *Tropos*

The mesh networks of Tropos (http://www.tropos.com/) distinguish MNs and end-user devices: the end-user device follows a conventional single-hop WLAN-access based on 802.11b/g. The WLAN AP is part of a mesh infrastructure. These wirelessly connected MNs form the extension to a fixed infrastructure network. Some MNs are connected to the wired backbone. From an architectural perspective, the mesh network can be classified as a closed network. However, as end UNs are not MNs, it can be considered as a wireless multihop infrastructure network extension that can be used practically for any type of WLAN-like network access.

Figure 6.3 shows the main components of network security that employs commonly accepted security technology [34]. It is a good engineering practice to use proved security technology when appropriate, instead of developing new specific security mechanisms. The user devices, named in the picture, UN employ the well-known WLAN access security mechanisms. Shown is network access following 802.1x/EAP-based authentication against a AAA-server (RADIUS). Other WLAN configurations employing PSK can be used as well. The WLAN-link is encrypted with WPA resp. WPA2 (TKIP or AES-based layer 2 link encryption), or even legacy WEP. The communication between the mesh nodes (MN and GW-MN), i.e., within the backhaul, is AES-encrypted. Both user traffic and mesh routing data exchanged between the MNs is protected. On top of that, a secure IPsec-based VPN can be established between the MS and a VPN-server located in the trusted part of the fixed infrastructure. This allows, e.g., to use the mesh network for different user classes as public access and for public safety. The network traffic is protected between the MS and the VPN-server located within the respective network, so that in the whole mesh network the traffic is not available in unprotected form.

Additional security measures that are commonly employed with WLAN APs are MAC address filters, packet filtering, suppression of broadcasting the WLAN network name (SSID), and remote administration using HTTPs. Also simple network management protocol (SNMP) is AES-encrypted.

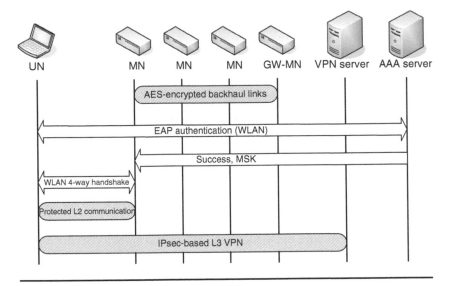

Figure 6.3 Mesh network multilayered security.

6.4.1.2 Meshdynamics

A similar basic approach is followed by Meshdynamics [31]. The user devices run WLAN WPA/WPA2 toward the first MN that looks to the user device like a WLAN AP. Authentication can be based on 802.1x/ EAP or PSK. Legacy WEP is supported. The backhaul links are AES-encrypted using on a common shared key from which link-specific keys are derived.

The MNs operate like a transparent layer-2 bridge so that DHCP, AAA, and VPN servers can operate with clients connected by mesh network without changes. So protocols such as IPsec-based VPNs can be used.

6.4.1.3 Research Proposals

MeshCluster is an infrastructure mesh network architecture [37]. Mesh relays forming the infrastructure extension are connected with the Internet by a GW node. For end-user devices, an MN operates as WLAN AP. Optionally, an MN performs initially a mutual EAP-based authentication with the GW node using a digital certificate or a secret key as credentials. The derived session key is used to protect the communication between the MN and the GW. A common dynamic mesh group key is provided to protect the mesh routing messages. This approach is interesting for larger mesh networks. The mesh network key can be easily updated as MNs obtain the updated mesh network key from the AAA infrastructure. It is not required to change the configuration of each MN separately, what would be required when each MN is configured directly with the mesh network secret key. As each MN is authenticated separately, a single-MN that shall be excluded from the mesh network can be removed easily from the authentication database.

Security for multihop relay extensions of an infrastructure network is investigated by the EU-funded Ambient Networks project [13]. After describing the reference model and security requirements, required security solution components are presented. These protect the access to a publicly accessible infrastructure network (e.g., to the Internet or to an operator network) over a multihop extension setup and operated by the public network operator. While secure access to the public network by users is based on the same preconditions as direct access to an AP, the relay extension has to be protected additionally to ensure its correct operation and to prevent that it is misused for other purposes than accessing the public network. Two variants for user network access authentication are presented: Either the user authenticates toward the GW, so that user traffic is protected the whole way through the mesh network. As second option, the user authentication is performed directly by the first-hop node. Here, already the first-hop MN terminates the user-specific SA, and the user traffic is protected with the mesh network key when forwarded to the GW. As in MeshCluster, a relay node authenticates first against the network infrastructure to obtain cryptographic keys allowing to set up a SA with neighbor relays. Filtering of user traffic within the mesh network intends to prevent overloading of the mesh network with traffic that will be discarded at the GW.

The mesh network as described by Microsoft research is a community network that is open to all [4]. However, users who contribute resources to the mesh network are prioritized. End devices do not participate in the mesh network, but they connect to an MN (mesh router, mesh box). Some MNs are connected to a GW that provides Internet connectivity. An assumption underlying this architecture is that an MN provides some physical security so that it cannot be hacked easily, i.e., it can be expected to behave as expected. So although being a community network, the security seems to be built on this assumption that only "officially provided" MNs participate in the mesh network, but not nodes under direct control by an end user. The security objectives are to protect against malicious users and freeloaders, to defend against faulty or hacked MNs, disruption (DoS) by malicious end devices, to ensure confidentiality of mesh traffic, and to protect access to network resources. MNs authenticate each other using EAP-TLS using a built-in digital certificate. Communication between MNs is encrypted. The certificate of a misbehaving MN is "blackballed" (revoked) to exclude it from participating in the mesh network. Also the authentication between clients and MN is EAP-based using PEAPv2 or EAP-TLS. The owner of a mesh router issues certificates to end devices using the mesh router certificate. The MN accepts certificates issued by any mesh box within range. The access to network resources, e.g., a file share offered by an end device is protected following the respective policy.

6.4.2 Authentication Protocols

Two novel authentication protocols are described to illustrate the provision of sufficient and efficient authentication for WMNs. The first protocol is the "WDAP" [53] for IEEE 802.11 WMNs. The WDAP provides dual authentication for a WS and its corresponding AP in a wireless network by an AS, and sets up a session key for confidential communications between the WS and the AP if the authentication is successful. The second protocol is the "SUMP" [21] that is originally designed for multihop ad hoc sensor networks. The SUMP requires the base station, whose function is like the GW in a WMN, to maintain network topology information and does not require individual node to store parent node information, thereby preventing an adversary from impersonating and limiting the amount of knowledge that an adversary gains by compromising a sensor node. Because of the multihop nature, SUMP is intrinsically applicable to WMNs. Note that one common characteristic of the two protocols is that the overhead incurred by authentication is largely centralized at the server side, either an AS or a GW, to reduce the burden on individual wireless nodes.

6.4.2.1 Wireless Dual Authentication Protocol

Before going into the details of WDAP, we first give a brief overview of the authentication scheme in 802.11i [19], the security standard used in 802.11 WMNs. As discussed in Section 6.1.3, 802.11i provides strong user-based authentication through the use of the 802.1x [20] standard and the EAP. The mutual authentication in 802.11i requires a UN to exchange a PMK with the AS before setting up a connection with an AP. (The negotiation request packets are forwarded to the AS by the AP using RADIUS, although no connection has yet been set up.)

The PMK is then made available to the AP after the AS authenticates the AP, such that the UN and AP can use the possession of the correct PMK to mutually prove their identity to each other. Moreover, as discussed above, re-authentication needs to be performed when a user moves out and back in the range of an AP.

Because the above authentication procedure is complicated and time-consuming, the 802.11i working group has come up with schemes aimed to mitigate the overhead. First, 802.11i provides pre-authentication for predictable roaming among APs in a network, such that when the UN arrives within the range of the next AP, the authentication is already completed. Second, the 802.11i standard offers a key caching option to allow the UN and the AP to remember the last used PMK, so that when the UN returns to this AP, the re-authentication can be done using the cached PMK, without the need to exchange a new PMK with the AS. However, adoption of the key caching option will render an 802.11 network vulnerable to an impersonation attack discussed in Section 6.3.2, in which a malicious AP uses previously cached PMKs to dupe UNs, whereas rejection of the key caching option will let the network still incur the expensive overhead of re-authentication.

WDAP is designed to provide sufficient and efficient authentication for IEEE 802.11 wireless networks during the initial connection stage and while roaming. The WDAP includes three sub-protocols: authentication protocol, deauthentication protocol, and roaming authentication protocol. Optionally the pre-authentication scheme described before can also be applied to WDAP. As in IEEE 802.11i, WDAP also involves three types of entities: WS, AP, and AS. For application of WDAP in WMNs, a WS can be regarded as a UN, and an AP can be regarded as an MN.

The authentication protocol is illustrated in Figure 6.4. In the authentication protocol, a WS broadcasts the WA-REQ message when it wants to connect to a wireless network. Normally, the AP that is closest to this WS will handle this message. It is assumed that each WS has to register with the AS beforehand to give the AS its MAC address and get from the AS a shared secret key and an initial sequence number. Therefore, the AS can use the MAC later as an index to find WSs shared secret key and sequence number. The sequence number is used to counter replay attacks, and should be incremented by one every time a new authentication request is sent. It is necessary to make the upper bound of the sequence number large enough (say 32 bits) to ensure that when the sequence number wraps around, it has been a very long period of time since the first sequence

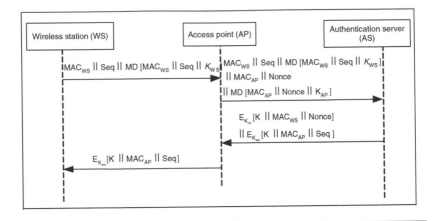

Figure 6.4   Authentication protocol in WDAP.

number was used and the WS's secret key has been updated at least once during this period. The WS's secret key can be updated using the current secret key or using some off-line schemes as discussed [45]. The message digest is used for integrity check and can be computed using a well-known hash function [24,35,40].

The AP that handles the WS's authentication request creates a similar message as in step 1 with a nonce instead of a sequence number, concatenates it with what was received from WS, and then sends it to the AS. This message is indeed where the "dual" part of the authentication protocol lies, since both WS and AP are assumed not to trust each other until the AS authenticates both of them. A session key (K) for the wireless communication between the WS and AP is generated and sent back to the AP if the dual authentication is successful. The two encryptions are used to make sure only the corresponding WS and AP can see the contents and extract the session key. The AP and WS will check the nonce and the sequence number, respectively, to verify the freshness of the message. The AP sends the session key to the WS. Only the legitimate WS can decrypt this message because it is encrypted using the WS's secret key. The session key shared between the AP and the WS can be used for their secure communications and secure deauthentication when the session is finished.

The deauthentication protocol provides secure deauthentication when a WS and its associated AP finish a session for three reasons: to prevent this connection from being exploited by an adversary, to prevent an adversary from spoofing deauthentication messages prematurely, and to stop the AP from transmitting any more frames.

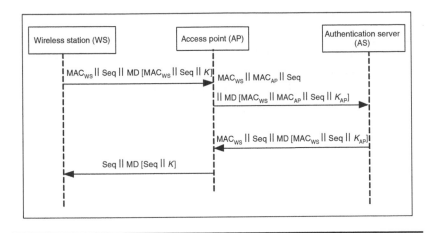

Figure 6.5 Wireless station deauthentication protocol in WDAP.

The 802.11 standards allow both the WS and the AP to send a deauthentication request, so there needs to be two versions of the deauthentication protocol: a wireless station deauthentication protocol that is illustrated in Figure 6.5, and a similar access point deauthentication protocol.

The roaming authentication protocol, provides authentication for a WS roaming within a basic service range (Figure 6.6). Before getting associated with a new AP, a WS needs to establish a dual authentication with the new AP. Note that the roaming authentication protocol only requires six messages, which is a saving of two messages compared to the case in which deauthentication with the old AP and authentication with the new AP are done separately. When the roaming WS needs to authenticate with the new AP, it sends out a roaming authentication request (the sequence number also needs to be incremented before used in the message), similar to the initial authentication. The new AP concatenates the WRA-REQ message and its own authentication message, and sends it to the AS to authenticate both the new AP and the roaming WS. After verifying the roaming dual authentication request, the AS uses WS's MAC address as an index to find the old AP and its secret key, generates and sends a session key revoke request message to old AP to invalidate the old session key used between the old AP and the roaming WS. The old AP notifies the AS that the old session key has been invalidated and stops using the old session key. The AS generates a session key for later communication between the roaming WS and the new AP, and sends a reply

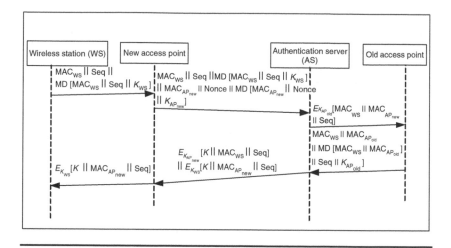

Figure 6.6   Roaming authentication protocol in WDAP.

back to the new AP. The new AP sends the new session key to the roaming WS.

It can be shown that WDAP is in conformance with the requirements of the IEEE 802.11i standard and effectively prevents an adversary from impersonating an AP. Moreover, results of a prototype implementation show that WDAP performs better with respect to communication time than IEEE 802.11i under the same security requirements. Detailed information about WDAP can be found [53].

6.4.2.2 Secure Unicast Messaging Protocol

SUMP is mainly designed to mitigate the threats of sinkhole and wormhole attacks for multihop ad hoc sensor networks. Most other authentication protocols for multihop ad hoc sensor networks, e.g., SNEP [36], are vulnerable to these attacks because they require nodes to maintain parent node information. The SUMP does not require a sensor node to store parent node information, thus mitigates sinkhole and wormhole attacks. In SUMP, the base station, which is equivalent to the GN in a WMN, incorporates path authentication information into every message it sends out. Additional benefits of SUMP include limiting the amount of knowledge that an adversary gains by compromising a node and providing a level-wise grouping of nodes in contrast to the locality-wise grouping approach used in other works.

The SUMP consists of two phases of operation: initialization and messaging. The purpose of the initialization phase is to provide the base station with the knowledge of individual node connectivity, density of distribution (with regards to connectivity, not locality), and establishment of paths. The messaging phase is the normal mode of operation used for instruction dissemination and data collection.

The base station and nodes maintain information regarding their view of the network structure. The base station's view of the network is global, and it maintains two primary structures: the node structure and the group structure. The node structure contains a list of all paths to each node, each node's hop count from the base station, ID, and individual key. The group structure maintains information about all groups in the network in a linked list of group elements. A group element contains information about the group including a listing of all node IDs, the distance from the base station to the group (called the level), and methods used for group membership authentication. A sensor node only maintains its own key information and group membership information.

The initialization phase in SUMP is divided into two steps: path establishment and verification. In the path establishment step, the base station initiates a breadth first search to discover the hop count and paths from the base station to each node. In the verification step, the base station updates nodes that received an incorrect hop count due to discrepancies in the communication range.

The base station initiates the path establishment step by issuing a hello message that contains a count of zero. The count corresponds to the current hop count from the base station. Nodes do not respond to any communications until the hello message is received. Once a node receives the hello message the node will record the hop count into its memory, increment the hop count in the hello message, and forward the message. The node then replies to the base station with a hello reply message containing the hop count recorded and the ID of the node. When the node receives a hello reply from another node it concatenates its own ID to the end of the message, and retransmits the message. If the node receives a hello reply message that contains its own ID, it will not respond. This avoids the formation of infinite routing loops that deplete resources. When the base station receives a hello reply from a node it finds the path based on the ID list in the reply message. The path derived from the first reply message received from a node is stored as the primary path to the node, and all paths derived from replies received after the first reply are stored as alternate paths, which are used to reduce packet loss and to enhance the survivability of the network when node death occurs. An example of path establishment sequence is shown in Figure 6.7.

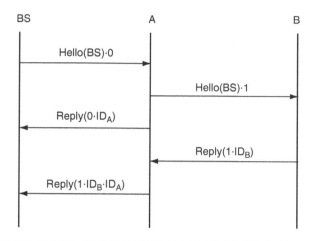

Figure 6.7 Message sequence diagram of path establishment.

After the expiration of the path establishment step, all nodes and the base station enter the verification step, in which the base station compares the node's recorded hop count with the length of the first path received in the path establishment step. If a discrepancy is found, the base station rectifies this by sending a hop count change request, which includes the ID of the destination node and the hop count value determined by the base station, and is encrypted with the individual key of that node. This ensures that the hop count is representative of a symmetric path between the base station and the destination node. If a node receives a hop count change request intended for it, the node updates its hop count accordingly.

Once verification of hop count is completed, the hop count is used by the base station to group nodes together. This is to implement a routing by level scheme in which only one node per hop on the primary path from the base station to the destination node will forward a received message to prevent arbitrary rebroadcast and malicious redirection of messages. All nodes of the same hop count are members of the group with the corresponding level value. For example, the group representing level one is comprised of all nodes with a hop count of one. The base station computes the group's key information and distributes it to all nodes in the group. To prevent malicious redirection of messages SUMP provides group authentication by Merkle hash trees [30], which uses a secure one-way hash function to generate a binary tree in which the members of the group are represented as the leaf values. The tree is formed by concatenating

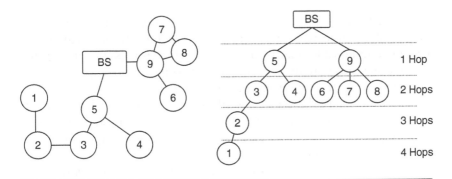

Figure 6.8 A sample network and its corresponding tree structure.

the sibling values and hashing the result to form a parent element of the tree. For example, in Figure 6.8, the network is divided into levels based on hop count, and the group that represents level 1 consists of nodes 5 and 9. Level 1 group is represented as a Merkle hash tree (Figure 6.9). In this representation, the hash values of nodes 5 and 9 form the leaves of the tree. The remainder of the tree is formed according to the following three rules. First, the tree is a balanced binary tree. Second, if an element is a leaf of the tree, then its value is the hash of a nodes ID. Third, if an element is not a leaf of the tree, then its value is the result of hashing the concatenation of its two children elements values. As a result of these rules the root value of the Merkle hash tree is a representative value of the entire group membership and can be used to authenticate a message. The base station maintains a representation of the entire tree of each level,

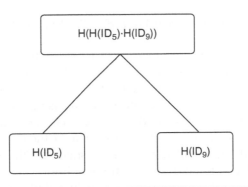

Figure 6.9   Merkle hash tree of level 1.

while each individual node only needs to store the root value and height of the tree of its level, and its own ID. The tree cannot be produced unless all node IDs are known. Since nodes do not store the IDs of other nodes in the network, an adversary cannot capture a node and reconstruct the tree from the root value.

After the establishment of Merkle hash tree at every level, begins the messaging phase that aims to securely unicast an outbound message hop-by-hop from the base station toward its destination. Since the path to every node is known and the nodes are grouped according to hop count, the base station can communicate with a destination node by encrypting messages according to the established primary path to that node. The base station concatenates the ID of the destination node to the message and encrypts the result with the key it shares with the destination node. Then the h authentication values, where h is the height of the tree at the destination node's level, are attached to the beginning of the message. These authentication values are the values needed by the destination node to reproduce the root value from already known information, i.e., the sibling values of elements in the path from the nodes hashed ID to the root of the tree. The base station then uses the next node on the stored primary path from the destination node to the base station to encrypt the message further. For an intermediate node between the base station and the destination node the base station encrypts the entire message with the individual key of the intermediary node and concatenates the authentication values of intermediate node with the resulting message. This results in the encapsulation of the original message, M, in a message to the intermediary node. For example, if the primary path from the base station to a node is {5, 3, 2}, the message produced is

$$E_{K5}(\{auth\,Values5\} \cdot E_{K3}(\{auth\,Values3\} \cdot E_{K2}(\{auth\,Values2\} \cdot ID_2 \cdot M)))$$

Upon receiving a message of this structure, a node uses its key to decrypt the message and attempts to authenticate the message to determine if the message is to be forwarded by the node. If the node successfully authenticates the message by using the authentication values in the message to reproduce the correct root value, it checks if the decrypted message begins with its ID. If so, the decrypted message is for the node and will not be retransmitted; otherwise the node forward the decrypted message to the next hop.

The structure of outbound messages in SUMP is shown in Figure 6.10. An individual node only uses as many nonzero authentication

8	16	24	32
Msg type	Seq#	Auth 1	
Auth 2		Auth 3	
Auth 4		Auth 5	
Auth 6		Auth 7	
Auth 8		Auth 9	
Auth 10		Dest ID	
Message			
Checksum			

Figure 6.10 Messaging phase outbound message.

values as necessary to authenticate a message for that hop, and unneeded or previously processed authentication values are zeroed out. For instance, if a given level has a Merkle hash tree of height 2, then only the first two nonzero authentication values are used in authentication. Once used the authentication values are replaced by zeros. If the message is authenticated, the remaining authentication values, message field, and checksum are decrypted. The resulting packet is forwarded to the next hop.

The sufficiency and efficiency strengths of SUMP are manifest in the following three regards [21]. First, SUMP is not susceptible to sinkhole and wormhole attacks. Second, very little storage is required by SUMP for sensor nodes to securely route outbound messages. Third, communication overhead is alleviated by avoiding arbitrary rebroadcast of messages.

6.5 SUMMARY AND OPEN ISSUES

This chapter discusses security for WMNs. After a brief overview of relevant security technology, different relevant usage scenarios with significantly different security requirements and their main distinguishing criteria have been described in Section 6.2. Specific WMN security issues are covered in Section 6.3, followed by a description of some concrete proposals in Section 6.4.

Despite many research work already spent on wireless mesh and ad hoc networks, still challenging open issues exist: A great number of security proposals exist for specific issues, but it is, in general, unclear which ones are useful for a certain usage scenario, and how those being proposed independently can be combined to arrive at a complete security solution. Individual countermeasures cover only a specific subset of relevant threats and are difficult to integrate [32]. Complete security solutions adapted for the differing security requirements of more challenging usage scenarios are missing. Other open issues are practically usable, commercially suitable security solutions for cooperative mesh networks between unknown and untrusted end that may or may not be cooperative, uncooperative, or even intentionally misbehaving. While proposals like granting incentives for enforcing cooperation, misbehavior detection, and robustness toward malicious nodes are described, it is unclear whether these solutions are adequate and sufficient for real-world applicability. A further open issue is the compensation of end-user mesh nodes providing access to an infrastructure network.

ACKNOWLEDGMENT

We would like to thank Dr. Guo Dai Fei for his discussions on MAC layer security.

REFERENCES

1. B. Aboba, L. Blunk, J. Vollbrecht, J. Carlson, and H. Levkowetz. Extensible authentication protocol (EAP). RFC 3748, June 2004.
2. I.F. Akyildiz, X. Wang, and W. Wang. Wireless mesh networks: A survey. *Computer Networks*, vol. 47, p. 445–487, 2005.
3. P.G. Argyroudis and D. O'Mahony. Secure routing for mobile ad hoc networks. *IEEE Communications Surveys and Tutorials*, vol. 7, no. 3, 2003.
4. V. Bahl. Mesh networking. In *Mesh Networking Summit 2004—Making Meshes Real*, 2004. http://research.microsoft.com/meshsummit/.
5. S. Buchegger and J. Le Boudec. Nodes bearing grudges: Towards routing, security, fairness, and robustness in mobile ad hoc networks. In *Proceedings of the Tenth Euromicro Workshop on Parallel, Distributed and Network-based Processing*, January 2002.
6. A. Burg. Ad hoc network specific attacks. Technical report, Technische Universität München, 2003. http://www13.informatik.tumuenchen.de/lehre/seminare/WS0304/UB-hs/burg-ad_hoc_specific_attacks-pres.pdf.
7. L. Buttyan and J. Hubaux. Enforcing service availability in mobile ad-hoc WANs. In *1st IEEE/ACM Workshop on Mobile ad hoc Networking and Computing (MobiHOC)*, 2000.
8. Air Defense. http://www.airdefense.net.
9. T. Dierks and C. Allen. The TLS protocol version 1.0. RFC 2246, January 1999.
10. P. Eronen and H. Tschofenig. Pre-shared key ciphersuites for transport layer security (TLS). RFC 4279, December 2005.
11. K. Sanzgiri et al. A secure routing protocol for ad hoc networks. In *Proceedings of 10th IEEE International Conerence on Network Protocols (ICNP02)*, p. 78–89, IEEE Press, 2002.

12. I.F. Akyildiz and X. Wang. A survey on wireless mesh networks. *IEEE Radio Communications*, pp. 23–30, September 2005.

13. R. Falk, H. Tschofenig, and A. Prasad. Secure access over multi-hop relay extensions of public networks. In *Wireless Personal Multimedia Communications*, Aalborg, Denmark, pp. 17–22, September 2005.

14. V. Gupta, S. Krishnamurthy, and M. Faloutsos. Denial of service attacks at the MAC layer in wireless ad hoc networks. In *Proceedings of IEEE MILCOM*, pp. 7–10, October 2002.

15. Y. Hu, A. Perrig, and D. Johnson. Ariadne, a secure on-demand routing protocol for ad hoc networks. In *8th ACM International Conference on Mobile Computing and Networking*, MobiCom, pp. 12–23, September 2002.

16. Y.-C. Hu, D.B. Johnson, and A. Perrig. Sead: Secure efficient distance vector routing for mobile wireless ad hoc networks. In *Proceedings of 4th IEEE Workshop on Mobile Computing Systems and Applications*, Callicoon, NY, p. 313, June 2002.

17. Y.-C. Hu, A. Perrig, and D.B. Johnson. Packet leashes: A defense against wormhole attacks in wireless networks. In *Proceedings of IEEE INFOCOM'03*, San Francisco, CA, April 2003.

18. Yih-Chun Hu, Adrian Perrig, and David Johnson. Rushing attacks and defense in wireless ad hoc network routing protocols. In *ACM Workshop on Wireless Security (WiSe 2003)*, San Diego, CA, September 2003.

19. IEEE standard for information technology telecommunications and information exchange between systems local and metropolitan area networks specific requirements, part 11, amendment 6: Medium access control (MAC) security enhancements. Std 802.11i-2004, 2004.

20. IEEE standards for local and metropolitan area networks—port based network access control. Std 802.1X-2001, 2001.

21. J. Janies, C.-T. Huang, and N.L. Johnson. SUMP: A secure unicast messaging protocol for wireless ad hoc sensor networks. In *Proceedings of 2006 IEEE International Conference on Communications (ICC'06)*, Istanbul, Turkey, 2006.

22. C. Kaufman. Internet key exchange (IKEv2) protocol. RFC 4306, December 2005.

23. J. Konorski, Multiple access in ad hoc wireless LANs with noncooperative stations. In *NETWORKING*, vol. LNCS 2345. Springer, 2002.

24. H. Krawczyk, M. Bellare, and R. Canetti. HMAC: Keyed-hashing for message authentication. RFC 2104, February 1997.

25. Pradeep Kyasanur and Nitin H. Vaidya. Detection and handling of MAC layer misbehavior in wireless networks. In *International Conference on Dependable Systems and Networks (DSN'03)*, 2003.

26. Janne Lundberg. Ad hoc routing security. Technical report, Seminar on Network Security, Telecommunications Software and Multimedia Laboratory, Helsinki University of Technology, 2000.

27. R. Mahajan, M. Rodrig, D. Wetherall, and J. Zahorjan. Sustaining cooperation in multihop wireless networks. In *Networked Systems Design and Implementation (NSDI)*, May 2005.

28. S. Marti, T.J. Giuli, K. Lai, and M. Baker. Mitigating routing misbehavior in mobile ad hoc networks. In *ACM International Conference on Mobile Computing and Networking (MobiCom)*, p. 255–265, 2000.

29. A.J. Menezes, P.C. Van Oorschot, and S.A. Vanstone. *Handbook of Applied Cryptography*, CRC Press, Boca Raton, FL, USA, 1996.

30. R. Merkle. Efficient distribution of key chain commitments for broadcast authentication in distributed sensor networks. In *Proceedings of the IEEE Symposium on Security and Privacy*, 1980.

31. MeshDynamics. Structured mesh security primer. http://www.meshdyna mics.com/Publications/MDSECURITYPRIMER.pdf, 2005.

32. P. Michiardi and R. Molva. *Mobile Ad hoc Networking*, Chapter 12: ad hoc Networks Security, pp. 329–354. IEEE, 2004.

33. R. Negi and A. Perrig. Jamming analysis of MAC protocols. Technical memo. Carnegie Mellon, February 2003.

34. Tropos Networks. Multi-layered security framework for metro-scale Wi-Fi networks—a security whitepaper. http://www.tropos.com/pdf/Tropos_Security_WP.pdf, 2005.

35. Secure hash standard. FIPS PUB 180–1, 1995.

36. A. Perrig, R. Szewczyk, V. Wen, D. Culler, and J.D. Tygar. SPINS: Security protocols for sensor networks. In *Seventh Annual International Conference on Mobile Computing and Networks (MobiCOM 2001)*, Rome, 2001.

37. K.N. Ramachandran, M.M. Buddhikot, G. Chandranmenon, S. Miller, E.M. Belding-Royer, and K.C. Almeroth. On the design and implementation of infrastructure mesh networks. In *First IEEE Workshop on Wireless Mesh Networks, WiMesh 2005*, Santa Clara, CA, 2005.

38. R. Ramanujan, A. Ahamad, and K. Thurber. Techniques for intrusion resistant ad hoc routing algorithms (TIARA). In *Proceedings Military Communications Conference (MILCOM 2000)*, Los Angeles, CA, pp. 660–664, October 2000.

39. M. Raya, J.-P. Hubaux, and I. Aad. Domino: A system to detect greedy behavior in IEEE 802.11 hotspots. In *Proceedings of the Second International Conference on Mobile Systems, Applications and Services (MobiSys2004)*, Boston, Massachussets, June 2004.

40. R. Rivest. The MD5 message digest algorithm. RFC 1321, April 1992.

41. N.B. Salem and J.-P. Hubaux. Securing wireless mesh networks. *Special Issue of IEEE Wireless Communications Magazine on Wireless Mesh Networking, Theories, Protocols, and Systems*, vol. 13, no. 2, April 2006.

42. B. Schneier. *Applied Cryptography* 2nd ed. Wiley, New York, 1995.

43. Siemens. Enterprise-grade wireless LAN security. HiPath Wireless Whitepaper, 2005.

44. F. Stajano and R. Anderson. The resurrecting duckling: Security issues for ad hoc wireless networks. In *Proceedings of 7th International Workshop on Security Protocols*, 1999. http://www-lce.eng.cam.ac.uk/~fms27/duckling/.

45. W. Stallings. *Cryptography and Network Security* 3rd ed. Prentice Hall, New Jersey, 2003.

46. N.H. Vaidya. Tutorial on security and misbehavior handling in wireless ad hoc networks. In *IEEE INFOCOM*, 2005.

47. Y. Xue and K. Nahrstedt. Providing fault-tolerant ad-hoc routing service in adversarial environments. *Wireless Personal Communications, Special Issue on Security for Next Generation Communications*, vol. 29, no. 3–4, pp. 367–388, 2004.

48. H. Yang, J. Shu, X.Q. Meng, and S. Lu. SCAN: Self-organized network-layer security in mobile ad hoc networks. *Selected Areas in Communications, IEEE Journal on, Special Issue on Security in Wireless Ad Hoc Networks*, vol. 24, no. 2, pp. 261–273, February 2006.

49. S. Yi, P. Naldurg, and R. Kravets. Security-aware ad hoc routing for wireless networks. In *Proceedings of the 2nd ACM Symposium Mobile Ad Hoc Networking and Computing (Mobihoc01)*, Long Beach, CA, pp. 299–302, October 2001.

50. S. Yi, P. Naldurg, and R. Kravets. Security-aware ad hoc routing for wireless networks. Technical report, University of Illinois at Urbana-Champaign, August 2001.

51. M.G. Zapata and N. Asokan. Secure ad hoc on-demand distance vector routing. *ACM Mobile Computing and Communication Review*, vol. 3, no. 6, pp. 106–107, July 2002.

52. Y. Zhang and W. Lee. Intrusion detection in wireless ad hoc networks. In *Proceedings of MOBICOM'00*, pp. 275–283, 2000.

53. X. Zheng, C. Chen, C.-T. Huang, M.M. Matthews, and N. Santhapuri. A dual authentication protocol for IEEE 802.11 wireless LANs. In *Proceedings of 2nd International Symposium on Wireless Communication Systems (ISWCS'05)*, Siena, Italy, 2005.

54. L. Zhou and Z.J. Haas. Securing ad hoc networks. *IEEE Network Magazine*, vol. 13, no. 6, pp. 24–30, 1999.

55. Y. Zhou, D. Wu, and S.M. Nettles. Analyzing and preventing MAC-layer denial of service attacks for stock 802.11 systems. In *Proceedings of IEEE/ACM 1st International Workshop on Broadband Wireless Services and Applications (BroadWISE 2004)*, San Jos, CA, USA, October 2004.

7

SCALABILITY IN WIRELESS
MESH NETWORKS

Jane-Hwa Huang, Li-Chun Wang,
and Chung-Ju Chang

CONTENTS

The wireless mesh network (WMN) is an economical solution to support the ubiquitous broadband services by low transmission power. However, multihop networking suffers from the scalability issue as coverage and users increase, since throughput enhancement and coverage extension are two contradictory goals in the multihop WMNs. Specifically, the multihop communications can indeed extend the coverage area to lower the total infrastructure cost. However, as the number of hops increases, the repeatedly relayed traffic will exhaust the radio resource. The throughput will also sharply degrade due to more collisions from a large number of users. This chapter addresses this key challenge of the WMNs from the network architecture perspective, aiming at maintaining the throughput while extending the coverage area.

We consider two typical application scenarios of WMNs, including the dense urban and wide-area scenarios. At first, we investigate the issue of deploying WMNs in the dense urban environment, where several adjacent access points (APs) will form a cluster. In a cluster, the APs are connected through wireless relays to ease deployment. The mixed integer nonlinear programming (MINLP) optimization approach is applied to determine the optimal number of APs in a cluster and the best separation distance between APs. The objective is to maximize the ratio of the total carried traffic load to the total cost for a cluster of APs connected by wireless relays.

We also present a scalable multichannel ring-based WMN for wide-area coverage. In the ring-based WMN, each cell is divided into several rings with different allocated channels. Without modifying the IEEE 802.11 medium access control (MAC) protocol, the simple ring-based frequency planning can make the system more scalable to the cell coverage. The MINLP optimization approach is also employed to determine the optimal number of rings in a cell and the associated ring widths, which aims to maximize the cell coverage with a guaranteed user throughput.

7.1 INTRODUCTION

Nowadays, the development of next-generation wireless systems aims to provide high data rates in excess of 1 Gbps. Thanks to the capability of enhancing coverage and capacity with low transmission power, WMNs play a significant role for broadband access with ubiquitous coverage [1–9].

In general, the advantages of wireless mesh networking technology can be summarized into four folds. First, it is well known that mesh networking technology can combat shadowing and severe path loss to extend service coverage. Second, WMN can be rapidly deployed in a large-scale area with minimal cabling engineering work so as to lower the infrastructure and deployment costs [1–4]. Third, WMN can concurrently support a variety of wireless radio and access technologies such as 802.16 (WiMAX), 802.11 (WiFi), and 802.15 (Bluetooth and Zigbee), thereby providing the flexibility to integrate different radio access networks [5–7]. Fourth, a WMN can be managed in a self-organization and self-recovery fashion [8,9]. Thus, if some nodes are down, the forwarded traffic can be delivered via other adjacent nodes. Due to these advantages, the WMN is believed to be a key enabling technology for 4G wireless systems.

However, multihop networking suffers from the scalability issue, especially when the coverage area or the number of contending users increases [9]. The scalability issue lies in the fact that throughput enhancement and coverage extension are two contradictory goals in the multihop WMNs. On the one hand, the multihop communications can indeed extend the coverage area to lower the total infrastructure cost. On the other hand, as the number of hops increases, the repeatedly relayed traffic will exhaust the radio resource. Meanwhile, the throughput will also sharply degrade due to the increase of collisions from a larger number of users. Therefore, maintaining the throughput while extending the coverage area becomes a difficult and important challenge for designing a scalable WMN.

This chapter addresses the scalability issue of the WMN from a network architecture perspective. We consider two most typical

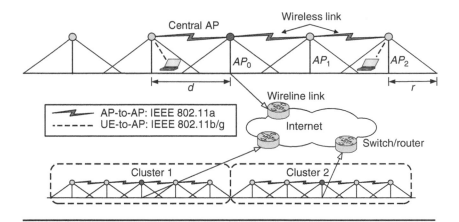

Figure 7.1 Clusters of APs in the wireless mesh network for the dense urban coverage.

application scenarios of WMNs [10,11], i.e., the dense urban and wide-area scenarios as shown in Figure 7.1 and Figure 7.2. First, we consider the WMNs in the dense urban area. Recently, deploying public outdoor WLANs in the metropolitan area has become a very hot topic. In the dense urban environment (e.g., the so-called Manhattan scenario discussed in [1] and universal mobile telecommunication systems (UMTS) [12]), heavy attenuation due to walls or buildings is expected. A number of APs will be deployed along the streets. Since connecting APs through cables is costly and difficult, connecting APs through wireless becomes an interesting option. Figure 7.1 illustrates an example of WMN for the dense urban area. In this network, several adjacent APs form a cluster and are connected to the backbone network through the same switch/router. In each cluster, only the central access point AP_0 connects to the backbone network through wires. Other APs are required to communicate with the neighboring APs via wireless links. Then the intermediate APs will relay the data to the central AP_0. By doing so, the network deployment in the urban area becomes easier because the cabling engineering work is minimized. We also propose a scalable multichannel ring-based WMN for wide-area coverage, as shown in Figure 7.2. Referring to the figure, the central gateway and stationary mesh nodes in the cell form a multihop WMN. The mesh cell is divided into several rings, which are allocated with different channels. In the same ring, the mesh nodes follow the legacy IEEE 802.11 MAC protocol to share the radio medium. In addition, mesh nodes in the inner rings will relay data for

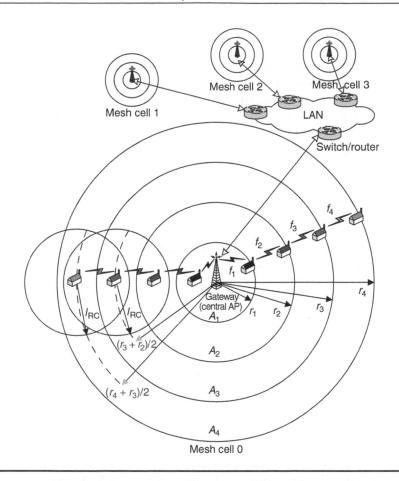

Figure 7.2 Ring-based cell architecture in the wireless mesh network for wide-area coverage, where each ring is allocated with different allocated channel. This is an example of WMN for community/campus networking.

nodes in the outer rings toward the central gateway. With this mesh cell architecture, the service coverage of the central gateway/AP can be significantly extended with less cost.

In this chapter, we also investigate the optimal trade-off between capacity and coverage for the scalable WMNs. Most traditional WMNs are not scalable to the coverage area, because the throughput is not guaranteed with increasing collisions. By contrast, the proposed WMNs are scalable in terms of coverage, since the frequency planning with multiple available channels is used to resolve the contention issue and thus the throughput can be ensured by properly designing the deployment parameters. The deployment parameters include the the number of APs in a cluster and the separation distance between APs for the considered WMN in the urban area (Figure 7.1); the number of rings in a cell and the associated ring widths for the ring-based WMN (Figure 7.2). The remaining important problem lies in the way to determine these deployment parameters so as to achieve the optimal trade-off between throughput and coverage performances. We will apply the MINLP optimization approach to determine the optimal deployment parameters.

The rest of this chapter is organized as follows. Section 7.2 discusses the related works of WMN. Section 7.3 investigates the issue of deploying APs for WMNs in the dense urban area. We formulate the AP placement problem as an optimization issue and discuss the proposed AP placement strategies. Section 7.4 describes the proposed ring-based WMN. We develop a cross-layer throughput model for the ring-based WMN in Section 7.5 and formulate an optimization problem to maximize the coverage and capacity in Section 7.6. Concluding remarks and open issues are given in Section 7.7 and Section 7.8.

7.2 RELATED WORKS

First, we discuss the issue of AP placement in WMNs for dense urban coverage. Most works were based on the architecture that all the APs are connected to the backbone network through cables [13–17]. An integer linear programming (ILP) optimization model was proposed for the AP placement problem, where the objective function was to maximize the signal level in the service area [13]. An optimization approach was proposed to minimize the areas with poor signal quality and improve the average signal quality in the service area [14]. The authors [15,16] proposed optimization algorithms to minimize average bit error rate (BER). The AP deployment problem was also formulated as an ILP optimization problem with the objective function of minimizing the maximum of channel utilization to achieve load balancing [17]. The concept of wireless multihop communication has not been considered [13–17].

Performance issues of WMNs have been studied mainly from two directions [1,2,18–21]. On the one hand, from a coverage viewpoint, authors [18] compared the coverage performance of a multihop WMN with that of a single-hop infrastructure-based network by simulations. On the other hand, from a capacity viewpoint, it was shown [19,20] that the throughput per node in a uniform multihop ad hoc network is scaled like $O(1/\sqrt{k \log k})$, where k is the total number of nodes. Moreover, the authors [2] showed that the achievable throughput per node in a multihop WMN will significantly decrease as $O(1/k)$ due to the bottleneck at the central gateway. To resolve the scalability issue of multihop network, authors [21] proposed a multichannel WMN to improve the network throughput. Fewer papers considered both the capacity and coverage performance issues for a WMN, except for [1]. However, the work [1] considered the single-user case. In addition, the scalability issue of WMNs has not been well addressed [1,2,18–21].

7.3 SCALABLE WIRELESS MESH NETWORK FOR DENSE URBAN COVERAGE

7.3.1 Architecture and Assumptions

In this section we consider the WMNs in the dense urban area as shown in Figure 7.1 [10]. In each cluster, only the central AP_0 has the wireline connection to the switch or router for accessing the Internet. Other APs (like AP_2) can access the Internet through wireless communications between AP_1 and AP_2 first and then through the wireline communications from AP_0 to the switch/router in the LAN. In this case, the function of AP_1 is referred as a relay. Indeed, many WLAN equipment vendors are developing the IEEE 802.11a/b/g multimode APs with wireless relays, where the IEEE 802.11b/g mode is mainly used to connect users to an AP and the IEEE 802.11a mode is mainly used for connecting two APs. In such a WMN architecture, the WLAN system can be deployed in the urban area with less cabling engineering work.

Specifically, we consider a WMN for the dense urban area, where the IEEE 802.11a WLAN standard is mainly used for data forwarding among APs, while the IEEE 802.11b/g is for data access between APs and user terminals. Recall that the IEEE 802.11a WLAN is assigned with eight nonoverlapping channels for outdoor applications in the spectrum of 5.25 to 5.35 GHz and 5.725 to 5.825 GHz, whereas the IEEE 802.11 b/g WLAN has three nonoverlapping channels in the spectrum of 2.4 to 2.4835 GHz. To avoid the cochannel interference, frequency planning is applied to ensure two buffer cells between the two cochannel APs. Thus, the intercell cochannel interference is reduced and will not be considered in this work.

To deploy a WMN in a dense urban environment, the coverage range of an AP is a key parameter. Table 7.1 shows the relationship

Table 7.1 Link Data Rates vs. Coverage Ranges for the IEEE 802.11a/b WLANs

(a) Transmission Performance of IEEE 802.11a

Data link rate (Mbps)	54	48	36	24	18	12	9	6
Indoor range [22][a] (m)	13	15	19	26	33	39	45	50
Outdoor range [22][a] (m)	30				180			304
Link capacity [23][b] (Mbps)	27.1	25.3	21.2	15.7	12.6	9.0	7.0	4.8

[a] 40 mW with 6 dBi gain patch antenna.
[b] packet error rate (PER) = 10% and packet length = 1500 octets.

(b) Transmission Performance of IEEE 802.11b

Data link rate (Mbps)	11	5.5	2	1
Indoor range [22][a] (m)	48	67	82	124
Outdoor range [22][a] (m)	304			610

[a] 100 mW with 2.2 dBi gain patch antenna.

between coverage range and link capacity for both the IEEE 802.11a/b WLANs [22].*

7.3.1.1 Throughput Model between Access Points

The throughput model between two APs follows the IEEE 802.11a WLAN specifications. Table 7.1a lists the coverage range and link capacity for the IEEE 802.11a WLAN [22,23]. By means of the curve-fitting method as illustrated in Figure 7.3, the radio link capacity $H(d)$ can be expressed as a function of the separation distance d:

$$H(d) = a_1 e^{a_2 d} + a_3 \text{ (Mbps)}, \quad d \le d_{\max} \qquad (7.1)$$

where

$$(a_1, a_2, a_3) = \begin{cases} (45.4, -2.8 \times 10^{-3}, -14.6), & \text{outdoor} \\ (47.8, -3.3 \times 10^{-2}, -3.96), & \text{indoor} \end{cases}$$

---

* Even if the coverage range varies depending on the environments, the proposed optimization approach for the access point placement can be applied by using the various coverage ranges as input parameters.

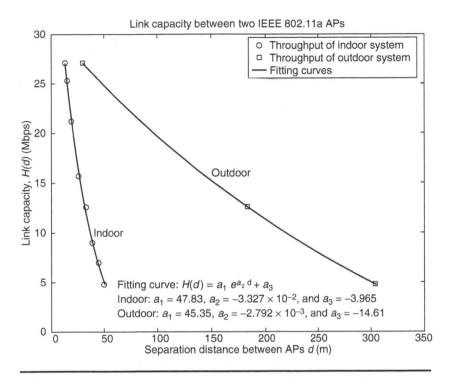

Figure 7.3 The outdoor/indoor 802.11a link capacity performance $H(d)$ at a separation distance d between access points.

Note that d_{max} is the maximum reception range of the IEEE 802.11a WLAN. In addition, since the APs are mounted on the streetlamps, the separation distance d between APs is written as $d = \Omega L_s$, where Ω is a positive integer and L_s is the separation distance between streetlamps.

7.3.1.2 Throughput Model between an Access Point and Users

The design of cell size in WMN for urban coverage can be considered from two folds. First, the cell radius should be less than r_{max} to maintain an acceptable data rate. Second, the cell radius should be larger than r_{min} to lower the handoff probability.

In each cell, users share the medium and employ the carrier sense multiple access with collision avoidance (CSMA/CA) MAC protocol to communicate with an AP. We assume that the users are uniformly distributed on the road with density D_M (Users/m). If the cell coverage (in radius) is r, the average number of users in a cell is $k = 2rD_M$. According to the method [24], the cell saturation throughput of

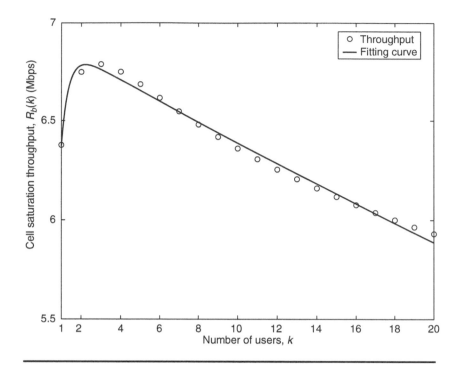

Figure 7.4 The cell saturation throughput vs. the number of users for the IEEE 802.11b WLAN.

the IEEE 802.11b WLAN for various numbers of users k is shown in Figure 7.4, where data rate is 11 Mbps and average packet payload is 1500 bytes. By using the curve-fitting method, the cell saturation throughput $R_b(k)$ for this particular case can be expressed by

$$R_b(k) = b_1 e^{b_3 k} + b_2 e^{b_4 k} \tag{7.2}$$

where $b_1 = 6.9$, $b_2 = -6.9$, $b_3 = -8.2 \times 10^{-3}$, and $b_4 = -2.6$.

7.3.2 Optimal Access Point Placement

7.3.2.1 Problem Formulation

Radio link throughput and coverage are two essential factors in placing APs in a WMN for dense urban coverage. From the viewpoint of coverage, a larger cell is preferred because less number of APs are required. From the standpoint of throughput, however, a smaller cell size will be better since it can achieve a higher data rate in the wireless

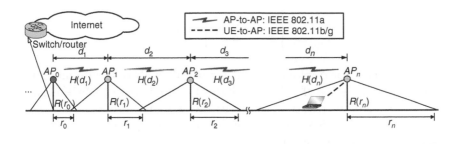

Figure 7.5   A cluster of APs in the dense urban environment. (This is an example for the increasing-spacing placement strategy, where $d_1 \leq d_2 \leq \cdots \leq d_n$.)

link. In this work, we formulate an optimization problem to determine the best separation distance for APs with consideration of these two factors.

Figure 7.5 illustrates a scenario where APs are deployed on the streetlamps. Since APs will be symmetrically deployed to the central AP_0 in a cluster, only one side of the cluster needs to be considered. The notations in Figure 7.5 are explained as follows:

n: the number of APs in the single side of the cluster;
d_i: the separation distance between AP_{i-1} and AP_i;
$H(d_i)$: the radio link capacity between AP_{i-1} and AP_i at a distance d_i, according to the IEEE 802.11a WLAN specification;
r_i: the cell radius of AP_i;
$R(r_i)$: the aggregated traffic load from all the users associated to AP_i, in which $R(r_i) = 2r_i D_M R_D$ and R_D is the average demanded traffic of each user.

Clearly, the separation distance can be written as:

$$d_i = r_i + r_{i-1}, \quad \text{for } i = 1, 2, \ldots, n \tag{7.3}$$

and the aggregated traffic load in a cell should be constrained by the cell saturation throughput, i.e.,

$$R(r_i) \leq R_b(k) \tag{7.4}$$

In the considered scenario as depicted in Figure 7.5, the total service area in a cluster of APs is $[2r_0 + 2\sum_{i=1}^{n} 2r_i]$. Therefore, the total carried traffic load of a cluster of APs through the wireline connection can be given as

$$2\left[r_0 + 2\sum_{i=1}^{n} r_i\right] D_M R_D$$

The total cost for deploying a cluster of APs with one wireline connection is $(2n + 1 + \rho)$, which includes the total cost of $(2n + 1)$ APs and the fixed overhead cost due to the wireline connection ρ. For convenience, in this work, the wireline overhead ρ has been normalized by the cost of one AP.

In this work, the AP placement problem will be formulated as an MINLP problem with the following decision variables: n and r_0, $r_1, \ldots, r_n$. The objective function is to maximize the ratio of the total carried traffic load to the cost for a cluster of APs. In the following, we discuss the two AP placement strategies: the increasing-spacing and the uniform-spacing placement strategies.

7.3.2.2 Increasing-Spacing Placement Strategy

Figure 7.5 illustrates an example for the proposed increasing-spacing placement strategy, where $d_1 \leq d_2 \leq \cdots \leq d_n$. In a cluster, the aggregated carried traffic load of the wireless link between AP_{i-1} and AP_i is a decreasing function of i. That is, the further the AP_i from the central AP_0, the less the carried traffic load in the wireless link between AP_{i-1} and AP_i. Accordingly, it is expected to deploy APs with increasing separation distance (i.e., $d_1 \leq d_2 \leq \cdots \leq d_n$) to deliver a higher traffic load for a cluster of APs. The system parameters according to the increasing-spacing AP placement strategy can be obtained by solving the following MINLP optimization problem:

$$\underset{n, r_0, r_1, \ldots, r_n}{\text{MAX}} \frac{\text{Total carried traffic load in a cluster of APs}}{\text{Total cost for deploying a cluster of APs}}$$

$$= \frac{2\left[r_0 + 2\sum_{i=1}^{n} r_i\right] D_M R_D}{(2n + 1 + \rho)} \tag{7.5}$$

subject to

$$2r_i D_M R_D \leq R_b(k), \quad i = 1, 2, \ldots, n \tag{7.6}$$

$$H(d_i) \geq \sum_{j=i}^{n} R(r_j) = \sum_{j=i}^{n} 2r_j D_M R_D, \quad i = 1, 2, \ldots, n \tag{7.7}$$

$$d_i = r_i + r_{i-1}, \quad i = 1, 2, \ldots, n \tag{7.8}$$

$$r_i \geq r_{min}, \quad i = 0, 1, \ldots, n \tag{7.9}$$

$$r_i \leq r_{max}, \quad i = 0, 1, \ldots, n \tag{7.10}$$

$$d_i \leq d_{max}, \quad i = 1, 2, \ldots, n \tag{7.11}$$

$$d_i = \Omega_i L_S, \quad i = 1, 2, \ldots, n \tag{7.12}$$

$$n, \Omega_i \in Z^+ \tag{7.13}$$

In the following, we will explain the above constrains. Equation 7.6 means that in each cell the total carried traffic load is constrained by the cell saturation throughput. Equation 7.7 states the condition that the radio link capacity $H(d_i)$ between AP_{i-1} and AP_i should be greater than the aggregate carried traffic load from the cells served by AP_i, $AP_{i+1}, \ldots$, and AP_n. Equation 7.8 is the relationship between the separation distance d_i and the cell radius r_i. Equation 7.9 and Equation 7.10 refer to the limits of cell radius, i.e., r_{min} and r_{max}. According to Equation 7.11, the maximum separation distance between two APs is limited to d_{max}. With respect to Equation 7.12, it is a limit on the separation distance d_i due to the distance between streetlamps. Equation 7.13 means that Ω_i and n (the number of APs in a cluster) are positive integers.

7.3.2.3 Uniform-Spacing Placement Strategy

Referring to Figure 7.1, the uniform-spacing placement strategy is to make all the cells in a cluster have the same radius, and thus the APs are uniformly deployed in the service area. Therefore, there are additional constraints for this placement, i.e., $r_i = r$, for $i = 0, \ldots, n$, and thus $d_i = d = 2r$, for $i = 1, \ldots, n$. Accordingly, $R(r_i) = R(r)$, for $i = 0, \ldots, n$, and $H(d_i) = H(d)$, for $i = 1, \ldots, n$. Then, the MINLP formulation of AP placement problem can be modified as

$$\underset{n,r}{\text{MAX}} \frac{(2n+1) \cdot 2r\, D_M\, R_D}{(2n+1+\rho)} \tag{7.14}$$

subject to

$$R_b(k) \geq R(r) = 2r\, D_M\, R_D \tag{7.15}$$

$$H(d) \geq nR(r) = n \cdot 2r\, D_M\, R_D \tag{7.16}$$

$$d = \Omega L_S \tag{7.17}$$

$$n, \Omega \in Z^+ \tag{7.18}$$

Here, Ω is a positive integer.

7.3.3 Numerical Examples of WMN for Dense Urban Coverage

We compare the performances of the increasing-spacing placement strategy and the uniform-spacing placement strategy. The system parameters in the numerical examples are summarized in Table 7.2.

Figure 7.6 compares the achieved profits of the objective function for the increasing-spacing and the uniform-spacing placement strategies with various wireline overheads ρ. Figure 7.6 demonstrates the advantage of the increasing-spacing placement strategy over the uniform-spacing placement strategy. The achieved profit of the objective function is a concave function of the number of APs, n, as depicted in Figure 7.6. Therefore, there exists an optimal solution of n to maximize the profit of the objective function. For example, when the wireline overhead $\rho = 4$, $n = 3$ will achieve the best performances for both placement strategies. The corresponding cell radii for the increasing-spacing placement strategy are $(r_0, r_1, r_2, r_3) = (113.3, 66.7, 143.3, 156.7$ m$)$ and that for the uniform-spacing placement strategy is $r = 105$ m, respectively. Accordingly, the corresponding separation distances for the increasing-spacing placement strategy are $(d_1, d_2, d_3) = (180, 210, 300$ m$)$ and that for the uniform-spacing placement strategy is $d = 210$ m, respectively. In this case, the increasing-spacing placement strategy can achieve 15% higher profit of the objective function than the uniform-spacing placement strategy. In Figure 7.6, we can also observe that the best number of APs in a cluster can vary for different strategies. When the wireline overhead $\rho = 2$, $n = 2$ will achieve the best performance for the increasing-spacing placement strategy, and $n = 1$ for the uniform-spacing placement strategy. In this case, the achieved profit of the objective function for the increasing-spacing placement strategy is about 6% better than that for the uniform-spacing placement strategy.

Table 7.2 System Parameters for Numerical Examples

Symbol	Item	Nominal Value
D_M	Road traffic density	0.08 Users/m
L_S	Distance between two streetlamps	30 m
R_D	Traffic demand of each user	0.2 Mbps
r_{min}	Minimum of cell radius	45 m
r_{max}	Maximum of cell radius	300 m
d_{max}	Maximum distance between APs	300 m

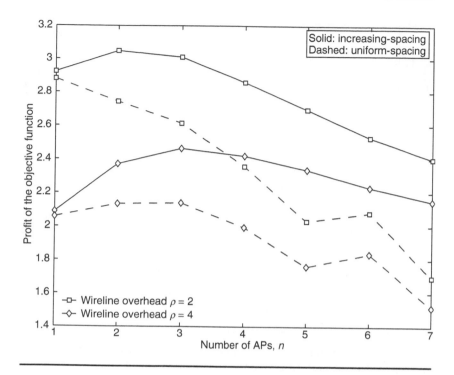

Figure 7.6   Comparison of the increasing-spacing and the uniform-spacing placement strategies in terms of the achieved profit of the objective function for different wireline overheads ρ.

Figure 7.7 shows the sum of carried traffic load and the total service area for a cluster of $(2n + 1)$ APs according to the increasing-spacing and the uniform-spacing placement strategies. One can observe that the total carried traffic load with the increasing-spacing placement strategy increases faster than that with the uniform-spacing placement approach as the number of APs in a cluster increases. Furthermore, the increment of the traffic load for the uniform-spacing strategy will gradually diminish (see $n = 6$ to $n = 7$). Since the profit of the objective function is proportional to the total carried traffic load and inversely proportional to the cost of a cluster of APs, the achieved profit of the objective function is therefore a concave function of n as shown in Figure 7.6.

In Figure 7.8, we show that the average carried traffic load per cell and the average cell radius for the increasing-spacing and the uniform-spacing placement strategies decrease as the number of APs increases.

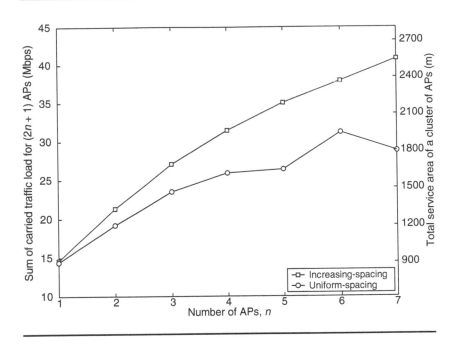

Figure 7.7   Performance comparison of the increasing-spacing and the uniform-spacing placement strategies, from the viewpoint of one cluster.

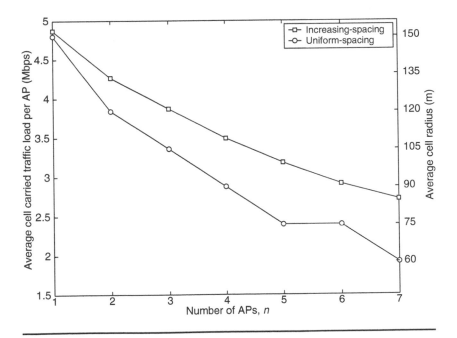

Figure 7.8   Comparisons of the average cell carried traffic loads and the average cell radii for the increasing-spacing and the uniform-spacing placement strategies.

In the considered WMN with wireless relays to forward data for the neighboring APs, the cluster size can not be too large due to the constraint of radio link capacity. Because of this reason, one can see that the average traffic load per cell decreases as n increases in the cluster. Nevertheless, the average traffic load per cell for the increasing-spacing placement strategy decreases slower than that for the uniform-spacing placement strategy. Moreover, one can observe an interesting phenomenon for the uniform-spacing placement strategy at $n = 5$ and $n = 6$, where both cases have the same traffic load and cell radius ($r = 75$ m). This explains why the profit of the objective function for the uniform-spacing strategy increases when n changes from 5 to 6 in Figure 7.6.

7.4 SCALABLE RING-BASED WMN FOR WIDE-AREA COVERAGE

7.4.1 Network Architecture and Assumptions

Figure 7.2 illustrates the scalable ring-based WMN for wide-area coverage [11]. In each mesh cell, all users are connected to the central gateway in a multihop fashion. Each intermediate node operates as a wireless relay to forward data traffic to the gateway. The gateway connects to the backbone network via a wired or wireless connection. Using this mesh architecture, the cabling engineering work for WMN deployment can be reduced.

In this work, we consider a multichannel WMN. In this WMN, each mesh cell is divided into several rings, denoted by $A_i, i = 1, 2, \ldots, n$. The user in the ring A_i will connect to the central gateway via an i-hop communication. We assume that each node can concurrently receive and deliver the forwarded traffic as [2,9,21]. That is, each node is equipped with two radio interfaces, and the users in ring A_i will communicate with the users in rings A_{i-1} and A_{i+1} at two different channels f_i and f_{i+1}, respectively. By doing so, the multihop mesh network becomes scalable to the number of users since the contention issue can be resolved by the multichannel arrangement in a ring-based network.

We assume that frequency planning is applied to avoid the cochannel interference, and thus the inter-ring cochannel interference will not be considered in this work. In a multichannel network [21], the dynamic frequency assignment can flexibly utilize the available channels, but it needs a multichannel MAC protocol that is sometimes complicated. In the considered ring-based WMN, however, the fixed frequency planning is simple because it only needs to consider the width of each ring to ensure a sufficient cochannel reuse distance.

The carried traffic load in each mesh node includes its own traffic and the forwarded traffic from other users. Assume that all the nodes in the inner ring A_i share the relayed traffic from the outer ring A_{i+1}. Suppose that the user density is ρ. The average number of nodes c_i in the ring A_i can be expressed as

$$c_i = \rho a_i = \begin{cases} \rho \pi r_i^2, & \text{for } i = 1 \\ \rho \pi (r_i^2 - r_{i-1}^2), & \text{for } 1 < i \le n \end{cases} \tag{7.19}$$

where a_i and $(r_i - r_{i-1})$ are the area and the width of ring A_i, respectively. Let R_D and R_i be traffic load generated by each node and the total carried traffic load per node in ring A_i, respectively. Then, it is followed that

$$R_i = \frac{c_{i+1}}{c_i} R_{i+1} + R_D$$

$$= \left[\frac{\sum_{j=i+1}^{n} c_j}{c_i} + 1 \right] R_D \tag{7.20}$$

For the outermost ring A_n, $R_n = R_D$.

7.4.2 Wireless Collision Domain and Sensing Region

The *wireless collision domain* is defined as a region where a number of users are contending for the same radio channel. Figure 7.9 shows a wireless collision domain in the ring A_i, which is an annulus sector with the central angle of $\theta_{W,i}$. Furthermore, we define the *sensing region* as the maximal area in which any two users can sense the activity of each other. In Figure 7.9, the sensing region in the ring A_i is an annulus sector with a central angle of $\theta_{S,i}$. Let l_{RC} be the distance between two opposite boundaries of the sensing region. As shown in the figure, the central angle $\theta_{W,i}$ of the wireless collision domain in the ring A_i is equal to the angle $\theta_{S,i-1}$ of the sensing region in the ring A_{i-1}. That is

$$\theta_{W,i} = \theta_{S,i-1}$$

$$= \begin{cases} 2\sin^{-1}\left(\frac{l_{RC}}{r_{i-1} + r_{i-2}} \right), & \text{for } l_{RC} \le (r_{i-1} + r_{i-2}) \\ \pi, & \text{otherwise} \end{cases} \tag{7.21}$$

Besides, the areas $A_{W,i}$ and $A_{S,i}$ of the wireless collision domain and the sensing region can be expressed respectively as

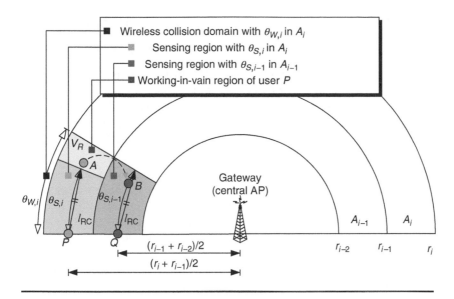

Figure 7.9 Examples of wireless collision domain and sensing region in the ring-based wireless mesh network.

$$A_{W,i} = \frac{\theta_{W,i}}{2}(r_i^2 - r_{i-1}^2) \tag{7.22}$$

$$A_{S,i} = \frac{\theta_{S,i}}{2}(r_i^2 - r_{i-1}^2) \tag{7.23}$$

Noteworthily, under the ring-based cell structure, the wireless collision domain is larger than the sensing region. As the example in Figure 7.9, users A and P in the ring A_i belong to the same wireless collision domain, but user A is out of the sensing region of user P. This phenomenon is due to the fact that the four-way handshaking request-to-send/clear-to-send (RTS/CTS) mechanism is employed to avoid the hidden node problem. When user A is sending data to user B, user P can send a RTS frame to user Q in the ring A_{i-1}. Nevertheless, user Q can not reply a CTS frame to user P because user B is inside the sensing region of user Q. In this case, the existence of transmitter in the region V_R invalidates the RTS request of P. Consequently, the region V_R with a central angle of $(\theta_{W,i} - \theta_{S,i})$ is defined as the *working-in-vain region* of P. Such impacts of ring structure on frame contention will be incorporated into the cross-layer throughput model.

7.5 CROSS-LAYER THROUGHPUT ANALYSIS

7.5.1 Channel Activity in Ring-based Multihop WMN

In a CSMA-based network, the activity of the radio medium can be logically described by a sequence of effective time slots [24–26]. Subject to the backoff procedures, an effective time slot is defined as the time interval between two consecutive backoff counter decrements of the considered user. Therefore, there are five types of effective time slots, including

1. Successful frame transmission from the considered user
2. Unsuccessful frame transmission from the considered user
3. Empty slot, when all the users are backlogged or idle
4. Successful frame transmission from other user
5. Unsuccessful frame transmission from other user

Their durations are defined as $T_1 = T_4 = T_S$, $T_2 = T_5 = T_C$, $T_3 = \sigma$, where σ is the duration of an empty slot, T_S and T_C detailed in Equation 7.35 and Equation 7.36 are the successful transmission time and collision duration, respectively. Thus, the average duration T_v of the effective time slot can be written as

$$T_v = \sum_{j=1}^{5} v_j T_j \qquad (7.24)$$

Here, v_j is the corresponding probability for the effective slot type and will be calculated in the following.

7.5.1.1 Successful/Unsuccessful Transmission from Considered User

The considered user P can successfully send data, as long as no other user is transmitting in the adjacent wireless collision domains of P, as shown in Figure 7.10. Consider the user P along with its adjacent wireless collision domains influenced by two neighboring transmitters P_L and P_R. Let ψ_L and ψ_R represent the positions of P_L and P_R, respectively. If the transmitter P_L (or P_R) is within the working-in-vain regions of user P (i.e., ψ_L or $\psi_R \in [\theta_{S,i}, \theta_{W,i}]$), the considered user P still can send the RTS request at the beginning of an effective slot. However, user Q will not reply the CTS acknowledgment since user Q_L (or Q_R) is inside the sensing region of user Q. Suppose that $Z_{W,i}$ is the channel utilization in a wireless collision domain as defined in Equation 7.42, which represents the average probability that in the adjacent wireless

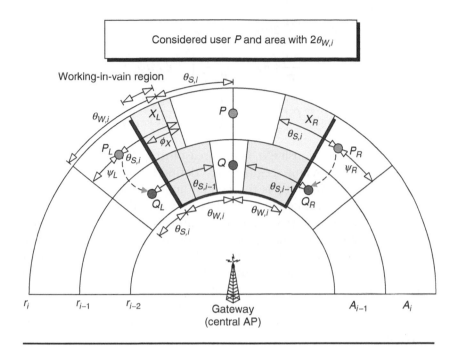

Figure 7.10 The considered user *P* and two adjacent wireless collision domains, where user *P* is contending for the radio channel.

collision domain one user is delivering its data traffic. Referring to Figure 7.10, the working-in-vain probability P_v of one user can be computed by

$$p_v = 1 - \Pr\{\psi_L, \psi_R \notin [\theta_{S,i}, \theta_{W,i}]\}$$

$$= 1 - \left[1 - Z_{W,i}\frac{\theta_{W,i} - \theta_{S,i}}{\theta_{W,i}}\right]^2 \qquad (7.25)$$

Now, we consider the case that both transmitters P_L and P_R are not in the working-in-vain regions of user P (i.e., $\psi_L, \psi_R \in [0, \theta_{S,i}]$). In the considered area, only the users in the area $\{2A_{W,i} - (X_L + X_R)\}$ are able to send RTS frames as shown in Figure 7.10. Those users in regions X_L and X_R cannot send their requests since the transmitters P_L and P_R are located in their sensing regions. Let $\overline{\phi}_X$ be the average central angle for the region X_L. Then, the average number of contending users in the considered area of angle $2\theta_{W,i}$ is equal to the number of users in the area of $\{2A_{W,i} - (X_L + X_R)\}$, i.e.,

$$c_{1,i} = \frac{\rho a_i}{2\pi} 2(\theta_{W,i} - Z_{W,i}\bar{\phi}_X)$$

$$= \frac{\rho a_i}{\pi} \left(\theta_{W,i} - \frac{Z_{W,i}}{\theta_{W,i}} \int_0^{\theta_{S,i}} \psi_L \, d\psi_L \right)$$

$$= \rho(r_i^2 - r_{r-1}^2) \left(\theta_{W,i} - \frac{Z_{W,i}\theta_{S,i}^2}{2\theta_{W,i}} \right) \tag{7.26}$$

where ρ is the user density; a_i is the area of ring A_i; $\theta_{S,i}$ is the central angle of the sensing region as defined in Equation 7.21; $\phi_X = (\psi_L + \theta_{S,i}) - \theta_{S,i} = \psi_L$ is the central angle of the region X_L and ψ_L is uniformly distributed in $[0, \theta_{W,i}]$ as shown in Figure 7.10. Subject to the RTS/CTS procedures, the frame collisions may occur only when the contending users concurrently deliver their RTS requests at the beginning of an effective slot. Let τ be the transmission probability of an active user, as detailed in Equation 7.38. Suppose that P_0 is the average probability of a user being idle due to empty queue, as defined in Equation 7.40. Incorporating the impacts of ring structure on frame contention, the unsuccessful transmission probability p_u can be computed by

$$p_u = p_v + (1 - p_v)[1 - (1 - \tau(1 - P_0))^{c_1, i-1}] \tag{7.27}$$

Here, the first term accounts for the probability that at least one transmitter is inside the working-in-vain regions of user P. That is, the considered user P will not receive the CTS acknowledgment. The second term represents the probability that the RTS request from the considered user is collided with other RTS frames.

In the result, given that the considered user has a nonempty queue, the probability that in an effective slot the considered user successfully/unsuccessfully transmits its traffic can be expressed as

$$\nu_1 = \tau(1 - p_u) \tag{7.28}$$

$$\nu_2 = \tau p_u \tag{7.29}$$

7.5.1.2 Empty Slot

As shown in Figure 7.11, the considered user P observes an empty slot, as long as all the users in the sensing regions of user P are silent. Within the sensing regions of user P, the users in regions Y_L and Y_R can not send their RTS requests due to the influence from the transmitters P_L, and P_R. Let $\bar{\phi}_Y$ be the average central angle of the region Y_L. In the

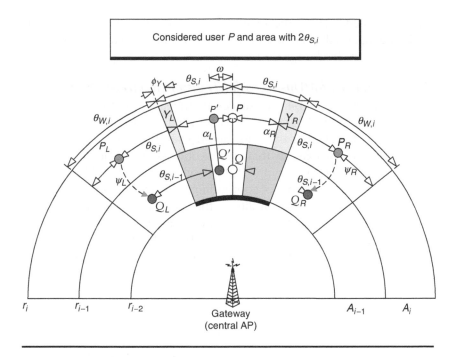

Figure 7.11 The considered user P and its two sensing regions, where user P is backlogged at the current slot.

considered area of angle $2\theta_{S,i}$, the average number of contending users is equal to the number of users in the area of $\{2A_{S,i} - (Y_L + Y_R)\}$, i.e.,

$$
\begin{aligned}
c_{2,i} &= \frac{\rho a_i}{2\pi} 2(\theta_{S,i} - Z_{W,i}\bar{\phi}_Y) \\
&= \frac{\rho a_i}{\pi}\left(\theta_{S,i} - \frac{Z_{W,i}}{\theta_{W,i}}\int_0^{\theta_{W,i}} \max(0,\ \psi_L + \theta_{S,i} - \theta_{w,i})d\psi_L\right) \\
&= \rho(r_i^2 - r_{r-1}^2)\left(\theta_{S,i} - \frac{Z_{W,i}\theta_{S,i}^2}{2\theta_{W,i}}\right)
\end{aligned}
\tag{7.30}
$$

where $\phi_Y = max\,(0,\ \psi_L + \theta_{S,i} - \theta_{w,i})$ is the central angle of the region Y_L, as shown in Figure 7.11. Therefore, from the viewpoint of the considered user P, the empty-slot probability can be computed by

$$
\nu_3 = (1 - \tau)[1 - \tau(1 - P_0)]^{c_2,\,i-1}
\tag{7.31}
$$

where the first term is the probability of the considered user being backlogged, and the second term represents the probability that all the other users are backlogged or idle.

7.5.1.3 Successful/Unsuccessful Transmission from Other User

To calculate the probability of successful transmission from other user, we consider user P and its two sensing regions, as shown in Figure 7.11. Given that the considered user P is backlogged at the current slot, the probability that at least one user sends its RTS request is equal to $p_{\text{otr}} = 1 - [1 - \tau (1 - P_0)]^{c_{2,i}-1}$, where $c_{2,i}$ derived in Equation 7.30 is the average number of contending users in the considered area. Suppose that X_j is the probability of the considered area being influenced by j neighboring transmitters. Consequently, in the considered area of angle $2\theta_{S,i}$, the conditional probability that there is at least one successful transmission from other user can be expressed as

$$
p_{\text{os}} = \frac{\sum_{j=0}^{2} (2s_{1,j} - s_{2,j})X_j}{p_{\text{otr}}} \tag{7.32}
$$

where $X_j = \binom{2}{j} Z_{W,i}^j (1 - Z_{W,i})^{2-j}$, $s_{1,j}$ is the probability that the left-side sensing region of user P has a successful transmission, and $s_{2,j}$ is the probability that each sensing region of P has a successful transmission. Then, from the viewpoint of the considered user P, the probability of an effective slot containing successful/unsuccessful transmission from other user can be expressed as

$$
\nu_4 = (1 - \tau)p_{\text{otr}}\, p_{\text{os}} \tag{7.33}
$$

$$
\nu_5 = (1 - \tau)p_{\text{otr}}\, (1 - p_{\text{os}}) \tag{7.34}
$$

where the first term accounts for the probability of the considered user being backlogged. Due to page limitation, the derivations for $s_{1,j}$ and $s_{2,j}$ are omitted.

7.5.2 MAC Throughput

First, we calculate the time durations of successful frame transmission and collision. Let l be the payload size of data frame, m_a and m_c be the transmission PHY mode for data frames and that for control frames, respectively. Subject to the IEEE 802.11 CSMA MAC protocol with RTS/CTS, the successful frame transmission time T_S and collision time T_C are expressed as follows:

$$T_S = T_{\text{RTS}}(m_c) + \delta + \text{SIFS} + T_{\text{CTS}}(m_c) + \delta + \text{SIFS}$$
$$+ T_{\text{DATA}}(l, m_a) + \delta + \text{SIFS}$$
$$+ T_{\text{ACK}}(m_c) + \delta + \text{DIFS} \qquad (7.35)$$

$$T_C = T_{\text{RTS}}(m_c) + \delta + \text{DIFS} \qquad (7.36)$$

where δ is the propagation delay, SIFS and DIFS stand for the durations of a short interframe space and a distributed interframe space. T_{DATA} (l, m_a) is the transmission time for a data frame with payload size l using PHY mode m_a. T_{RTS} (m_c), T_{CTS} (m_c), and T_{ACK} (m_c) are the transmission durations for RTS, CTS, and acknowledgment (ACK) control frames using PHY mode m_c, respectively. According to the IEEE 802.11a WLAN standard [27], the values of T_{DATA} (l, m_a), T_{RTS} (m_c), T_{CTS} (m_c), and T_{ACK} (m_c) can be specified.

To evaluate the MAC throughput in the ring-based WMN, we should consider the impacts of the physical layer ring structure on frame contention. Consider a binary exponential backoff procedure with the initial backoff window size of W. Let m_{bk} be the maximum backoff stage. Therefore, the average backoff time can be calculated by

$$\overline{B_k} = (1 - p_u)\frac{W-1}{2} + p_u(1-p_u)\frac{2W-1}{2} + \cdots$$
$$+ p_u^{m_{bk}}(1-p_u)\frac{2^{m_{bk}}W - 1}{2}$$
$$+ p_u^{(m_{bk}+1)}(1-p_u)\frac{2^{m_{bk}}W - 1}{2} + \cdots$$
$$= \frac{[1 - p_u - p_u(2p_u)^{m_{bk}}]W - (1 - 2p_u)}{2(1 - 2p_u)} \qquad (7.37)$$

where p_u is the unsuccessful transmission probability with considering the impacts of ring structure in the physical layer, as defined in Equation 7.27. Since a user transmits frames every $(\overline{B_k} + 1)$ slots [28], the transmission probability τ for a user can be written as

$$\tau = \frac{1}{\overline{B_k} + 1} = \frac{2}{1 + W + p_u W \sum_{i=0}^{m_{bk}-1}(2p_u)^i} \qquad (7.38)$$

From Equation 7.27 and Equation 7.38, we can obtain the unique solution of τ and p_u for a given idle probability P_0 of a user. The idle probability P_0 will be derived by the following queueing model.

Figure 7.12 illustrates the proposed discrete-time queueing model for a user in the ring A_i, where the state variable k_i represents the

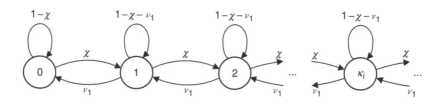

Figure 7.12 State transition diagram for the considered user, where the state variable k_i is the number of frames queued at the considered user.

number of frames queued in the user. As defined in Section 7.5.1, in each effective time slot one user can successfully transmit its data frame with probability ν_1. Consequently, the total contention delay spent for a frame (i.e., the frame service time) will be a geometric random variable with the mean of $1/\nu_1$ effective slots. In a multihop network, this phenomenon means that the arrival process of relayed traffic is also Markovian since the interarrival time of relayed traffic is geometrically distributed. Let l be the payload size of data frame. Then, it is reasonable to assume that the frame arrivals at one user follow a Poisson process with a rate of $\lambda = R_i/l$ frames/s. Here, R_i is the total carried traffic load of a user in the ring A_i, including the local traffic of user and the forwarded traffic from others, as defined in Equation 7.20. From above considerations, the state-transition probabilities for the queue model can be defined as

$$p_{k,k+1} = \chi = \lambda T_v$$
$$p_{k,k-1} = \nu_1$$
$$p_{k,k} = 1 - \chi - \nu_1 \qquad (7.39)$$

Then, we can derive the state probability [29]

$$P_k = u_c^k(1 - u_c) \qquad (7.40)$$

where $u_c = \chi/\nu_1$ and the idle probability of a user can be given as $P_0 = (1 - u_c)$.

Now, we evaluate the MAC throughput of one user. On top of the effective slot concept, the average busy probability $Z_{O,i}$ of one user and the channel utilization $Z_{W,i}$ in a wireless collision domain are expressed as

$$Z_{O,i} = \frac{\nu_1 T_1(1 - P_0)}{T_v} \qquad (7.41)$$

$$Z_{W,i} = \frac{\rho a_i}{2\pi} \theta_{w,i} \, Z_{O,i} \qquad (7.42)$$

where ν_1 is the probability that one user successfully sends a frame in an effective slot, $T_1 = T_S$ is the time duration for successful frame transmission, T_v is the average duration of the effective slot, and $\frac{\rho a_i}{2\pi} \theta_{w,i}$ is the number of users in a wireless collision domain. From Equation 7.24, Equation 7.28, and Equation 7.40 through Equation 7.42, ν_1, T_v, and P_k can be calculated by using an iterative method. Then, the capacity $H_t(d)$ of a mesh link between two nodes at a separation distance d can be calculated by

$$H_i(d) = \frac{\nu_1 T_1}{T_v} \cdot \frac{l}{T_S} = \frac{\nu_1 l}{T_v} \qquad (7.43)$$

where l is the payload size of data frame. It is noteworthy that the payload size l of data frame will be affected by the separation distance d and the PHY mode m_a, which will be discussed in the following.

In the multihop WMN, the throughput of wireless link is also affected by the hop distance. Generally, the radio signal will not only suffer from the path loss, but also from shadowing as well as multipath fading. With consideration of these radio channel effects, we assume that the average reception ranges for eight PHY modes are respectively d_j, $j = 1, 2, \ldots, 8$, where $d_1 > d_2 > \cdots > d_8$. In principle, two users with a shorter separation distance can transmit at a higher data rate. Therefore, the transmission PHY mode m_a is determined according to the separation distance d between two users, i.e.,

$$m_a = j, \text{ if } d_{j+1} < d \le d_j \qquad (7.44)$$

Furthermore, we suggest that all data frames have the same transmission time $T_{\text{DATA}} (l, m_a)$. That is, the payload size l of data frame is determined by the adopted PHY mode m_a. The same transmission time for each data frame can achieve fairness and avoid throughput degradation due to low-rate transmission [30,31].

7.6 OPTIMAL COVERAGE AND CAPACITY OF RING-BASED MESH CELL

7.6.1 Problem Formulation

Coverage and throughput are both essential performance issues in a WMN. From the viewpoint of deployment cost, a larger coverage area per cell is better since it requires less APs. From the standpoint of link throughput, however, a smaller cell is preferred since less number of users will share the radio channel resource. In the following, we formulate an optimization problem to determine the optimal width of each ring subject to the trade-off between coverage and capacity.

To begin with, we discuss the constraints in the considered optimization problem:

■ It is obvious that the capacity $H_C(i)$ of the lowest-rate link in ring A_i should be greater than the traffic load carried at each node R_i (defined in Equation 7.20), i.e., $H_C(i) = H_i(r_i - r_{i-1}) \geq R_i$, where $(r_i - r_{i-1})$ is the width of ring A_i. This constraint guarantees the minimum throughput for each user. As shown in Figure 7.13, the lowest-rate link in ring A_i is the link between nodes $P_{C,i}$ and $Q_{C,i}$ at a separation distance $d = (r_i - r_{i-1})$.
■ The maximum reception range should be larger than the ring width $(r_i - r_{i-1})$, i.e., $(r_i - r_{i-1}) \leq d_{\max} = d_1$.
■ The ring width should be greater than the average distance $d_{\min}$ between two neighboring nodes, i.e., $(r_i - r_{i-1}) \geq d_{\min}$ where $d_{\min} = 1/\sqrt{\rho}$ (m) is dependent on the user node density ρ.

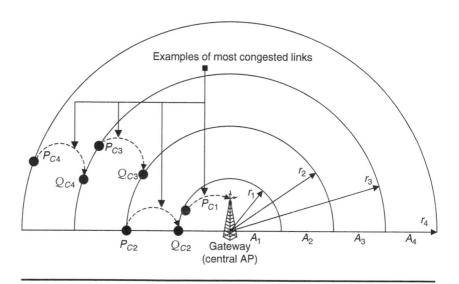

Figure 7.13 Examples for the lowest-rate links for a ring-based mesh cell with $n = 4$.

7.6.2 MINLP Optimization Approach

From the above considerations, the optimal coverage issue in a WMN can be formulated as an MINLP problem with the following decision variables: n (the number of rings in a mesh cell) and $r_1, r_2, \ldots, r_n$. The objective function is to maximize the coverage of a mesh cell as follows.

$$\underset{n, r_1, r_2, \ldots, r_n}{\text{MAX}} \quad r_n (\text{coverage of a mesh cell}) \qquad (7.45)$$

subject to

$$H_C(i) \geq R_i \qquad (7.46)$$

$$d_{\max} \geq (r_i - r_{i-1}) \geq d_{\min} \qquad (7.47)$$

$$n \in Z^+ \qquad (7.48)$$

7.6.3 Numerical Examples of Ring-based WMN

The system parameters are summarized in Table 7.3 and Table 7.4. We will consider a simple case where all the ring widths in a cell are the same, i.e., $(r_i - r_{i-1}) = r$. The control frames (RTS/CTS/ACK frames) are transmitted with PHY mode $m_c = 1$ for reliability. The mesh nodes are uniformly distributed with density $\rho = (100)^{-2}$ (nodes/m^2). We assume the sensing range $l_{\text{RC}} = \gamma_I d_{\max}$, where γ_I is 1.5. The chosen data frame payload sizes for eight PHY modes are {425, 653, 881, 1337, 1793, 2705, 3617, 4067 ($4095 - MAC_{\text{hdr}} - MAC_{\text{FCS}}$)} bytes [30]. Referring to the measured results [22], the corresponding average reception ranges are $d_j = \{300, 263, 224, 183, 146, 107, 68, 30\}$ m. It is true that these reception ranges vary for different environments. However, the proposed optimization approach is general enough to evaluate

Table 7.3 System Parameters for Numerical Examples

Symbol	Item	Nominal Value
ρ	User node density	$(100)^{-2}$ m^{-2}
R_D	Demanded traffic of each user node	0.5 Mbps
$d_{\min}$	Minimum of ring width, i.e., $(1/\sqrt{\rho})$	100 m
$d_{\max}$	Maximum reception range	300 m
l_{RC}	Sensing range $(\gamma_I d_{\max})$	450 m

Table 7.4   Relevant Network Parameters for an IEEE 802.11a WLAN

PHY Mode for Data Frame, m_a	$1 \sim 8$
PHY Mode for Control Frame, m_c	1 (6 Mbps)
Propagation delay, δ	1 μs
SIFS	16 μs
DIFS	34 μs
Empty slot time, σ	9 μs
m_{bk}	6
Initial contention window, W	16

the performances of different WMNs by adopting various reception ranges.

In Figure 7.14, the achieved cell coverage against the number of rings in a mesh cell for $R_D = 0.5$ Mbps is shown. One can observe that

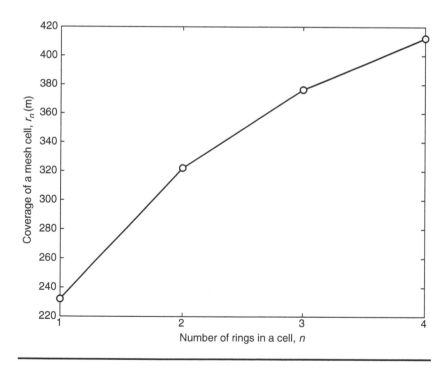

Figure 7.14   Cell coverage vs. the number of rings *n* in a mesh cell, where the demanded traffic per user is $R_D = 0.5$ Mbps.

the optimal achieved cell coverage is 412 (m) with $n = 4$. Compared with the coverage of the single-hop network ($n = 1$), the multihop mesh network improves the coverage by 77%. Figure 7.15 illustrates the capacity performance against the number of rings in a cell, for $R_D = 0.5$ Mbps. In this example, the corresponding optimal cell throughput is 26.7 Mbps with $n = 4$. Compared with $n = 1$, the multihop mesh network improves the cell throughput by 215%.

Figure 7.14 and Figure 7.15 show that the proposed ring-based WMN can enhance the cell coverage and throughput compared with the single-hop network. More importantly, we find that the optimal number of rings is equal to $n = 4$ for $R_D = 0.5$ Mbps. In these figures, it is shown that the more the number of rings in a mesh cell, the better the coverage and capacity. However, the constraints on the mesh link throughput and the separation distance between the mesh nodes determine the optimal solution.

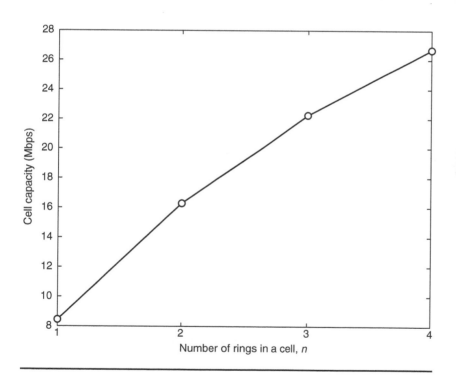

Figure 7.15 Achieved cell capacity vs. the number of rings n in a mesh cell, where $R_D = 0.5$ Mbps.

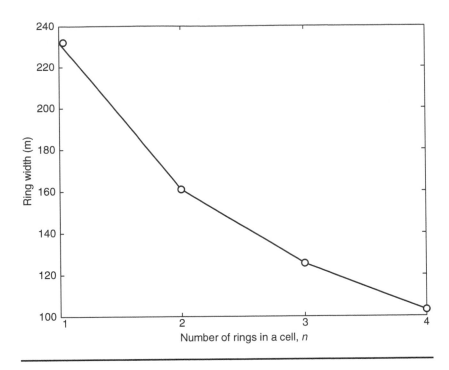

Figure 7.16   Ring width *r* vs. the number of rings *n* in a cell, where $R_D = 0.5$ Mbps.

Figure 7.16 shows the ring width for various number of rings *n* in a cell. Referring to this figure, when the number of rings increases, the ring width decreases. In general, when the number of rings *n* in a cell increases, the cell coverage also increases as shown in Figure 7.14. For handling the increment of relay traffic as *n* increases, each ring width will decrease to shorten the hop distance and thus improve the link capacity. However, since the ring width should be larger than the average distance between two neighboring nodes as discussed in Section 7.6.1, there exists a maximum value of *n*. In this example, the maximum allowable number of rings in a mesh cell is $n = 4$.

7.7   SUMMARY

WMNs are the promising solution in the next-generation communication system to support the ubiquitous broadband services by low transmission power. However, multihop networking suffers from the scalability issue. This chapter addresses this key challenge of the WMN from a network architecture perspective. We consider two main application scenarios of WMNs, including the dense urban and wide-area scenarios. The proposed WMNs are scalable in terms of coverage, since the frequency planning with multiple available channels can effectively resolve the contention issue and thus the throughput can be ensured by properly designing the deployment parameters. We investigate the optimal trade-off between capacity and coverage for the scalable WMNs. In addition, we apply the MINLP optimization approach to determine the optimal deployment parameters, subject to the trade-offs between throughput and coverage.

7.8   OPEN ISSUES

Scalability is a quite desirable feature of WMNs. With the support of this feature, the network performances including user throughput and even the quality of service (QoS) (e.g., delay, jitter, and packet loss rate) can still be maintained as network coverage and users increase. In the following, we will discuss some interesting issues in WMNs, from the viewpoint of scalability.

7.8.1   Quality of Service (QoS)

It goes without saying that the multihop communications can extend the coverage of an AP with more hops and/or longer hop distance. However, the repeatedly relayed traffic with more hops will easily exhaust the radio resource and degrade the quality of service, e.g., higher delay and jitter. Meanwhile, longer hop distance will also lead to lower data rate in the relay link between nodes and then higher delay. Therefore, maintaining the throughput and QoS while extending the coverage area of a multihop network is an essential issue.

Another interesting issue is to support differentiated services with different priorities. In the literature, there are some delay guarantee mechanisms proposed for WLAN. The point coordination function (PCF) of IEEE 802.11 WLAN standard is used to send delay-sensitive services in the contention free periods [32]. Based on the distributed coordination function (DCF), the work [33] supports delay differentiation by adjusting the interframe space and the contention window size. Another well-known QoS mechanism is the enhanced DCF (EDCF) in IEEE 801.11e [34], which groups services into eight categories with different priorities. By EDCF lower delay can be achieved by using the higher access priority. However, these works [32–34] mainly focus on the single-hop wireless networks.

In the WMN, mesh nodes are connected to the central gateway in a multihop fashion. At each hop, the packet may be delivered at different transmission rate with different packet loss rate, thereby experiencing different queuing delay and contention delay. Due to lack of central control, as network size increases, providing end-to-end QoS guarantees for different service types in multihop WMNs is still a challenge.

7.8.2 Cross-Layer Design

Cross-layer design can improve the network performance and scalability. For example, this chapter exploits the physical layer system architecture with a simple multichannel frequency planning to improve MAC throughput as network size increases, thereby making the WMN more scalable. Indeed, in the wireless networks, there are many interactions among the transport, routing, MAC, and physical layer protocols. For example, the transmission power and rate in the physical layer will influence MAC throughput and routing decisions. The link selection in the routing layer will affect the MAC layer contention level. In addition, the MAC protocol can adapt the backoff window size according to the end-to-end delay information provided by the transport layer. How to exploit cross-layer interactions in the design of high-performance scalable WMNs and how to optimize the performances of WMNs are very interesting issues.

However, it should be noted that cross-layer design will face the loss of design abstraction and the incompatibility with existing protocols [35]. Furthermore, any protocol modification may result in unforeseen impact on the whole system, and difficulty in management. To avoid these potential problems, some design principles have been suggested [35].

7.8.3 Cooperative Communications

Differing from the conventional WMN, in cooperative communication systems different users in the wireless network will collaborate to deliver traffic, through distributed transmission and processing [36,37]. Specifically, user information is delivered not only by the source node, but also by the cooperative relays. Then, the destination node combines these signals transmitted from different nodes. By doing so, several single-antenna cooperative relays can form a virtual antenna array system to combat server path loss and fading, thereby improving link reliability and capacity. In addition, cooperative communications also help achieve a low-power communication system and realize seamless networking.

Obviously, low transmit power and high reliability are two appealing features of cooperative communications. Low transmit power will reduce contention collisions and improve the efficiency of spatial frequency reuse. High communication reliability will further improve relay link capacity and reduce retransmission delay. Therefore, these features can be employed to improve the network scalability.

To conclude, in the design of a practical scalable cooperative communication system, many important issues still need to be addressed, including system architecture design; capacity, performance analysis and optimization; QoS, resource management and scheduling; MAC and routing protocol design.

ACKNOWLEDGMENT

This work was supported jointly by the MoE ATU program, the National Science Council and the Program for Promoting Academic Excellence of Universities under the grand numbers, EX-91-E-FA06-4-4, NSC 94-2213-E-009-030, and NSC 94-2213-E-009-060.

REFERENCES

1. R. Pabst et al., "Relay-based deployment concepts for wireless and mobile broadband radio," *IEEE Commun. Mag.*, vol. 42, no. 9, pp. 80–89, September 2004.
2. J. Jun and M. Sichitiu, "The nominal capacity of wireless mesh networks," *IEEE Wireless Common. Mag.*, vol. 10, no. 5, pp. 8–14, October 2003.
3. M. Zhang and R. Wolff, "Crossing the digital divide: cost-effective broadband wireless access for rural and remote areas," *IEEE Commun. Mag.*, vol. 42, no. 2, pp. 99–105, February 2004.
4. T. Fowler, "Mesh networks for broadband access," *IEE Rev.*, vol. 47, no. 1, pp. 17–22, January 2001.
5. *MeshNetworks website*, http://www.meshnetworks.com.
6. *MeshDynamics website*, http://www.meshdynamics.com.
7. B. Lewis, "Mesh Networks in fixed broadband wireless access," July 2003. [Online]. Available: http://grouper.ieee.org/groups/802/16/docs/03/80216-03_10r1.pdf.
8. L. Qiu et al., "Troubleshooting multihop wireless networks," Microsoft Research Tech. Report, MSR-TR-2004-11, November 2004. [Online]. Available: ftp://ftp.research.microsoft.com/pub/tr/TR-2004-11.pdf.

9. I. Akyildiz, X. Wang, and W. Wang, "Wireless mesh networks: a survey," *Computer Networks*, vol. 47, pp. 445–487, March 2005.

10. J.-H. Huang, L.-C. Wang, and C.-J. Chang, "Deployment strategies of access points for outdoor wireless local area networks," in *Proc. IEEE VTC'05 Spring*, May 2005.

11. J.-H. Huang, L.-C. Wang, and C.-J. Chang, "Coverage enhancement for a multi-channel ring-based wireless mesh network with guaranteed throughput and delay," in *Proc. IEEE ICC'06*, June 2006.

12. 3rd Generation Partnership Project, "Technical specification group radio access networks; rf system scenarios (release 1999)," *Report 3GPP TR 25.942, V3.0.0, 3GPP*, March 2001.

13. R.C. Rogrigues, G.R. Mateus, and A.A.F. Loureiro, "On the design and capacity planning of a wireless local area network," in *Proc. IEEE/IFIP NOMS'00*, pp. 335–348, April 2000.

14. M. Amenetsky and M. Unbehaun, "Coverage planning for outdoor wireless LAN systems," in *Proc. IEEE Zurich Seminar on Broadband Communications*, pp. 49.1–6, February 2002.

15. M. Kobayashi et al., "Optimal access point placement in simultaneous broadcast system using OFDM for indoor wireless LAN," in *Proc. IEEE PIMRC'00*, pp. 200–204, September 2000.

16. T. Jiang and G. Zhu, "Uniform design simulated annealing for optimal access point placement of high data rate indoor wireless LAN using OFDM," in *Proc. IEEE PIMRC'03*, pp. 2302–2306, September 2003.

17. Y. Lee, K. Kim, and Y. Choi, "Optimization of AP placement and channel assignment in wireless LANs," in *Proc. IEEE LCN'02*, pp. 831–836, November 2002.

18. S. Naghian and J. Tervonen, "Semi-infrastructured mobile ad-hoc mesh networking," in *Proc. IEEE PIMRC'03*, pp. 1069–1073, September 2003.

19. P. Gupta and P.R. Kumar, "The capacity of wireless networks," *IEEE Trans. Inform. Theory*, vol. 46, pp. 388–404, March 2000.

20. J. Li et al., "Capacity of ad hoc wireless networks," in *Proc. ACM MobiCom'01*, July 2001.

21. A. Raniwala and T.-C. Chiueh, "Architecture and algorithms for an IEEE 802.11-based multi-channel wireless mesh network," in *Proc. IEEE INFOCOM'05*, March 2005.

22. CISCO, *Cisco Aironet 1200 Series Access Point*. [Online]. Available: http://www.cisco.com/en/US/products/hw/wireless/ps430/index.html.

23. J.C. Chen, "Measured performance of 5 GHz 802.11a wireless LAN systems." [Online]. Available: http://epsfiles.intermec.com/esp_files/eps_wp/Atheros-RangeCapacity- Paper.pdf.

24. G. Bianchi, "Performance analysis of the IEEE 802.11 distributed coordination function," *IEEE J. Select. Areas Commun.*, vol. 18, no. 3, pp. 535–547, March 2000.

25. P. Chatzimisios, A.C. Boucouvalas, and V. Vitsas, "Packet delay analysis of the IEEE MAC 802.11 protocol," *IEE Electron. Lett.*, vol. 39, pp. 1358–1359, September 2003.

26. X.J. Dong and P. Variya, "Saturation throughput analysis of IEEE 802.11 wireless LANs for a lossy channel," *IEEE Commun. Lett.*, vol. 9, no. 2, pp. 100–102, February 2005.

27. IEEE 802.11a, *Part 11: Wireless LAN, Medium Access Control (MAC) and Physical Layer (PHY) Specifications: High-Speed Physical Layer in the 5 GHz Band*, supplemnet to IEEE 802.11 Standard, September 1999.

28. Y.C. Yay and K.C. Chua, "A capacity analysis for the IEEE 802.11 MAC protocol," *Wireless Network*, pp. 159–171, July 2001.

29. D. Gross and C.M. Harris, *Fundamentals of Queueing Theory*, 3rd ed. New York: John Wiley & Sons, 1998.

30. L.J. Cimini Jr. *et al., Packet Shaping for Mixed Rate 802.11 Wireless Networks*, United States Patent Application Number: US 20030133427, July 2003.

31. M. Heusse et al., "Performance anomaly of 802.11b," in *Proc. IEEE INFO-COM'03*, pp. 836–843, March 2003.

32. J.-Y. Yeh and C. Chen, "Support of multimedia services with the IEEE 802.11 MAC protocol," in *Proc. IEEE ICC'02*, pp. 600–604, 2002.

33. S.-T. Sheu and T.-F. Sheu, "A bandwidth allocation/sharing/extension protocol for multimedia over IEEE 802.11 ad hoc wireless LANs," *IEEE J. Select. Areas Commun.*, vol. 19, no. 10, pp. 2065–2080, October 2001.

34. IEEE Std. 802.11e, *Wireless LAN Medium Access Control (MAC) and Physical Layer (PHY) Specifications, Medium Access Control (MAC) Enhancement for Quality of Service (QoS)*, November 2005.

35. V. Kawadia and P.R. Kumar, "A cautionary perspective on cross layer design," *IEEE Wireless Commun. Mag.*, vol. 12, no. 1, pp. 3–11, February 2005.

36. G. Kramer, M. Gastpar, and P. Gupta, "Cooperative strategies and capacity theorems for relay networks," *IEEE Trans. Inform. Theory*, vol. 51, no. 9, pp. 3037–3063, September 2005.

37. A. Nosratinia, T.E. Hunter, and A. Hedayat, "Cooperative communication in wireless networks," *IEEE Commun. Mag.*, pp. 74–80, October 2004.

8

LOAD BALANCING IN WIRELESS MESH NETWORKS

B.S. Manoj and Ramesh R. Rao

CONTENTS

In a wireless mesh network (WMN), load balancing is critical to utilize the network capacity efficiently. The effects of unbalanced load include gateway loading, center loading, and bottleneck node formation. The gateway nodes connect the WMN to the external Internet. The traffic aggregation at the gateway nodes creates load imbalance at certain gateways which in turn results in congestion, packet loss, and buffer overflow, at the gateway nodes. In addition, the gateway's backhaul connection to the external network may be bandwidth constrained. Hence, load balancing across gateways in a WMN is critical for improving the bandwidth utilization and network scalability.

Another issue arising out of unbalanced load in a WMN even with uniform node density and uniform traffic density, is the center loading. This is primarily due to shortest path routing. The shortest path routing, i.e., routing in which a path is chosen as close to the straight-line path between the source and the destination as possible, seems to be an attractive alternative to any nonlinear path. However, in shortest path routing schemes, WMN nodes closer to the center of the network lie on more such shortest paths than nodes farther away and hence become points of contention. The WMN nodes that lie on too many paths exhaust their resources such as bandwidth, processing power, and memory storage in static WMNs. Moreover, in portable WMN nodes with limited battery energy reserve, load balancing assumes even bigger significance. Nonuniform loading of a WMN prevents effective data forwarding and thus reduces the bandwidth achieved by the nodes. Further, the number of collisions during transmission increases with the increase in contention for a particular mobile node, leading to a degradation in throughput. Also, the bandwidth usage at any part of the network depends upon the number of nodes contending for the bandwidth in a region and the traffic on each of them.

The region of maximum bandwidth utilization becomes the bottleneck for the load on the entire network.

In addition to the gateway nodes and the center of the network, certain nodes that are located at critical positions in the WMN form network bottlenecks. Such nodes are also likely to get heavily loaded other than the normal nodes, gateway nodes, and the nodes at the center of the network. Therefore, load balancing is an essential ingredient in improving the achievable throughput and also for improving the scalability of the WMN.

Gateway-based load balancing solutions discussed in this chapter attempt to relive or balance the load that may be present in a WMN with multiple gateway nodes. Ring-based routing schemes distribute the load, close to the center of the network, over to the periphery of the network. This chapter discusses the issues related to nonuniform load distribution in a WMN and their effects on the performance of WMNs. Followed by the issues related to the nonuniform load, this chapter presents solutions for gateway load balancing, solutions for avoiding center loading, theoretical formulations for estimating the load distribution when different load balanced routing strategies are employed, and presents a few other solutions for load balancing.

8.1 INTRODUCTION

With the use of multihop wireless relaying, increasing number of users, and the very limited electromagnetic spectrum, the WMN is limited by two main resources: bandwidth and network capacity. While bandwidth refers to the data rate achievable through the radio interface, capacity refers to the data transport capacity available for each node in the network. Many attempts have been made to optimally utilize the available bandwidth and load balancing is one of them. The chief motivation behind the exploration of load balancing techniques is their ability to provide a higher network capacity and more balanced resource usage in order to achieve a higher throughput for nodes. For example, load balanced routing addresses the problem of contention for some nodes that lie on many shortest paths, which once again leads to the problem of unfairness in terms of opportunity to relay. In a system where mobility is high this problem may not be as significant as in WMNs, because mobility may cause mobile nodes located in a congested location to move to less congested locations and vice versa, thus on an average each node will relay the same amount of data.

Unlike an autonomous ad hoc wireless network, a WMN is primarily designed to provide connectivity to the wired network infrastructure, i.e., the Internet. Therefore, in order to improve the bandwidth available to the backbone network, it is a usual practice to add multiple gateways to the WMN. The important problems resulting from load imbalance in WMNs are: (a) relay induced capacity degradation, (b) overloading of certain gateway nodes, (c) overloading of WMN nodes closer to the center of the network, and (d) formation and overloading of bottleneck nodes. The major challenges in solving each of the above issues and existing solutions which attempt to solve them are presented.

The use of WMNs provides a fast and low-cost access to the Internet for the residential and mobile users. In WMNs, traffic is routed using multiple hops to a gateway which is connected to the Internet. In a given network, there may be multiple gateways and the regions close to the gateways and the gateways themselves may become bottleneck in the network resulting in larger packet loss due to congestion. Therefore, load balancing across the gateways is very important in providing a better service in WMNs.

The use of multihop relaying in WMNs imposes certain issues specific to these networks. The prominent among them is the relay induced load and interference. The relaying process when applied on single channel systems with carrier sense multiple access protocols prevents simultaneous transmission across neighborhood hops. In a linear string topology WMN, during the transmission on the first hop, at least the subsequent two hops cannot proceed with transmission. This almost limits the capacity achieved by a multihop relayed path in comparison to the physical layer data rate. Therefore, in an ideal case, the multihop throughput achieved is less than one-third [1] of the physical layer data rate. With the presence of interference, this capacity further degrades. Therefore, the throughput degradation associated with the relaying process prevents efficient use of the channel and one method to improve the system capacity is to use load balancing approaches, the focus of this chapter. This chapter mainly focuses on the following: (a) solutions for gateway loading, (b) solutions for center loading, and (c) other load balancing routing solutions. In addition to the above, the theoretical formulations for analyzing the load distribution and path length are provided.

The rest of this chapter is organized as follows: Section 8.2 provides a detailed explanation of solutions for gateway load balancing and Section 8.3 explains the issues and solutions for avoiding the center loading problem in WMNs. Section 8.4 briefs other prominent issues in load balancing in WMNs and Section 8.6 summarizes the chapter.

8.2 GATEWAY LOAD BALANCING IN WIRELESS MESH NETWORKS

Gateway nodes connect a WMN to the wired network, generally to the Internet. Therefore, traffic aggregation happens at gateway nodes which essentially limits the WMN's capacity. In addition to the limited capacity, the gateway node particularly expends much more energy for handling large number of packets it forwards, and this high energy consumption leads to quicker failure of gateway nodes in energy constrained WMNs. Therefore, gateway load balancing assumes significance in order to achieve the following:

- Efficient traffic allocation
- Efficient use of backhaul links
- Maximal use of network capacity
- Minimizing the resource consumption at the gateway nodes
- To counter the effects of traffic imbalance due to node mobility

In a WMN with multiple gateways, load balancing is particularly important when there exists low bandwidth backhaul links. The backhaul networks may be heterogeneous types. For example, a WMN may use a wide variety of backhaul links such as wired network links, wireless local area network (LAN) links, cellular network links such as GPRS, UMTS, and CDMA 2000, data CDMA networks such as $1 \times$ RTT, and $1 \times$ EVDO, WiMAX networks, and satellite links. The type and characteristics of the network affect the bandwidth and latency of the backhaul link of the gateways. For example, a WLAN backhaul may support, depending on the technology used, data rates from 1 Mbps (IEEE 802.11) to 54 Mbps (IEEE 802.11 a/g) whereas cellular networks may provide data rates from 100 Kbps to 2 Mbps.

Since traffic load and load variation across a set of gateways are dynamic factors, quantifying the imbalance of load requires parameters reflecting the load variation. One measure for quantifying the load gradient across gateways is the index of load balance (ILB) [2] which is defined as

$$ILB = \frac{\max\{LI_i\} - \min\{LI_i\}}{\max\{LI_i\}} \qquad (8.1)$$

where LI_i refers to the load index, of a particular gateway i, which is defined as $LI_i = \frac{\sum_{k \in N} \beta k_i \times T_k}{C_i}$. Here the factor βk_i is the fraction of node

8.2.1.2 Load Index-Based Moving Boundary Approach

The load index-based moving boundary approach [2] provides for a more dynamic load balancing solution compared to the shortest path-based moving boundary approach. As per this scheme, each gateway advertises its current load index (LI) through periodic gateway announcement packets. The WMN nodes choose their gateway nodes according to the LI and therefore, the lightly loaded gateway will serve more nodes and the heavily loaded nodes will serve fewer nodes.

The WMN nodes that receive these gateway advertisement packets, accepts the gateway as its own gateway if the node does not already have a dominator gateway or if the LI of the advertisement packet is lesser than the LI of its current dominator gateway by a particular threshold value, switching load threshold. Upon reception of a gateway advertisement packet, every node (N_i) in the network runs a gateway selection algorithm. When node i receives a gateway advertisement packet from gateway G_i, the gateway selection algorithm executes the following steps: (a) if G_i is the current dominator gateway of N_i, it just updates the LI_i, load index of gateway i and forwards the advertisement packet to its neighbors, (b) if N_i is currently served by a different dominator gateway, k, it switches over to gateway i with a gateway switching probability P_s if the load threshold check function, given in Equation 8.2 satisfies, and (c) if the node N_i is not currently served by any gateway node, it accepts the gateway G_i as its dominator gateway, updates the LI value, and forwards the advertisement. Every node uses a load threshold check function to make a decision on whether to choose another gateway as its dominator gateway and the threshold check function is given by:

$$(\text{Service duration of } G_k > T_{\text{limit}}) \text{ AND} \left(\left[\frac{LI_k - \frac{T_{N_i}}{C_{G_k}}}{LI_i + \frac{T_{N_i}}{C_{G_i}}} \right] > \text{LFT} \right) \quad (8.2)$$

Where LI_z, T_{N_i}, C_{G_z}, and LFT represents the LI at gateway Z, traffic generated by node N_i, Capacity of gateway Z, and load fraction threshold, respectively. The first part of Equation 8.2 avoids frequent gateway switching by checking if the old gateway has been serving for a time duration greater than T_{limit}. The second part checks if the switching process leads to creation of bandwidth bottlenecks at the new gateway. The gateway switching probability is also designed to avoid frequent switching of gateways even if all other conditions satisfy. In addition, there is a chance that a lightly loaded gateway may be flooded with traffic because of the sudden decision by every

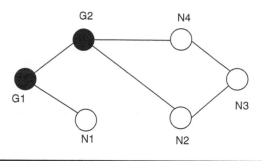

Figure 8.2 Disadvantage of load index-based moving boundary approach.

node in the network. The probability of switching factor is designed to slow down the gateway switching process.

This scheme provides a much more load balanced situation than the shortest path-based moving boundary solution. On the other hand, this approach also suffers from certain disadvantages. Figure 8.2 illustrates a simple topology with two gateways G1 and G2 with four nodes. Here the gateway G2 needs to forward the advertisement packet for gateway G1 also. In such cases, the gateway node G2 will experience additional traffic which is supposed to be directed to G1. Therefore, gateway G2 will almost experience the sum of its own traffic and the traffic for G1 which is forwarded by G2.

Now the switching decision taken by any of the nodes N2, N3, and N4 to use G1 as its gateway will, in fact, affect gateway G2 also. Another problem with this approach is rather practical where the gateway G2 will need additional routing information to route the packets to gateway G1. This is because, normally nodes have a default gateway information which directs a gateway to forward any packet without specific routing instruction. In this example, the default gateway information in gateway G2 will direct all externally addressed packets outside the network and therefore it is likely to forward all the packets to the backhaul network. In order to route a set of packets to gateway G1, the gateway G2 requires additional routing entries which, in practice, is difficult and complex.

8.2.2 Partitioned Host-Based Load Balancing

The PHLB approach classifies the set of nodes into a number of groups, each designated to be served by a particular gateway node. Although this approach is similar to the MBLB methods, the major difference stems from the fact that there are no clear boundaries in this method. Therefore, members of each group may be randomly present within the network. The two main approaches in this category are: (a) centralized host partitioning (CHP) approach and (b) distributed host partitioning (DHP) approach.

8.2.2.1 Centralized Host Partitioning-Based Load Balancing

In order to implement this scheme, a central server which has the complete load information about all the gateway nodes and the traffic requirements of all the nodes in the network, is required. The centralized server assumes the responsibility of assigning a particular node to a gateway in such a way that all the gateway nodes are equally loaded. This node assignment is proven to be NP complete [2] and is computationally very expensive. Therefore, a heuristic algorithm based on greedy approach is used to achieve the host partitioning. According to this heuristic host partitioning algorithm, the nodes are ordered in descending order according to the traffic requirements and a sequential assignment of each node N_i to a particular gateway G_i is carried out based on the minimum $LG_{i_{new}}$ where $LG_{i_{new}}$ is estimated as $LG_{i_{new}} = LG_i + T_{N_i}/CG_i$ where LG_i is the current LI of a gateway i, $LG_{i_{new}}$ is the estimated LI of gateway node when the node N_i is assigned to the gateway G_i, T_{N_i} is the traffic requirement of the node, and CG_i is the capacity of the gateway G_i.

The advantage of this scheme is its capability to provide a near optimal solution as the allocation is primarily carried out by a centralized server. However, this scheme suffers from several disadvantages such as high information exchange overhead in collecting the traffic and load information from nodes and gateways and the requirement for high processing capability for the allocation process.

8.2.2.2 Distributed Host Partitioning-Based Load Balancing

The DHP approach solves some of the disadvantages of the centralized approach. In this case, a logical network is formed by the gateway nodes in order to exchange load and traffic related information. In order to form the logical network, the gateway may use the wireless links between WMN nodes. The WMN nodes periodically update their dominating (serving) gateway nodes about its traffic demands. Every gateway, periodically exchanges its load and capacity information to its neighbor gateways through the logical network. A given gateway G_i estimates load imbalance and if it experiences high load, it decides to handover (the process of delegating a node under a loaded gateway to another lightly loaded gateway) a node, currently served by G_i and the removal of which would lead to the maximum reduction in load, to other gateway which is lightly loaded. In this process, in order to avoid rapid node handover and associated frequent load fluctuation, a given node is handed over only if it has been served by the gateway for sufficient amount of time. In addition, a gateway can handover only one node during the load estimation interval, a periodic time interval over which the gateways run the load check. While handing over a node to a lightly loaded target gateway G_t, the heavily loaded current gateway G_i chooses a set of potential target gateways if they satisfy the condition $LG_i/LG_t >$ LT where LG_i, LG_t, and LT refer to the load indices of the current gateway and a potential target gateway, and the load threshold, respectively. Here the LT is used to avoid frequent fluctuations arising from minor differences in the load across the gateways. Among the potential target gateway nodes, the current gateway selects the one target gateway, to which the handover of the chosen node will lead to the largest amount of reduction of the ILB. Once the node to be handed over and target gateway are identified, the current gateway sends a delegate (G_t) message for instructing the node to start using the new gateway G_t.

Although this approach is better than the centralized approach, it also has a few disadvantages. The main disadvantage is that this scheme depends on the accuracy of information exchanged by the gateway nodes over the logical network. In addition, the logical network must be connected and formation of a connected topology may result in significant control overhead.

8.2.3 Probabilistic Striping-Based Load Balancing

In the above-mentioned MBLB and PHLB approaches, every WMN node utilizes only one gateway for its external communication. Therefore, achieving perfect load balancing is very difficult. In PSLB [2] approach, every node may utilize multiple gateway nodes simultaneously and therefore a perfect gateway load balancing is, theoretically, possible. The two primary approaches in this category are: (a) all node probabilistic striping and (b) boundary node probabilistic striping approach.

8.2.3.1 All Node Probabilistic Striping Approach

According to this scheme, every WMN node in the network identifies all the gateway nodes in the network and attempts to send a fraction of its traffic through every gateway. Here striping refers to splitting traffic across to multiple gateways. Every gateway node advertises its backhaul capacity throughout the network in a way similar to the shortest path-based moving boundary approach discussed in Section 8.2.1. The capacity information from all the gateway nodes are collected by every WMN node. Based on the capacity of each gateway node, a given WMN node estimates what fraction of his traffic can be sent through each gateway. For example, given that G is the set of gateway nodes in the network, CG_i is the capacity of a given gateway node i, then a fraction of its traffic equivalent to $CG_i/\sum_{j \in G} CG_j$ is sent over the gateway CG_i. Theoretically this scheme can achieve perfect load balancing, but in practice, this scheme also suffers from high control overhead. The traffic striping leads to additional overhead and inefficiency in the end-to-end connection. For example, consider an end-to-end TCP connection which undergoes the striping process across several gateways. Each gateway's backhaul may experience a different delay performance and therefore, the end-to-end connection may experience high packet reordering resulting in large number of retransmissions.

8.2.3.2 *Boundary Node Probabilistic Striping Approach*

In comparison with the above-mentioned scheme, only a selected subset of the WMN nodes choose to enforce the probabilistic striping. This scheme is also called partially probabilistic routing [2] and it is a combination of the shortest path-based moving boundary load balancing approach and probabilistic striping approach. In this case, only the boundary nodes, nodes which are at the boundary similar to that of the gateway coverage regions in MBLB scheme, employ probabilistic striping. In addition, each boundary WMN node can utilize only a limited subset of available gateway nodes. The number of available gateways are decided by the number of gateway advertisements that reach every node. Compared to the previous approaches such as MBLB, here a filtering is employed on the advertisements from lightly loaded nodes. Every gateway node broadcasts its capacity information in gateway advertisements and each WMN node has a primary gateway which serves the node. The primary gateway can either be manually assigned or dynamically obtained by using any of the above-mentioned schemes. When a WMN node receives a gateway advertisement from its primary gateway, the node broadcasts it further. However, when a WMN node receives a gateway advertisement from a gateway which is not the node's primary gateway, then it forwards the advertisements only if $LG_i/LG_p > \text{LT}$ where LG_i is the ILB of the new gateway and LG_p is the ILB of the primary gateway. The load threshold, LT, helps in broadcasting the advertisements from lightly loaded gateway nodes much farther than from highly loaded gateway nodes. Every node strips its traffic to a set of gateway nodes G' from which it receives gateway advertisement packets. The cardinality of the set G' ranges from 1 to G where G is the set of all gateways in the network. A WMN node will strip the traffic among the available gateways in a probabilistic approach. That is, whenever a packet needs to be routed, the WMN node routes it to the primary gateway with a traffic distribution probability P_d and with probability $1 - P_d$, it employs the traffic striping approach across gateways in set G'. While striping the traffic across all gateways in set G', a node employs a probability value proportional to the capacity of the available gateway capacities. For example, assuming CG_i represents the capacity of given gateway i, then packets will be routed to the gateway k with the probability $\frac{CG_k}{\sum_{j \in G'} CG_j}$ where G' is the subset of gateways from which gateway advertisements reach the given node. Therefore, while the probability of using the primary gateway for a packet under consideration is P_d, the probability of using a gateway $k \in G'$ is $(1 - P_d) \times \frac{CG_k}{\sum_{j \in G'} CG_j}$. While this solution, though good for better load balancing, compared to the above-mentioned approaches, it may not be useful for incoming connections. This is because, incoming connections may not get striped under this scheme, as the gateway themselves are not aware of the load balance situation.

8.3 CENTER LOADING IN WIRELESS MESH NETWORKS

Similar to the gateway loading, another important issue in WMN is the center loading which refers to the issue of nodes, located near the geographical center of the network, getting overloaded in comparison to other nodes in the network. Such overloading is present in every kind of multihop wireless network and the main reasons behind center loading are: (i) nodes near the center of the network lie on many shortest path routes than other nodes in the network, (ii) the use of multihop relaying, and (iii) the relatively static nature of WMNs. Though multihop relaying is used in several kinds of networks such as ad hoc wireless networks, center loading is not a serious issue there because of the highly dynamic topology of ad hoc wireless networks. In WMNs, overloading of the WMN nodes closer to the center of the network still remains a challenging issue as the nodes lie on many shortest paths, because it is static or with low mobility. The solutions which attempt to resolve this problem by using schemes that dynamically determine the load on WMN nodes in the system and thus route new data through less loaded nodes. Such schemes can be classified broadly into the following types:

1. Networks in which the traffic and load information at each WMN node is communicated to all its neighbor nodes. In such cases, the neighbor nodes may decide an alternate next hop which may be less loaded. This approach is more suitable for a hop-to-hop routing system.
2. Networks in which the traffic and load information at each WMN node is communicated to all other nodes in the WMN so that a source routing with load balanced can be utilized.
3. Networks in which the traffic and load information at each WMN node is communicated to a central node which helps all the WMN nodes in finding less loaded routes. This approach is better suited for hybrid WMN where a central node is available. Example for such architectures includes throughput enhanced wireless in local loop (TWiLL) [3] and multihop cellular networks (MCNs) [4,5].

All the schemes mentioned above have their own advantages and disadvantages. The first approach has an advantage; the load information need not be propagated over the entire network and thus they avoid flooding caused by the latter. At the same time, the first approach may be disadvantageous when compared to the second approach when the number of hops taken by a packet to reach the destination may not be bounded. Moreover, in some cases the packet may even end up in a loop if not handled explicitly in the scheme. Further, there is a possibility of oscillation of load due to the dynamic load-based routing approach. This is because when nodes are informed that some nodes and/or gateways are lightly loaded while others are heavily loaded all the nodes may end up sending data to the lightly loaded nodes/gateways resulting in rapid fluctuations in the load between the lightly loaded and the heavily loaded nodes and vice versa. The third category may not always be feasible in a WMN network as it demands a centralized node for aiding the load balancing activity. In addition, communicating the load and traffic information by every node in the network may lead to scalability and performance issues with the centralized node in a large WMN.

The center loading problem is also called a ring formation problem in ad hoc wireless networks where the power constrained nodes fail much earlier at the center of the network thus leading to a ring-shaped network [7]. This problem causes, in the center region, increased MAC contention delay, packet loss, and collision leading to performance degradation. The primary reason behind the center loading is the use of routing protocols that find shortest path routes or that use routing metrics which in turn obtain a near shortest path route. In this section, a number of load distribution schemes which provide ring-based routing solutions are listed by which the nodes follow a set of simple rules so that the traffic load gets distributed over the network.

8.3.1 Shortest Path Routing and Center Loading

The primary reason for the center loading, as mentioned in Section 8.3, is the use of shortest path routing. In this section, the load distribution for circular network of radius R units which has uniform node distribution is presented. A ring is defined by its smaller and bigger radii r_1 and r_2, respectively. For example, the range of a ring is defined as $[r_1, r_2]$ where r_1 and r_2 correspond to the minimum and the maximum radius from the center to any circles forming this ring. As shown in Figure 8.3, a given ring i divides the network into three regions: X, Y, and Z. Region X corresponding to the nodes at a distance less than r_i from the center, region Y corresponding to ring i itself and region Z corresponding to the nodes at a distance greater than r_{i+1} from the center. In order to simplify the analysis, a continuous model is considered in the following sections. Also, in the analysis, it is assumed that all nodes generate data at the same rate to random destinations in the network.

An impact of the traffic density increase near the center of the network in a WMN is the saturation of buffers and bandwidth

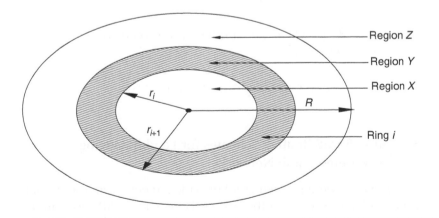

Figure 8.3   Representation of a ring.

resources that happens much earlier in those nodes compared to other nodes in the network. Increase in the hop count bears a direct impact on the transmission delay associated with the data. Additionally, another angle to the increase in the traffic load is the increase in the path length. For example, if the average path length (hop count) in the network is high, that would lead to higher relay load and hence the overall load on the network increases. We considered the following two node distributions: (i) radial nonuniform distribution and (ii) uniform node distribution. First, the radial nonuniform node distribution is considered where the transformation from the polar coordinates results in $x = r \cos \phi$ and $y = r \sin \phi$ where $r \in [0, R]$ and $\phi \in [0, 2\pi)$. The average path length analysis is given below.

Assuming an ideal situation where the average path length is the straight-line distance between the source and the destination, the path length analysis is presented here. For a simpler analysis, the polar coordinates are used to represent the source and the destination as (r_1, ϕ_1) and (r_2, ϕ_2), respectively. More than the absolute values of ϕ_1 and ϕ_2, the relation is more important in this case and therefore ϕ is defined as

$$\phi = |\phi_1 - \phi_2| \quad \text{if } 0 \leq |\phi_1 - \phi_2| \leq \pi$$
$$\phi = 2\pi - |\phi_1 - \phi_2| \quad \text{if } |\phi_1 - \phi_2| > \pi \qquad (8.3)$$

The average path length in shortest path-based routing with radially nonuniform distribution is given by

$$\frac{\int_0^R \int_0^R \int_0^\pi \theta(r_1, r_2, \phi) \, dr_1 dr_2 d\phi}{\int_0^R \int_0^R \int_0^\pi dr_1 \, dr_2 \, d\phi} \approx 0.726R$$

where $\theta(r_1, r_2, \phi)$ provides the path length between source and destination and is given by

$$\theta(r_1, r_2, \phi) = \sqrt{r_1^2 + r_2^2 - 2r_1 r_2 \cos \phi} \qquad (8.4)$$

8.3.2 Ring-Based Routing Schemes for Load Balancing in Wireless Mesh Networks

The ring-based routing schemes [6] provide a mechanism for alleviating the load at the center of the WMNs. Here, each WMN node in the network belongs to a ring. A ring is an imaginary division of the network into concentric rings about the center of the network. An arbitrary ring, ith ring, which as a minimum radius of r_i and maximum radius of r_{i+1} is denoted by $ring_i$ (r_i, r_{i+1}). The thickness of such a ring is given by $r_{i+1} - r_i$. In WMNs, it is measured in terms of hops from the center of the network where the center of the network is defined as the node for which $max_{\forall x}$ $(Hops(C, x)) \leq min_{\forall y}$ $(max_{\forall z}$ $(Hops(y, z)))$ where $Hops(x, y)$ refers to the number of hops along the shortest path from node x to node y where x, y, and z are nodes in the WMN.

Figure 8.4 shows an example of center of a network where the vertexes represent the nodes and the links indicate connections. The numbers indicate the value of max $(Hops(node, x))$ for each node and based on the above definitions nodes C and D qualify to be center of the network.

In order to apply the ring-based routing approaches, it is essential to determine the center of the network. Using the network topology information, each WMN node builds the connectivity matrix for the network and using Warshall's algorithm the center can be determined in $O(n^3)$ time where n is the number of nodes in the system. There exists the possibility of a tie which can be resolved in favor of lesser node id or IP address. Once the center is determined, the rings to which the nodes belong can be determined. The distance from the center of the node referred to as distance, in this case, refers to the hop count from the center (Figure 8.5). The thickness of the ring is again

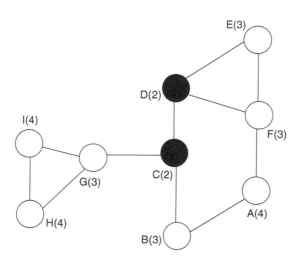

Figure 8.4 Determination of the center of network in an example topology.

Edge between Scheme	Nodes in other rings (both the nodes in the same ring)	Nodes in source ring	Nodes in destination ring	Nodes in different rings (edge crosses the ring boundary)
PORS	MAX	MAX	MIN	MAX
		MIN	MAX	

Figure 8.7 Edge weights to be used in Dijkstra's algorithm for finding a route according to PORS.

destination ring larger than the source ring and the lower for destination ring smaller than the source ring.

The downside of PORS is the increase in path length. The purpose of these ring-based routing schemes is to move the load from closer to the center towards the periphery of the network. This is better accomplished by the PORS though at the cost of a greater average path length.

8.3.2.1.1 Theoretical Analysis of Load Distribution in PORS. In the PORS the nodes follow the rule that no traffic generated by nodes must be passed to higher or lower rings than the source and the destination rings unless it is inevitable. Further, the packet must be circulated around in the outer of two rings (ring containing the source node and ring containing the destination node).

We estimate some of the parameters which are essential for the analysis of load distribution in this section. The average distance, through region Y, between a random source and the destination pair when both nodes are located in region Y can be approximated as

$$D_{YY} = \pi \frac{r_{i+1} + r_i}{4} + \frac{r_{i+1} - r_i}{2} \qquad (8.5)$$

Based on the above rules, in a network with radius R, the traffic that contributes to load in ring i can be divided into the following four categories:

1. Traffic load experienced in region Y (see Figure 8.3 for a definition of regions) by the communication between nodes in region Y is proportional to probability that the source and

the destination are in region $Y \times$ average distance between source and the destination due to the use of PORS.

$$= \frac{(\pi r_{i+1}^2 - \pi r_i^2)^2}{(\pi R^2)^2} \times (D_{YY}) \qquad (8.6)$$

2. Traffic load experienced in region Y due to the communication between nodes in regions X and Y is proportional to probability that the source lies in region X or Y and the destination in region Y or X, respectively $\times$ average distance for transmission in region Y. This can be approximated as the following.

$$= 2 \times \frac{(\pi r_{i+1}^2 - \pi r_i^2) \times \pi r_i^2}{(\pi R^2)^2} \times D_{YY} \qquad (8.7)$$

3. Traffic load experienced in region Y due to the communication between nodes in regions Y and Z is proportional to probability that the source lies in region Z or Y and the destination in regions Y or Z, respectively $\times$ average distance for transmission in region Y

$$= 2 \times \frac{(\pi r_{i+1}^2 - \pi r_i^2) \times (\pi R^2 - \pi r_i^2)}{(\pi R^2)^2} \times \left(\frac{r_{i+1} - r_i}{2} \right) \qquad (8.8)$$

4. Traffic load experienced in region Y due to the communication between nodes in regions X and Z is proportional to probability that the source lies in regions X or Z and the destination in regions Z or X, respectively $\times$ average distance for transmission in region Y alone

$$= 2 \times \frac{\pi r_i^2 \times (\pi R^2 - \pi r_{i+1}^2)}{(\pi R^2)^2} \times \delta \qquad (8.9)$$

where δ is the ring thickness and can be obtained as

$$\delta = (r_{i+1} - r_i) \qquad (8.10)$$

Thus, the total load on nodes in ring i is given by the sum of Equation 8.6 through Equation 8.9. The average per node traffic load can be approximated by dividing the above total load by the area occupied by ring i.

The term r is defined as the mean distance of ring i from the center and δ as the thickness of ring i. Thus,

$$r = \frac{r_i + r_{i+1}}{2} \tag{8.11}$$

$$r_{i+1} = r + \delta/2 \tag{8.12}$$

$$r_i = r - \delta/2 \tag{8.13}$$

Hence the total load on individual nodes is proportional to:

$$\frac{1}{\pi r_{i+1}^2 - \pi r_i^2} \times \left(\begin{array}{c} \text{sum of Equation} \\ \text{8.6 through Equation 8.9} \end{array} \right) \tag{8.14}$$

$$\Rightarrow \left(\frac{\left(r + \frac{\delta}{2}\right)^2 + \left(r - \frac{\delta}{2}\right)^2}{\pi^2 R^4} \right) \times \left(\pi r + \delta - \frac{R^2}{r} - \frac{\left(r + \frac{\delta}{2}\right)^2}{r} \right) \tag{8.15}$$

8.3.2.1.2 Average Path Length Analysis for PORS. The average path length for PORS in a WMN with radially nonuniform node distribution can be approximated as

$$\frac{\int_0^R \int_0^R \int_0^\pi \theta(r_1, r_2, \phi) \, dr_1 \, dr_2 \, d\phi}{\int_0^R \int_0^R \int_0^\pi \, dr_1 \, dr_2 \, d\phi} = \frac{(\pi + 1)R}{3} = 1.38R \tag{8.16}$$

where $\theta(r_1, r_2, \phi)$ represents the path length between any two nodes in PORS and can be approximated as

$$\theta(r_1, r_2, \phi) = \max(r_1, r_2)\phi + |r_1 - r_2| \tag{8.17}$$

8.3.2.2 Preferred Inner Ring Routing Scheme

According to PIRS, the traffic generated by ring i for a destination node in ring j must not be forwarded through nodes which are located in rings beyond area enclosed by rings i and j. Further, the packet must be preferably routed through in the inner of the sender's or receiver's rings. In other words, for the nodes belonging to the same ring, the packet must be transmitted in the same ring. For nodes belonging to different rings, the packet must go across the rings in radial direction and the radial transmissions must be restricted to the inner of the two rings (of the source or the destination). For example, path III in Figure 8.6, the source lies in ring 5 while the destination resides in ring 2. Hence no transmission must go through ring 0 (the center node) or ring 1 unless inevitable. Further, in the path traced, transmission reaches ring 2 from ring 5 and after reaching ring 2, the radial direction of the path begins (see path III in Figure 8.6). Similar to PORS, tracing a route in this scheme requires changing the weights of the edges in the adjacency matrix of the graph (Figure 8.8) before running Dijkstra's shortest path algorithm. Similar to PORS, the overall complexity of finding the route thus remains the same as in the shortest path routing. Figure 8.8 shows the edge weights to be used with Dijkstra's shortest path algorithm in order to obtain a PIRS path. Here also, MAX represents highest edge weight possible for a connected link. Similarly, MIN refers to the lowest possible edge weight for a connected link. In Figure 8.8, some entries are divided into two. The upper represents the case of destination ring larger than the source ring and the lower for destination ring smaller than the source ring.

8.3.2.1.3 Theoretical Analysis of Load Distribution in PIRS.   Like the PORS, in the PIRS, the nodes follow the rule that no traffic generated by nodes must be passed to higher or lower rings than the source and the destination rings unless it is inevitable. However, the packet must be forwarded around in the inner of two rings (the ring containing the source node and the ring containing the destination node). The same four categories as that of PORS, are used for the analysis:

1. The traffic load experienced in region Y by the communication between nodes in region Y is proportional to

$$\frac{\left(\pi r_{i+1}^2 - \pi r_i^2\right)^2}{\left(\pi R^2\right)^2} \times D_{YY} \qquad (8.18)$$

where D_{YY} can be obtained from Equation 8.5.

Edge between / Scheme	Nodes in other rings (both the nodes in the same ring)	Nodes in source ring	Nodes in destination ring	Nodes in different rings (edge crosses the ring boundary)
PIRS	MAX	MIN	MAX	MAX
		MAX	MIN	

Figure 8.8   Edge weights to be used in Dijkstra's algorithm for finding a route as per PIRS.

and PSRS is the same path as that of PIRS (path III in Figure 8.6) and PORS (path II in Figure 8.6), respectively.

In different kinds of WMN applications, each of these schemes performs differently. These schemes essentially make a trade-off between the increase in hop count and the ability to move the load away from the center and/or balance the load in a more uniform way. In this case, the PORS is found to be performing better as it does the maximum to move the traffic away from the center to the periphery. But the hop-count increase due to PORS may also cause increased delay, which may not be tolerable to some applications. The PSRS and the PDRS which appear very much similar may give varied performance in the case of the presence of gateways in WMNs where the destination may not be decided at random. Here our assumption that the heavy load region is the center of the network may not hold. This is because the gateway nodes may experience heavy load and these gateway nodes may be present at multiple places in the network.

8.3.2.4.1 Theoretical Analysis of Load Distribution in PDRS/PSRS. The rule followed by the nodes in this scheme is that packets are sent around in the destination (source) ring only (unless inevitable). Though the scheme emphasizes the destination ring, the analysis remains the same for the source ring case as well. Considering the four categories as that of the previous analysis:

1. Traffic load experienced in region Y due to the communication between nodes in region Y is proportional to

$$\frac{\left(\pi r_{i+1}^2 - \pi r_i^2\right)^2}{(\pi R^2)^2} \times D_{YY} \tag{8.25}$$

2. Traffic load experienced in region Y due to the communication between nodes in regions X and Y is proportional to

$$\frac{\left(\pi r_{i+1}^2 - \pi r_i^2\right) \times \pi r_i^2}{(\pi R^2)^2} \times \left(\underbrace{\frac{\delta}{2}}_{Y \text{ to } X} + \underbrace{\frac{\delta}{2} + \frac{\delta}{2} + \pi \frac{r_{i+1} + r_i}{4}}_{X \text{ to } Y}\right) \tag{8.26}$$

3. Traffic load experienced in region Y due to the communication between nodes in regions X and Z is proportional to

$$\frac{\left(\pi r_{i+1}^2 - \pi r_i^2\right) \times \left(\pi R^2 - \pi r_i^2\right)}{(\pi R^2)^2} \times \left(\pi \frac{r_{i+1} + r_i}{4} + 2 \times \frac{\delta}{2}\right) \tag{8.27}$$

4. Traffic load experienced in region Y due to the communication between nodes in regions X and Z is proportional to

$$2 \times \frac{\pi r_i^2 \times (\pi R^2 - \pi r_{i+1}^2)}{(\pi R^2)^2} \times (\delta) \qquad (8.28)$$

The load per node at a distance r from the center of the network can be obtained by summing up Equation 8.25 through Equation 8.28 and is proportional to

$$\frac{1}{\pi R^2}\left(\frac{\pi r}{2}+\delta\right) + \frac{\left(r - \frac{\alpha}{2}\right)^2 \left(\pi R^2 - \pi \left(r + \frac{\delta}{2}\right)^2\right)}{\pi^2 R^4 r} \qquad (8.29)$$

The traffic load in the PDRS, as obtained in Equation 8.29, can provide load balancing for high values of δ.

8.3.2.4.2 Average Path Length Analysis for PDRS and PSRS. The average path length for PIRS in a WMN with radially nonuniform node distribution can be approximated as

$$\frac{\int_0^R \int_0^R \int_0^\pi \theta(r_1, r_2, \phi)\, dr_1\, dr_2\, d\phi}{\int_0^R \int_0^R \int_0^\pi dr_1\, dr_2\, d\phi} = R\left(\frac{1}{6}+\frac{\pi}{4}\right) = 0.952R \qquad (8.30)$$

where $\theta(r_1, r_2, \phi)$ is the path length between two nodes in the network which can be approximated as $\theta(r_1, r_2, \phi) = r_2\phi + |r_1 - r_2|$.

8.3.3 Average Path Length Analysis with Uniform Node Density

The uniform node distribution is considered where the transformation from the polar coordinates results in $x = \sqrt{r}\cos\phi$ and $y = \sqrt{r}\sin\phi$ where $r \in [0, R]$ and $\phi \in [0, 2\pi)$. The average path length for the shortest path routing is given by

$$\frac{\int_0^R \int_0^R \int_0^\pi \sqrt{(r_1 + r_2 - 2\sqrt{r_1 r_2}\cos\phi)}\, dr_1\, dr_2\, d\phi}{\int_0^R \int_0^R \int_0^\pi dr_1\, dr_2\, d\phi} \approx 0.905\sqrt{R}$$

Similarly extending the analysis for PORS, PIRS, and PDRS, the average path lengths can be obtained as $1.52\sqrt{R}$, $1.1044\sqrt{R}$, and $1.313\sqrt{R}$, respectively.

8.4 OTHER LOAD BALANCING SOLUTIONS IN WIRELESS MESH NETWORKS

In addition to the gateway loading and center loading, there are other issues that arise out of load imbalance in a much more local sense in a WMN. For example, certain critical nodes can be on a bottleneck position, similar to a bridge node, the removal of which will create a network partition. This section discusses rather general load balancing solutions which avoid traffic through loaded WMN nodes. This is necessary because, in certain cases, locally high load imbalance can result in performance degradation from congestion, packet loss, and buffer overflow. Therefore, two general load balancing solutions which use on-demand load balancing routing discussed in this section, i.e., load balanced ad hoc routing (LBAR) and load balanced AODV.

8.4.1 Load Balanced Ad Hoc Routing

The LBAR [8] protocol is a routing protocol with a routing metric, degree of nodal activity, that inherently represents the path load. Therefore, LBAR attempts to find a path which avoids congested locations within the network. The path finding process consists of two phases: a forward phase during which a route-setup packet is propagated by the source node to search for a new path towards a destination node and a backward phase in which the route-ACK packet is propagated back from the destination to the source node. In the forward phase, a source node broadcasts a route-setup packet containing the routing cost (RC) (degree of nodal activity) to all its neighbors and the traversed path field. The routing cost contained in the route-setup packet at any intermediate node reflects the degree of nodal activity, a representation of traffic load, from the source node to the current intermediate node.

The RC is estimated along the path from source node towards the destination and by the following approach. The routing cost (RC_p) for a candidate path p is given by $RC_p = \sum_{j \in p} (P_j + PI_j)$ where P_j reflects the traffic load experienced by the node j, excluding source and destination, and is the number of active path through a given node j and PI_j represents the number of interfering paths at the node j. The increasing value for P_j reflects increasing traffic load at node j. The number of interfering paths for a given node j is estimated by $PI_j = \sum_{\forall k} (P_k^j)$ where k refers to the neighbor nodes of node j. The path interference reflects the number of paths through the neighbor nodes which interfere with a given node j's traffic. As the route-setup packet propagates towards the destination node, the RC field and the traversed path field are updated by each forwarding intermediate node. The traversed path (also known as route record) information contains the nodes traversed by the packet. The traversed path information is used for loop-free routing by ensuring that the nodes that are already present in the traversed path field avoid forwarding the packets further. An intermediate node upon receiving a route-setup packet, forwards it further with updated RC value and the traversed path field. The destination node collects all the route-setup messages for a certain period called route-select-waiting-period and chooses the path with the minimum routing cost. The chosen path, with minimum RC, represents the path with minimum traffic and interference from neighbors' traffic and hence the purpose of load balancing is met implicitly by the routing scheme. The backward phase of path finding begins immediately after the destination node chooses a path and this phase is marked by the initiation of a route-ACK packet which propagates towards the source node. This route-ACK packet follows the chosen path in the reverse direction until it reaches the source node of the path. Once the source node receives the route-ACK packet, it starts transmitting packets through the newly obtained path. The LBAR protocol performs better in terms of packets delivery ratio, average end-to-end delay, and normalized routing load compared to normal routing protocols such as AODV [9] and DSR [10]. The impact of load balancing is evident in the increase in performance of LBAR with increased network load and number of nodes.

8.4.2 Load Balancing Ad Hoc On-Demand Distance Vector

The load balancing ad hoc on-demand distance vector (LB-AODV) routing [11] protocol is on-demand, load balancing routing protocol for WMNs and ad hoc wireless networks with gateway nodes. In this solution, the load balancing is controlled by the gateway nodes by grouping the source nodes into a number of nonoverlapping groups. Here the gateway determines what is the optimal number of groups which can provide the maximum value of load balanced index (LBI) which is defined as

$$\mathrm{LBI} = \frac{\left[\sum_{k=1}^{N_C} N_k\right]^2}{N_G \times \sum_{k=1}^{N_G} N_k^2} \qquad (8.31)$$

where N_k represents number of source nodes that belongs to group k and the total number of groups is represented by N_G. The maximization

8.5 OPEN ISSUES

Although there exist several solutions for load balancing in WMNs as discussed in the above sections, there exist several open issues. First is, e.g., the issue of traffic stability arising out of the rapid fluctuations due to the use of load balancing schemes. The measure of traffic stability and their impact of routing is not well studied yet. Second issue that is not studied well is the bandwidth striping for load balancing and its implications on the performance of higher layer protocols such as TCP. Since the transport layer protocols such as TCP suffer heavily from the misordered packets or the packets arriving out-of-order, it becomes very important to study the impact on higher layer protocol performance when load balancing schemes that use bandwidth striping are employed. Another issue that needs research focus is the quantification of the relation between routing strategy and the load imbalance. For example, though the ring-based routing schemes are helpful, the packet level traffic load as a function of routing schemes is not yet accurately estimated. In addition to the above-mentioned issues, load balancing can be achieved even by using call admission control. Even empowering the routing protocols with a call admission control may help in load balancing. Finally, the most important open issue is the unknown relation between the achievable capacity and load balancing. The theoretical capacity limit provided by a WMN cannot be achieved without having an ideally load balanced system. It still remains an open issue to identify the relation between achievable capacity and load balancing. In conclusion, there exists a large number of open problems that needs soulutions for achieving a balanced load in a WMN.

8.6 SUMMARY

This chapter presents the issues arising out of load imbalance in WMNs and the existing solutions for balancing the load. The main issues in this case are: (a) gateway loading, (b) center loading, and (c) issues arising out of local load imbalance. Load imbalance in a WMN occurs even in the case of uniform data generation rate and uniform node distribution.

The first major issue arising out of load imbalance is the gateway loading, i.e., gateways get heavily loaded due to traffic aggregation from nonuniform number of nodes in the network. Without proper load balancing, the backhaul links may be congested or underused resulting in low scalability and network capacity. This chapter presented three major solutions for solving the gateway loading problem. The second major issue arising out of load imbalance in WMNs is the center loading and this is primarily due to the property that the nodes in the center region lie on more shortest paths than in other regions. This chapter also presents a number of solutions based on ring-based routing in order to alleviate this problem. The ring-based routing schemes are designed to distribute the load over the entire network from hot center areas in a WMN networks. Finally, the chapter concludes the discussion on load balancing with the LBAR and LB-AODV, two on-demand routing protocols that use load balancing principles in routing for multihop wireless networks such as WMNs and ad hoc wireless networks.

ACKNOWLEDGMENTS

This chapter is prepared with the support of the NSF project *Responding to Crises and the Unexpected* (NSF award numbers 0331707 and 0331690). We sincerely thank Professor C. Siva Ram Murthy and G.R. Bhaya for their help in preparing this chapter.

REFERENCES

1. J. Li, C. Blake, D.S.J. De Couto, H.I. Lee, and R. Morris, "Capacity of Ad Hoc Wireless Networks," *Proceedings of ACM Mobicom 2001*, pp. 61–69, July 2001.
2. C.F. Huang, H.W. Lee, and Y.C. Tseng, "A Two-Tier Heterogeneous Mobile Ad Hoc Network Architecture and Its Load-Balance Routing Problem," *Mobile Networks and Applications*, Vol. 9, pp. 379–391, 2004.
3. D. Christo Frank, B.S. Manoj, and C. Siva Ram Murthy, "Throughput Enhanced Wireless in Local Loop (TWiLL)—The Architecture, Protocols, and Pricing Schemes," *Proceedings of IEEE LCN 2002*, November 2002.
4. Y.D. Lin and Y.C. Hsu, "Multi-Hop Cellular: A New Architecture for Wireless Communications," *Proceedings of IEEE INFOCOM 2000*, March 2000.
5. R. Ananthapadmanabha, B.S. Manoj, and C. Siva Ram Murthy, "Multi-Hop Cellular Networks: The Architecture and Routing Protocols," *Proceedings of IEEE PIMRC 2001*, October 2001.
6. G. Bhaya, B.S. Manoj, and C. Siva Ram Murthy, "Ring-Based Routing Schemes for Load Distribution and Throughput Improvement in Multi Hop Cellular, Ad Hoc, and Mesh Networks," *Proceedings of HiPC 2003*, LNCS 2913, pp. 152–161, December 2003.
7. C. Siva Ram Murthy and B.S. Manoj, *Ad hoc Wireless Networks: Architectures and Protocols*, Prentice Hall PTR, New Jersey, May 2004.
8. H. Hassanein and A. Zhou, "Routing with Load Balancing in Wireless Ad Hoc Networks," *Proceedings of ACM MSWiM 2001*, pp. 89–96, July 2001.

9. C.E. Perkins and E.M. Royer, "Ad Hoc On-Demand Distance Vector Routing," *Proceedings of IEEE Workshop on Mobile Computing Systems and Applications 1999*, pp. 90–100, February 1999.

10. D.B. Johnson and D.A. Maltz, "Dynamic Source Routing in Ad Hoc Wireless Networks," *Mobile Computing*, Kluwer Academic Publishers, Vol. 353, pp. 153–181, 1996.

11. J.H. Song, V.W.S. Wong, and V.C.M Leung, "Efficient On-Demand Routing for Mobile Ad Hoc Wireless Access Networks," *IEEE Journal on Selected Areas in Communications*, Vol. 22, No. 7, September 2004.

12. L. Kleinrock and J. Silvester, "Optimum Transmission Radii for Packet Radio Networks or Why Six is a Magic Number," *Proceedings of IEEE National Telecommunications Conference 1978*, pp. 431–435, 1978.

13. E.M. Royer, P.M. Mellier-Smith, and L.E. Moser, "An Analysis of the Optimum Node Density for Ad Hoc Mobile Networks," *Proceedings of IEEE ICC 2001*, pp. 857–861, June 2001.

14. E. Fratkin, V. Vijayaraghavan, Y. Liu, D. Gutierrez, T.M. Li, and M. Baker, "Participation Incentives for Ad Hoc Networks," *Technical Report*, Department of Computer Science, Stanford University, 2001.

15. H. Wu, C. Qiao, S. De, and O. Tonguz, "Integrated Cellular and Ad Hoc Relaying Systems: iCAR," *IEEE Journal on Selected Areas in Communications*, Vol. 19, No. 10, pp. 2105–2115, October 2001.

16. M.R. Pearlman, Z.J. Haas, P. Sholander, and S.S. Tabrizi, "On the Impact of Alternate Path Routing for Load Balancing in Mobile Ad Hoc Networks," *Proceedings of ACM MOBIHOC 2000*, August 2000.

17. C. Perkins and P. Bhagwat, "Highly Dynamic Destination-Sequenced Distance-Vector Routing (DSDV) for Mobile Computers," *Proceedings of ACM SIGCOMM 1994*, August–September 1994.

18. G. Pei, M. Gerla, and T.W. Chen, "Fisheye State Routing in Mobile Ad Hoc Networks," *Proceedings of ICDCS Workshop 2000*, April 2000.

19. X. Zeng, R. Bagrodia, and M. Gerla, "Glomosim: A Library for Parallel Simulation of Large-Scale Wireless Networks," *Proceedings of PADS-98*, May 1998.

9

CROSS-LAYER
OPTIMIZATION FOR
SCHEDULING IN WIRELESS
MESH NETWORKS

*Vasilis Friderikos, Katerina Papadaki, David Wisely,
and Hamid Aghvami*

CONTENTS

Wireless mesh networks (WMNs) can be considered as a new paradigm for multihop wireless communication. Due to their inherent flexibility for providing first mile connectivity to the Internet and/or acting as a wireless backhaul network segment, they have recently attracted a significant research interest. The aim of this chapter is twofold. Firstly, to discuss resource management schemes for WMNs from a cross-layer optimization approach and secondly, to formulate the corresponding mathematical programs that utilize such cross-layer information. The mixed integer linear programs that are investigated focus on collision-free link scheduling for WMNs based on spatial time division multiple access (STDMA). Numerical investigations and computational complexity aspects of these schemes for different network sizes and number of active links are also reported. These networks are still in the early stages of the technology learning curve concerning both deployment issues and aspects related to the optimal interaction between information in different layers of the protocol stack. To this end, we close the chapter by identifying and discussing open-ended research challenges.

9.1 INTRODUCTION

9.1.1 Overview of Wireless Mesh Networks

Over the last few years we have witnessed an increased quest for revolutionizing traditional concepts in wireless communications. Without doubt the most prominent example is the effort towards the extension of the successful paradigm of single wireless hop cellular networks to multihop wireless communications. WMNs have naturally emerged as a result of this momentum and quickly become an intensive research topic [1,2]. The efforts for the realization of mesh networks span over a broad set of research activities: from theoretical studies on system capacity [3] to standardization fora such as the IEEE 802.16 standard [4]. WMNs are aiming to fulfil a number of different operational roles, which can vary from rapid deployable, low cost backhaul support to 3G/IEEE 802.11 "x" networks to first mile wireless connectivity to the Internet, or even to transient wireless networking.

WMNs, can be loosely defined, and this will be our fundamental assumption through-out this chapter, as wireless networks where nodes can act both as clients and routers [5]. We can distinguish two different types of WMNs: client-based and infrastructure-based. The main characteristic of client-based mesh networks is that portable mobile devices participate in the store and forward process. Client-based mesh networks operate in a rather autonomous way without the need of a central administration entity. From this perspective, these mesh networks resemble ad hoc networks, which are mainly characterized by energy constraint nodes, i.e., limited battery lifetime, and stochastic mobility. On the other hand, infrastructure-based mesh networks are characterized by nodes that are administered and controlled by a single entity and do not encounter energy constraints. It is noteworthy that this type of mesh networks flag a different set of research challenges compared to client-based mesh networks. Infrastructure-based mesh networks are characterized by low mobility (or no mobility at all) and by nodes that do not encounter energy constraints. In that respect, these two types of mesh networks have different operational domains, and in light of these differences the fitness of current proposed solutions for routing, scheduling, and rate control that emerged for ad hoc networks comes into question when applied to infrastructure-based mesh networks. In this chapter, we primarily focus on the second type of mesh networks.

One approach for multiple access in WMNs is to use contention-based (random) access schemes such as ALOHA or different flavors of CSMA. The benefits of random access schemes, such as carrier sense multiple access/collision avoidance (CSMA/CA), is that by using a combination of carrier sensing and back-off algorithms to prevent further conflicts, the nodes can operate in a rather autonomous way. This is a desirable feature for client-based mesh networks, where resource management procedures should be distributed in nature and quality of service (QoS) support is an add-on rather than a prerequisite feature. For infrastructure-based WMNs that are designed to provide, for example, last-mile broadband Internet access, QoS support, such as throughput and latency, evolves as a rather mandatory requirement. Thus, medium access control (MAC) schemes, such as TDMA, that allow more deterministic performance guarantees are highly desirable in such scenarios.

In this study, we address the issue of resource allocation in WMNs that are based on STDMA utilizing cross-layer information. In a carefully designed STDMA MAC protocol, collisions can be eliminated, fairness can be guaranteed, and bounds on per-hop latency can be provided. The principle operation of an STDMA MAC protocol is that nodes which are sufficiently far apart can utilize the same timeslot.

9.1.2 Chapter Structure

The chapter is organized as follows. In Section 9.2, we discuss architectural aspects of all Internet protocol (IP)-based WMNs. Based on the assumption of all IP networks, Section 9.3 discusses the cross-layer information plane that will be utilized for STDMA scheduling. In Section 9.4, the STDMA scheduling problem is formulated as a mathematical program and in Section 9.5 we extend the formulation to include power control and multirate allocation. Section 9.6 discusses approximation algorithms based on linear programming (LP)-relaxation and randomized rounding, while Section 9.7 provides the related Chernoff bounds. In Section 9.8, a methodology for increasing the performance of independent randomized rounding is presented. Furthermore, Section 9.9 discusses an alternative formulation for the link scheduling problem based on continuous optimization. In Section 9.10, numerical investigations using computer simulations are provided. Open problems and avenues for future research are discussed in Section 9.11, and finally, the conclusions end the chapter in Section 9.12.

9.2 ARCHITECTURAL CONSIDERATIONS—ALL IP WIRELESS NETWORKS

In this section, main architectural elements and the meaning of the notion all IP-based wireless networks will be briefly discussed together with the intrinsic advantages for moving towards all IP-based solutions. The architectural principles discussed here aim to create the generic framework upon where the cross-layer information plane and optimization problems will be developed.

From an architectural perspective, we believe that WMNs should not be considered in isolation but rather as one of many different wireless access networks that a mobile operator may have to manage in the near future. If seen from that perspective, an IP-based network provides a feature-rich convergence infrastructure to efficiently manage interworking aspects in such heterogeneous wireless environment. The above argument is supported by the intensive standardization activities within 3GPP towards the evolution and migration of both core and access segments to an all IP-based network. An important aspect of this migration process is how mobility functionalities will be implemented, where these functionalities will reside, and how to optimize network layer mobility processes [12]. Table 9.1 shows a nonexhaustive list of the main incentives for embracing all IP-based architectures.

From a cross-layer optimization perspective, which is our main focus point, an all IP-based wireless network can potentially propel developments towards protocols that embed information across the layers. This is because IP technology provides the required unification level across different access technology networks so that algorithms based on cross-layer information can be developed. Therefore, migration to a harmonious coexistence of different access networks based on IP technology will provide a common platform for the development of cross-layer optimization schemes.

9.2.1 Main Aspects of Resource Management in Mesh Networks and Focus of the Work

In this chapter we formulate a family of (mixed) integer optimization programs that are infused with cross-layer information. Integrating cross-layer information emanates from the need to provide, firstly efficient utilization of radio network resources and secondly to treat in a differential way traffic with inherently diverse requirements.

9.3 CROSS-LAYER INFORMATION PLANE

9.3.1 Benefits and the Need for Utilizing Cross-Layer Information

Traditional network engineering is based on a strict layering approach on the protocol stack. In this modular approach, fundamental

Table 9.1   Benefits of All IP Wireless Networks

Rational	Benefits
Deployment	Cost reduction by: – utilizing the economy of scale in IP technology
Operation	Cost reduction by: – capitalizing on existing administration tools – core moving to IP over WDM – shared lower cost transmission in the access network – capex lower for IP infrastructure (softswitches and media gateways)
New business opportunities	Current and emerging WLAN systems for covering hot spot areas Multimode terminals Intersystem handover
Heterogeneity	Efficient interworking of different wireless access networks
Network management and diagnosis	IP autoconfiguration and uniform tools across different access networks
Data services	Ease of deploying existing and emerging Internet-like data services
VoIP	Integration of VoIP services, capitalizing the strong momentum on the Internet
Multicasting	Leveraging existing multicast technologies over the Internet
Quality of service	Uniform application of IP-based QoS techniques (forwarding and control plane) in the core and different wireless access networks
IP routing	Increased network reliability and availability
Cross-layer designs	Improved network performance

functionalities such as access to the medium (link layer), routing (network layer), and congestion control (transport layer) are performed independently. This has proven to be a very powerful and successful paradigm for wireline networks, since the infrastructure is based on links with constant capacity and high degree of reliability.

A cross-layer design approach on the other hand, is one that utilizes information across different layers of the protocol stack. A number of studies over recent years highlighted that cross-layer designs that support information exchange between layers can yield significant performance gains [8,9]. The performance gains that can be attained should also be reflected on the robustness and longevity of

Table 9.2 Reasoning Behind the Utilization of Cross-Layer Information

	Motivation for Cross-Layer Designs
Link layer	Shared and interference limited nature of wireless channel
Network layer	Varying link capacities influence the optimality of routing decisions
Transport layer	Fundamental assumption of traditional congestion control algorithms is that end-to-end paths are predefined and link capacities are time invariant

the architecture among other design criteria [10]. The penalty that has to be paid for increasing system performance by deploying cross-layer designs is the complexity and communication overhead. Therefore, these two aspects should be carefully taken into account. Table 9.2 lists the main reasons towards the shift to cross-layer designs in wireless networks. As can be inferred from the table, the key reason for deploying cross-layer designs is that in wireless networks the capacities of the links can not be considered as constants, which was the fundamental assumption behind traditional layer designing. Therefore, in an optimal setting, scheduling, routing, and end-to-end congestion control in wireless networks should operate jointly under the awareness of the "soft" capacities of the link layer.

9.3.2 Incorporating Cross-Layer Information for Link Scheduling

To achieve increased system performance and efficient utilization of the scarce radio resources a linear integer mathematical programme is formulated that embeds information from the physical, link and network layer to optimize link scheduling decisions. Information from different layers is exchanged through the cross-layer information plane that span across the protocol stack. A representation of this concept is shown in Figure 9.1. The white boxes in the diagram show the information used to formulate the mathematical programs discussed in Section 9.4. Grey boxes represent information regarding routing and flow control that is not explicitly used, but we address this issue in Section 9.11. More specifically, at the network layer the parameters used are the bitrate (or bandwidth) requests from the nodes in the mesh network. At the link layer, which is based on STDMA, the link together with rate and power information is utilized to schedule packet transmission. Finally, at the physical layer

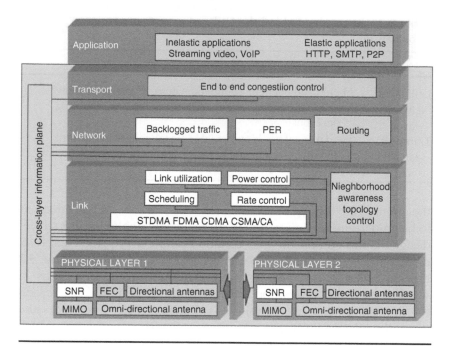

Figure 9.1 Functional diagram of the cross-layer information plane.

the signal-to-interference noise ratio (SIR) is used to characterize the quality of the link.

9.4 OPTIMIZATION ALGORITHMS FOR LINK SCHEDULING

9.4.1 Background Information

A first and rather naive approach to the problem of designing a feasible timeslot allocation is to assign a unique timeslot to each pair of active links in the mesh network. Such a scheme would clearly eliminate potential collisions of transmitted packets. The penalty to be paid is that a large number of timeslots per timeframe would be required and this number will increase linearly with the number of active links. Thus, despite the simplicity of this approach it is highly undesirable since the latency of the end-to-end communication will increase and network resources will be poorly utilized. To increase the performance of the system it is possible to allow coordinated concurrent transmissions from spatially distributed nodes within the mesh network, under the constraint that the underlying interference level is below a predefined SIR threshold. This spatial reuse of timeslots, which has been formally defined in the seminal work of Kleinrock [6], is called STDMA and can be considered as a collision-free multihop access scheme. The benefits inherent by adopting STDMA as access scheme is that it allows flexible management of uplink/downlink traffic in single frequency band based on dynamic adjustment of transmit/receive timeslot ratios to accommodate IP-based bursty traffic (i.e., accommodating asymmetric traffic). The optimization problems discussed in Section 9.4.2 allows such flexible allocation of uplink/downlink ratios by incorporating different number of timeslots per link.

The key operating issue in STDMA multihop networks boils down to the following acute question: How to maximize the spatial reuse of timeslots by ensuring at the same time that all SIR thresholds of active communication links are satisfied. Using graph theory previous work has shown that optimal scheduling policies turn to be $\mathcal{NP}$-hard problems [7,11]. Despite the inherent computational complexity of these optimization problems, the optimal scheduling decisions are of significant importance since they define the fundamental physical limitations as they pertain to the task of resource management. In other words, these optimal solutions provide the performance bounds of the system and thus ultimately dictate what can or cannot be accomplished.

Algorithms that perform optimal (or near-optimal) allocation of timeslots in an STDMA framework are undoubtedly the most versatile building block for such mesh networks. A number of previous work in this area studied both optimal allocations [13] but also sub-optimal solutions [14–16] under specific load and node density scenarios. In this work, we study near-optimal solutions under different pressing sets of scenarios. In contrast to numerous previous works that focused on minimum frame length we stand back and take a more holistic perspective on the problem by studying both minimum number of slots and constant frame length allocations. This is because we are interested in the embedded trade-offs on computational requirements that comes from these two problems. In addition, we also formulate multirate allocations with embedded continuous or discrete power control.

9.4.2 Problem Formulation

The discussion presented here sets forth the basic notation and the required system level assumptions to formulate the problem of

STDMA scheduling as a mathematical programme. Let U be the set of active links in the WMN, and U^T, U^R be the set of transmitting and receiving nodes respectively. By b_i ($i \in U^T$) we express the backlog pressure (number of packets waiting to be transmitted) in transmitting node i. For the first model, constant transmitted power is assumed for each transmitting node. This power level is equal to the required transmitted power to mitigate the path loss without considering any interference plus a slack value to allow for concurrent transmissions. The transmitted power level is incorporated within the path loss model and denoted as $\bar{g}_{ij}$ for the link $(i, j) \in U$. We relax the assumption of a constant power transmission per link in Section 9.5. Finally, we assume a constant frame length equal to T timeslots and by N we denote the lumpsum power of background and thermal noise. To formulate the problem we also introduce the Boolean variable x_{ij}^t, which is defined as

$$x_{ij}^t = \begin{cases} 1 & \text{if link } (i, j) \text{ transmits at timeslot } t \\ 0 & \text{otherwise} \end{cases} \tag{9.1}$$

Based on the above assumptions, the linear integer program with the objective of maximizing the weighted sum of allocated rates per link (i.e., throughput) can be written as

$$\max_x \sum_{t=1}^{T} \sum_{(i,j) \in U} \frac{b_{ij}}{\sum_{(k,m) \in U} b_{km}} x_{ij}^t$$

$$\text{s.t.} \sum_{t=1}^{T} x_{ij}^t \leq m_i \quad \forall\, (i,j) \in U \quad \text{(i)}$$

$$\sum_{t=1}^{T} x_{ij}^t \geq 1 \quad \forall\, (i,j) \in U \quad \text{(ii)}$$

$$\sum_{k \in U^T} x_{kj}^t \leq 1 \quad \forall\, j \in U^R, \quad \forall\, t \quad \text{(iii)}$$

$$\sum_{k \in U^R} x_{ik}^t \leq 1 \quad \forall\, i \in U^T, \quad \forall\, t \quad \text{(iv)}$$

$$\sum_{j \in U^R} x_{ij}^t + \sum_{k \in U^T} x_{ki}^t \leq 1 \quad \forall\, i \in U^T \cap U^R, \quad \forall\, t \quad \text{(v)}$$

$$\frac{\bar{g}_{ij} x_{ij}^t + (1 - x_{ij}^t)\Lambda}{\sum_{(k,m)/(i,j) \in U} \bar{g}_{kj} x_{km}^t + N} \geq \gamma \quad \forall\, (i,j) \in U, \quad \forall\, t \quad \text{(vi)}$$

$$x_{ij}^t \in \{0, 1\} \quad \forall\, (i,j) \in U, \quad \forall\, t \quad \text{(vii)} \tag{9.2}$$

The first set of constraints, (i), corresponds to a maximum allowable timeslot allocation per link. This allows the flexibility of dynamically assigning different upper bounds on rate allocation per link. Constraint (ii) guarantees a minimum rate allocation for every active link in the network (henceforth we assume that the minimum rate allocation is one timeslot per frame). The third set of constraints, (iii), ensures that only one node can send traffic to the same receiving node in each timeslot (indegree constraint). Likewise, constraint (iv) ensures that a transmitting node can only send traffic to one receiving node per timeslot (outdegree constraint). By the same token, constraint (v) binds the previous two constraints by ensuring that a node cannot transmit and receive at the same timeslot (degree constraint). This set of constraints is active only for the nodes that are both receivers and transmitters, i.e., $i \in U^T \cap U^R$. The SIR constraints, that should be satisfied for every link that transmits at a specific timeslot, are shown in (vi). In the SIR constraints, γ denotes the target SIR threshold. The parameter Λ is introduced so that the constraint is satisfied when a link is not transmitting on a specific timeslot. This parameter should be greater than the total aggregate interference produced in the mesh network, and therefore a suitable value could be

$$\Lambda = \alpha \cdot \gamma \left(\sum_{(i,j) \in U} \bar{g}_{ij} + N \right) \tag{9.3}$$

where α is a small integer. All of the foregoing analysis can be extended to include multirate decisions per timeslot and power controlled transmission. This is the focus of Section 9.5.

9.5 JOINT POWER CONTROL AND MULTIRATE ALLOCATION

Let us assume that the system is equipped with a predefined set of modulation schemes allowing multirate transmission. In such a setting, the focus point is how to optimally explore the relationship between power allocation and corresponding link capacity. Let $r \in R$ be the set of allowable transmission rates per timeslot. Contrary to Section 9.4.2, here we additionally perform power control by introducing the variable p_{ij}^t that expresses the transmitted power by node i in link (i, j) under the constraint that $0 \leq p_{ij}^t \leq P_{max}$, $\forall t \in \{1, \ldots, T\}$. The variable P_{max} expresses the power ceiling at the transmitting node. Under the assumption of a constant target bit error rate, i.e., $E_b/N_0 = \Gamma$, different transmission rates can be translated into different SIR requirements, which we will denote as γ_r. The rational of utilizing different thresholds is that if an SIR value of γ_r can be supported in a specific link, then the transmitting node of this link can utilize rate r at the allocated timeslot. To be able to express now the problem in a mathematical programming setting we introduce the Boolean variables x^t_{ijr}, which are defined as

$$x^t_{ijr} = \begin{cases} 1 & \text{if } r \in R/\{0\} \\ 0 & \text{otherwise} \end{cases} \qquad (9.4)$$

We therefore propose to solve the following mixed integer linear program for joint power and multirate control

$$\max_{x,p} \sum_{r \in R} \sum_{t=1}^{T} \sum_{(i,j) \in U} \frac{b_{ij}}{\sum_{(k,m) \in U} b_{km}} x^t_{ijr}$$

subject to

$$\sum_{r \in R} \sum_{t=1}^{T} x^t_{ijr} \le m_i \quad \forall\, (i,j) \in U \quad \text{(i)}$$

$$\sum_{r \in R} \sum_{t=1}^{T} x^t_{ijr} \ge 1 \quad \forall\, (i,j) \in U \quad \text{(ii)}$$

$$\sum_{k \in U^T} x^t_{kjr} \le 1 \,\forall\, j \in U^R \quad \forall\, t \in T, \quad \forall\, r \in R \quad \text{(iii)}$$

$$\sum_{k \in U^R} x^t_{ikr} \le 1 \,\forall\, i \in U^T \quad \forall\, t \in T, \quad \forall\, r \in R \quad \text{(iv)}$$

$$\sum_{j \in U^R} x^t_{ijr} + \sum_{k \in U^T} x^t_{ijr} \le 1 \quad \forall\, i \in U^T \cap U^R, \quad \forall\, t \in T, \quad \forall\, r \in R \quad \text{(v)}$$

$$\sum_{r \in R} x^t_{ijr} \le 1 \,\forall\, (i,j) \in U \quad \forall\, t \in T \quad \text{(vi)}$$

$$\sum_{r \in R} x^t_{ijr} \le \frac{p^t_{ij} \bar{g}_{ij}}{N \gamma r} \quad \forall\, (i,j) \in U \quad \forall\, t \in T, \quad \forall\, r \in R \quad \text{(vii)}$$

$$\frac{p^t_{ij} \bar{g}_{ij} + (1 - x^t_{ijr})\Lambda}{\sum_{(k,m)/(i,j) \in U} \bar{g}_{kj} p^t_{km} + N} \ge \gamma_r \quad \forall\, (i,j) \in U, \quad \forall\, t \in T, \quad \forall\, r \in R \quad \text{(viii)}$$

$$0 \le p^t_{ij} \le P_{\max} \quad \forall\, (i,j) \in U, \quad \forall\, r \in R \quad \text{(ix)}$$

$$x^t_{ijr} \in \{0, 1\} \quad \text{(x)} \qquad (9.5)$$

In order to couple the Boolean variables with the transmitted power variables, an extra constraint has been imposed that forces the Boolean variable x^t_{ijr} to zero when the corresponding transmitted power level p^t_{ij} is zero. Since there are more than one possible transmission rates, constraint (vi) ensures that only one rate can be used at any timeslot.

9.5.1 Incorporating Discrete Transmission Power Levels

A more realistic scenario would be to consider discrete power transmission levels of the nodes rather than continuous ones. The optimization problems defined so far can be extended to include a set of of discrete transmission power levels L_k $1 \leq k \leq K$ per link, i.e., p^t_{ijk} $(i, j) \in U, k \in L, t \in \{1 \cdots T\}$. By introducing the Boolean variables $\delta_k \in \{0,1\}$ the discrete transmission power level can be expressed as

$$p_{ijk} = \delta_k L_k \; (i,j) \in U, \quad k \in L \tag{9.6}$$

The following new constraint need to be added to the optimization problem to ensure that only one discrete power level is selected per timeslot.

$$\sum_{k \in L} p^t_{ijk} = \sum_{k \in L} \delta_k L_k \leq 1 \quad \forall \, (i,j) \in U, \quad t \in \{1 \cdots T\} \tag{9.7}$$

All previously mentioned constraints are also valid for the formulation of the problem with discrete levels of transmission power, with additional $|U| \cdot K$ further constraints.

9.5.2 Composite Functions: Maximizing Throughput While Minimizing Power

We can extend the previous formulated linear integer programs to include multiobjectives. One interesting scenario is a biobjective program that aims to maximize throughput with the minimum possible transmission power. This biobjective mixed integer program can be formulated as

$$\max_{x,p} \left[\sum_{r \in R} \sum_{t=1}^{T} \sum_{(i,j) \in U} \frac{b_{ij}}{\sum_{(k,m) \in U} b_{km}} x^t_{ijr} \right], \quad \left[\sum_{t=1}^{T} \sum_{(i,j) \in U} \sum_{k \in L} p^t_{ijk} \right] \tag{9.8}$$

constraints in Equation 9.5

From this biobjective program we can derive a single-objective mathematical program by taking a nonnegative linear combination of the objective functions [22]. Each selection of the weight will produce a different single-objective problem, and optimizing the resulting problem produces a Pareto outcome. For the above defined biobjective problem, the combined criterion is parameterized by the single scalar parameter $0 \leq \xi \leq 1$. These two weighted sums are in general non-conflicting objectives, since minimizing the consumed power within the network can potentially lead to an increase of the aggregate throughput. We should note that this may not be always the case. It has been shown that the optimal transmission power is determined by the number and density of nodes in the network together with the aggregate load and the minimal transmission power may not lead to maximum throughput [24]. Therefore by scalarizing the biobjective problem allows us to study these trade-offs and synergies between throughput and transmission power under different network topologies and number of active links. For a fixed weight ξ, the solution of the optimization problem would balance between aggregate transmission rate and power consumption in the network. The single-objective optimization program can be formulated as

$$\max_{x,p} (1-\xi) \sum_{r \in R} \sum_{t=1}^{T} \sum_{(i,j) \in U} \frac{b_{ij}}{\sum_{(k,m) \in U} b_{km}} x_{ijr}^{t} - \xi \sum_{t=1}^{T} \sum_{(i,j) \in U} \sum_{k \in L} p_{ijk}^{t}$$

$$\text{subject to } \sum_{k \in L} p_{ijk}^{t} = \sum_{k \in L} \delta_k L_k \leq 1 \quad \forall \, (i,j) \in U \quad \forall \, t \in \{1 \cdots T\}$$

(9.9)

constraints in Equation 9.5.

If we consider the first term in the objective as a reward function and the second as a penalty function, then the parameter ξ controls the relative "weight" of these two terms. The above problem formulation represents a weighted sum of two objective functions and since the weighting variable can vary between zero and one there are typically infinite number of optimal solutions (Pareto frontier). By inspecting the biobjective function we can provide a smaller range of desirable values for the weighting parameter ξ. Note that in order to allow a timeslot allocation to a specific link, the award acquired (first term in the objective) should be larger than the maximum penalty from the cost function (second term in the objective). Since the minimum reward is related to the link with the fewer number of backlogged packets the following constraint should be imposed to the choice of weight ξ.

$$\xi P_{\max} \leq (1 - \xi) \, \min \left(\frac{b_{ij}}{\sum\limits_{(k,m) \in U} b_{km}} \right) \tag{9.10}$$

or

$$\xi \leq \frac{\min (b_{ij})}{\min (b_{ij}) + P_{\max} \sum\limits_{(k,m) \in U} b_{km}} \tag{9.11}$$

9.6 LP-RELAXATION WITH ROUNDING

Since there is no efficient algorithm known for solving integer programs in bounded time (and there cannot exist any unless $P = NP$), the above formulation does not help us solve the scheduling problem. It does however guide us to a natural relaxation which helps us find a good approximation algorithm. In LP-relaxation, each of the variables x_{ij}^t is in the range from zero to one and therefore converted from integer variables to fractional ones. The LP-relaxation, problem can be solved in polynomial time, either by using Khaciyan's ellipsoid method [17] or some interior point method based on the ideas of Karmarkar [18]. In practice, the algorithms most commonly used are variations of Dantzig's simplex algorithm [19], even though this requires exponential computational time in the worst case scenario. The optimal solution of the LP-relaxation problem can be treated as a probability vector with which we choose to include a link to a specific timeslot. In that sense, if x_{ij}^t is near to one, it is likely that this link will be included in the current timeslot. If, on the other hand, the optimal solution of the LP-relaxation problem is near zero, it will probably not be included in the current timeslot. We denote the optimal solution of the linear relaxation problem as $\hat{x}_{ij}^t$.

9.6.1 Deterministic Rounding

Let us first assume the following rounding scheme:

$$x_{ij}^t = \begin{cases} 1 & \text{if } \hat{x}_{ij}^t \geq 0.5 \\ 0 & \text{if } \hat{x}_{ij}^t < 0.5 \end{cases} \tag{9.12}$$

We will now examine how the above deterministic rounding scheme affects the feasibility of the optimization problem (Equation 9.2).

On the constraint regarding the minimum allowable allocation per link, the optimal fractional solution satisfies the inequality and therefore we can write

$$\sum_{t=1}^{T} \hat{x}_{ij}^{t} \geq 1 \quad \forall\, (i, j) \in U \tag{9.13}$$

This implies that for link (i,j), either two different timeslots have been allocated fractionally with $\hat{x}_{ij}^{t} \geq 0.5$ or one timeslot has been allocated and the constraint is satisfied with equality, i.e., $\hat{x}_{ij}^{t} = 1$. Therefore, in both cases, the integral solution will also satisfy the inequality.

The fractional solution also satisfies the indegree constraint, i.e.,

$$\sum_{i \in U^{T}} \hat{x}_{ij}^{t} \leq 1 \quad \forall\, j \in U^{R}, \quad \forall\, t \tag{9.14}$$

The above inequality infers that either all $\hat{x}_{ij}^{t} < 0.5$ and therefore the integral inequality is satisfied or only one $\hat{x}_{ij}^{t} \geq 0.5$ and therefore the integral constraint is satisfied with equality. With the same argument we can show that also the outdegree and degree integral constraints are satisfied when the rounding scheme is applied on the fractional solution.

The fractional solution of the LP-relaxation problem satisfies the SIR constraint (Equation 9.5 (viii)), therefore we can write

$$\bar{g}_{ij}\hat{x}_{ij}^{t} + (1 - \hat{x}_{ij}^{t})\Lambda - \gamma \left(\sum_{(k,m)/(i,j)\in U} \bar{g}_{kj}\hat{x}_{km}^{t} + N \right) \geq 0 \tag{9.15}$$

If $\hat{x}_{ij}^{t} < 0.5$ then $x_{ij}^{t} = 0$ therefore the integral constraint is also satisfied. If we denote by M the set of links where $\hat{x}_{km}^{t} \geq 0.5$ then the integral constraint that should be satisfied can be written as

$$\bar{g}_{ij} \geq \gamma \left(\sum_{(k,m)/(i,j)\in M} \bar{g}_{km} + N \right) \quad \forall\, (i, j) \in M \tag{9.16}$$

9.6.2 Randomized Rounding

The randomized rounding approach to find efficient and near-optimal solutions is due to Raghavan and Thompson [20] and Raghavan [21], where each variable is rounded independently with probability depending on the fractional part of its value. This allows the use of Chernoff-type large deviation inequalities, showing that a sum of independent random variables is highly concentrated around its expectation. In that respect, randomized rounding has been applied to numerous combinatorial optimization problems that can be formulated as integer linear programs. In randomized rounding, the solution of the linear program relaxation is used as probability to round the variable towards zero or one. More specifically, the rounding procedure can be more formally described as

$$\bar{x}_{ij}^t = \begin{cases} 1 & \text{with probability } \hat{x}_{ij}^t \\ 0 & \text{with probability } 1 - \hat{x}_{ij}^t \end{cases} \tag{9.17}$$

An example of the operation of the randomized rounding procedure is summarized in Figure 9.2 and Figure 9.3. Figure 9.2 shows the solution of the linear program relaxation in the case of 10 active links that need to be scheduled in a frame with 5 timeslots. Figure 9.3 shows a feasible

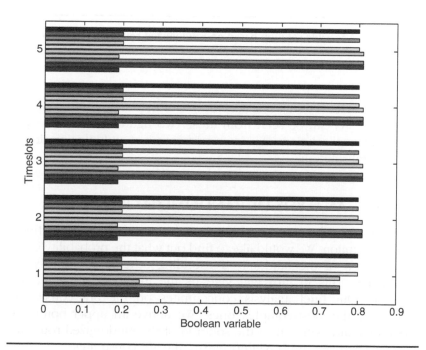

Figure 9.2 Fractional solution based on linear program relaxation [10 active links, 5 timeslots].

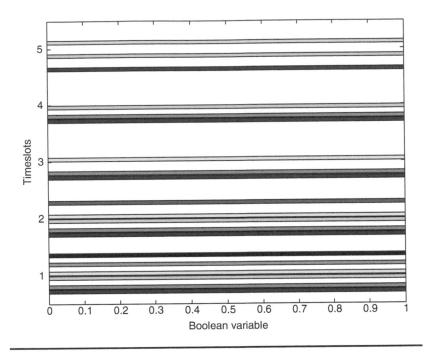

Figure 9.3 A feasible solution based on randomized rounding [10 active links, 5 timeslots].

allocation achieved in just few iterations of the randomized rounding procedure. As can be seen from these two figures, active links that have fractional solutions close to one, transmit at multiple timeslots in the integral solution.

9.7 ANALYSIS OF RANDOMIZED ROUNDING

The randomized rounding scheme needs several iterations to find a feasible solution. We would like to find out what the probability is that the solution of the randomized rounding scheme will satisfy the constraints. This in turn will give us the likelihood of a feasible solution. Our brief analysis concentrates on the SIR constraints, which are the most complex ones. We derive an upper bound on the probability with which the solution of the randomized rounding scheme (Equation 9.17) will not satisfy the SIR constraints.

We use the following well known Chernoff bound-type result.

Theorem 9.7.1 (HOEFFDING-CHERNOFF BOUND) *Let $u_k \in [0,1]$, $k \in K$ for a finite set K, be independent random variables. Let $V = \sum_{k \in K} u_k$ and $\mu = E\{\sum_{k \in K} u_k\}$, then we have:*

$$\Pr(V \leq (1 - \delta)\mu) \leq \exp\left(-\frac{\delta^2 \mu}{2}\right) \qquad (9.18)$$

$$\Pr(V \geq (1 + \delta)\mu) \leq \exp\left(-\frac{\delta^2 \mu}{3}\right) \qquad (9.19)$$

We apply Theorem 9.7.1 to find the desired upper bound. The SIR constraint can be written as follows for each link (i, j) and each timeslot t:

$$(\Lambda - \gamma N) \geq (\Lambda - \bar{g}_{ij})x_{ij}^t + \gamma \sum_{(k,m)/(i,j)\in U} \bar{g}_{kj} x_{km}^t \qquad (9.20)$$

Let us define by $W = \Lambda - \gamma N$, $w_{ij} = \Lambda - \bar{g}_{ij}$ and $w_{km} = \gamma \bar{g}_{kj}$ for $(k, m)/(i, j) \in U$. Thus we can rewrite the SIR constraint of Equation 9.20 as

$$\sum_{(i,j)\in U} w_{ij} x_{ij}^t \leq W \qquad (9.21)$$

Since transmission power levels are chosen so that the SIR constraint is satisfied in a pure TDMA transmission, i.e., $\bar{g}_{ij} \geq \gamma N$, we have

$$\frac{w_{ij}}{W} \leq 1 \quad \forall\, (i,j) \in U \qquad (9.22)$$

Since $\bar{x}_{ij}^t$ is the integral solution that results from the randomized rounding procedure defined in Equation 9.17, we note that $E(\bar{x}_{ij}^t) = \hat{x}_{ij}^t$. We define the random variable, $u_{ij} = \frac{w_{ij}}{W} \cdot \bar{x}_{ij}^t$. Due to the fundamental problem setting assumption of Equation 9.22, we have $u_{ij} \in [0,1]$. Further, the variables u_{ij} are independent since they are scalar multiples of the variables $\bar{x}_{ij}^t$, which are independently sampled using the randomized rounding scheme. The expected value of u_{ij} is

$$E\{v_{ij}\} = E\left\{\frac{w_{ij}}{W} \cdot \bar{x}_{ij}^t\right\} = \frac{w_{ij}}{W} E\{\bar{x}_{ij}^t\} = \frac{w_{ij}}{W} \cdot \hat{x}_{ij}^t \qquad (9.23)$$

Regarding the μ value, we can see that

$$\mu = E\left\{\sum_{(i,j)\in U} u_{ij}\right\} = \sum_{(i,j)\in U} \frac{w_{ij}}{W} \cdot \hat{x}_{ij}^t \qquad (9.24)$$

Since the solution of the LP-relaxation satisfies the SIR constraint, from Equation 9.21 we have

$$\sum_{(i,j)\in U} w_{ij}\hat{x}_{ij}^t \leq W \tag{9.25}$$

Thus the probability that the SIR constraints are not satisfied within a range of δ is

$$\Pr\left(\sum_{(i,j)\in U} w_{ij}\bar{x}_{ij}^t > (1+\delta)W\right)$$

$$\leq \Pr\left(\sum_{(i,j)\in U} w_{ij}\bar{x}_{ij}^t > (1+\delta)\sum_{(i,j)\in U} w_{ij}\cdot\hat{x}_{ij}^t\right)$$

$$= \Pr\left(\sum_{(i,j)\in U} \frac{w_{ij}}{W}\bar{x}_{ij}^t > (1+\delta)\sum_{(i,j)\in U} \frac{w_{ij}}{W}\cdot\hat{x}_{ij}^t\right)$$

$$= \Pr\left(\sum_{(i,j)\in U} u_{ij} > (1+\delta)\mu\right) \leq \exp\left(-\frac{\delta^2\mu}{3}\right)$$

where we used inequality (Equation 9.25), equality (Equation 9.24) and Theorem 9.7.1 for variables v_{ij} in this order.

The above analysis regarding the SIR constraint provides insights on the hardness of finding feasible solutions using the randomized rounding based on the fractional solutions provided by the relaxation algorithm. Thus, for a given WMN topology, average channel gains and received signal quality thresholds (γ) the probability of satisfying the SIR constraint for every active link can be calculated.

9.8 BOOSTING PERFORMANCE BY NONINDEPENDENT RANDOMIZED ROUNDING

So far, the basic independent randomized rounding scheme has been discussed that allows approximate solutions to the linear integer program to be found. However, timeslot allocations attained by independent randomized rounding may fail to satisfy a subset of the constraints. For example, with an increased number of links and time slots, fractional solutions such as $\hat{x}_{i,j}^t \approx 1/T$, which could arise from the constraint $\sum_t x_{ij}^t \geq 1$, result in a low probability that the corresponding rounding integer values $\bar{x}_{i,j}^t$ will satisfy this constraint. Nonindependent randomized rounding on the other hand allows flexibility on the rounding procedure that can potentially increase the performance of the algorithm [23].

The first level of dependence that can be introduced on the rounding decision is based on the indegree, outdegree, and degree constraints. When a link is rounded to one, the corresponding "conflicting" links as given by these constraints are automatically rounded to zero. This ensures that the indegree, outdegree, and degree constraints are always satisfied. Further, to avoid similar rounding in different iterations of the randomized rounding scheme, the links are picked randomly for performing rounding.

Based on the fractional solution provided by the LP-relaxation a number of different rounding strategies can be envisioned to provide a trade-off between feasibility and optimality of the integer solution. By taking a closer look into the structure of the fractional solutions provided by the linear program, two important properties can be drawn. Depending on the aggregate rate (bandwidth) requests and channel conditions, the fractional allocations that some links are receiving satisfies tightly the constraint of the minimum rate requirement (one timeslot per frame). On the other hand, for some other links the fractional solution will be close to one and therefore it is highly probable that they will receive multiple timeslots per frame after the rounding procedure. In light of these observations we taxonomize the active links in three different sets.

The first set, denoted by U_1, includes all the links which satisfy tightly the constraint regarding the allocation of one timeslot per frame. The probability that these links will be actually allocated a timeslot after the rounding procedure, is linearly decreased with the length of the frame (T). The rounding procedure works as follows: each rounding iteration starts with the first timeslot, then a link is randomly picked and the corresponding fractional solution from the linear programme is rounded into an integral variable, and then the algorithm moves to the next timeslot. For links in the set U_1, the algorithm monitors whether they have been allocated a timeslot so far (within the current timeframe) and if they have not, the probability of rounding towards to one is gradually increased over the next timeslots. This continues until an allocation occurs and then their probability of rounding towards one is setback to the original probability. For the links in U_1, the probability in each timeslot is increased according to the coefficient ω_1. By increasing these probabilities the number of feasible solutions is accordingly expected to increase.

The second set, denoted by U_2, includes the links with fractional solutions close to one. These links have a high probability of attaining multiple timeslots per frame in the integer solution. With an increased number of timeslots allocated to these links, the objective function is maximized but the probability of acquiring feasible solutions is decreased. This is due to the fact that these links will be allocated a timeslot with probability close to one and therefore the possible permutations of allocating timeslots among the active links are decreased. One way to increase the number of feasible solutions is to decrease the probability of the links in U_2 to be allocated a timeslot. The reduction of the probability is done with the coefficient ω_2. The penalty that has to be paid for increasing the number of feasible solutions is that the objective value is decreased.

In order to allow a trade-off between optimality and feasibility, a third set of links is defined, U_3, that correspond to the links with intermediate values of fractional solutions. The probabilities of the links in U_3 are decreased, once a slot allocation occurs, with a factor ω_3, which is larger than the one used for the links in the second set, i.e., $\omega_3 \leq \omega_2$. In this way the number of feasible solutions located by the randomized rounding algorithm is increased, while at the same time allows more allocations of links from the set U_2, that make a major contribution to the maximization of the objective function.

The parameters ω_1, ω_2, ω_3 can be adjusted depending on the desired preference of feasibility vs. optimality.

9.9 USING CONTINUOUS OPTIMIZATION FOR THE INTEGER TIMESLOT ALLOCATION PROBLEM

In this section the (mixed) linear integer programme is reformulated into a continuous quadratic nonconvex optimization problem. There are close connections between nonconvex quadratic optimization problems and combinatorial optimization. Many integer linear and integer quadratic optimization problems can be reformulated as quadratic programming problems in continuous variables. Such a reformulation usually may not lead to more effective algorithms, but it provides some insight into the structure of these problems. The previously formulated linear integer programs can be formulated as quadratic continuous programs as shown below:

$$\min_{x} \sum_{r \in R} \sum_{t=1}^{T} \sum_{(i,j) \in U} \mu\, x_{ijr}^{t}(1 - x_{ijr}^{t}) - \sum_{r \in R} \sum_{t=1}^{T} \sum_{(i,j) \in U} w_{ijr} x_{ijr}^{t} \qquad (9.26)$$

subject to constraints in (Equation 9.5)

$$x_{ijr}^{t} \in [0,1]$$

The parameter μ should be chosen large enough so that the quadratic term in the objective function becomes the dominant one for integral solutions. We should note that the continuous version of the problem still encounters the same level of computational complexity as the integer version. This is because the continuous version of the optimization problem has local minima at every vertex of the rectangular polytope defined by the upper and lower bounds of the x_{ijr}^t variables. It can be easily seen that the above problem has $2^{TR|U|}$ local minima. Therefore, in the worst case scenario an exponential amount of time would be required to find an optimal solution. Reformulating the problem of link scheduling as a continuous optimization problem allows the introduction of nonlinear terms in the objective function and therefore the utilization of nonlinear programming tools.

9.10 NUMERICAL INVESTIGATIONS

9.10.1 Experimental Design

The numerical investigations are based on grid topologies aiming to evaluate the effect of the number of active links on different performance aspects of the system. Even though grid topologies may not represent realistic mesh network configurations, they facilitate a detailed and systematic discussion on the performance of different algorithms. As an example a topology with 64 nodes and 40 active links in a 8×8 grid is shown in Figure 9.4. The coefficients of the objective function are normalized fractions which are uniformly randomly allocated to each active link. The link gains take into account only path loss and the SIR threshold, γ, is the same for all links. The reported simulation results have been conducted using ILOG AMPL/CPLEX 9.0 and MATLAB 7.0.

9.10.2 Simulation Results

The computational complexity of mainly all of the optimization problems discussed is characterized by a complex synergy between the number of active links (Boolean variables in the problem) and the number of constraints which is mainly affected by the number of available timeslots. Therefore, it is expected that the computational complexity will not decrease as the number of available timeslots increases. Figure 9.5 shows the computational complexity of LP-relaxation for different number of active links. Here the complexity does not increase linearly with the number of active links. Therefore, for very large number of active links, a possible different approach would be to solve the

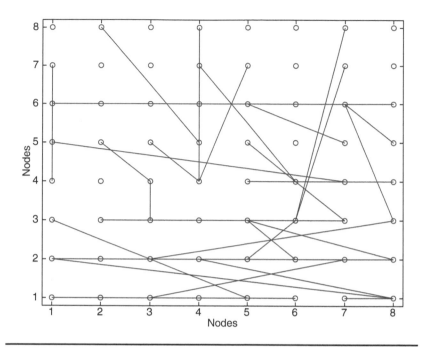

Figure 9.4   40 Active links in an 8 × 8 grid network.

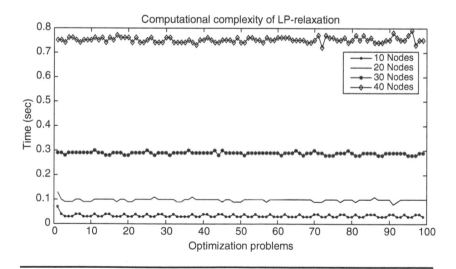

Figure 9.5   Computational time for the linear programming relaxation problem.

relaxation problem with a greedy algorithm before feeding this solution into the randomized rounding engine. We should note that for the same set of problems, CPLEX managed to find the optimal solution for integer programs with instances of up to 20 active nodes.

Figure 9.6 shows the number of feasible solutions that have been found by the randomized rounding procedure with respect to the number of timeslots per frame (for a fixed number of iterations). It is interesting to note that the number of feasible solutions decreases as the number of timeslots approaches the minimum frame length; this is because it becomes harder to pack these links within a small time frame and thus there are fewer feasible solutions. However, the number of feasible solutions also decreases after a certain number of time slots; this is due to the large size of the problem, which increases with T, and even though there are many feasible solutions the randomized rounding scheme needs more iterations to find the same number of feasible solutions.

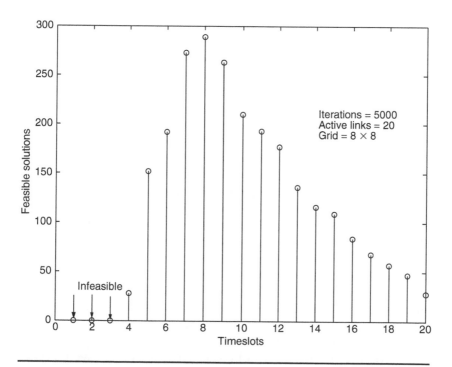

Figure 9.6 Number of feasible solutions found by the randomized rounding algorithm with respect to the number of available timeslots.

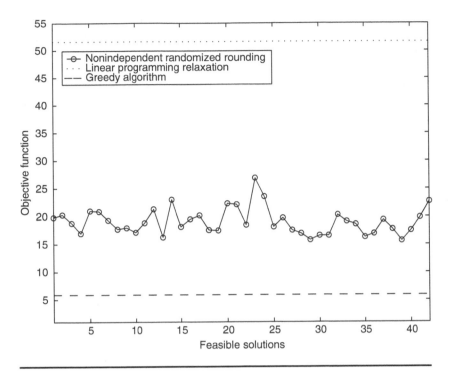

Figure 9.7   Comparison between the upper bound on the solution, randomized rounding and a baseline greedy algorithm.

Figure 9.7 depicts the solutions obtained by the randomized rounding algorithm in contrast with the upper bound, which is the fractional solution. The number of active links used is 40 and T equals to 20. The figure also includes the performance of the following rudimentary greedy algorithm. We sort the values of the objective function and we satisfy the T largest between them. The greedy algorithm then tries to pack the remaining links in any of the T time-slots but with only one passage through the timeslots. This greedy algorithm does not produce in general feasible solutions but was included in the figure to show the margins on the solution.

Finally, the box-plot of Figure 9.8 shows the trade-off between the number of feasible solutions that can be found and the objective value of these solutions, by adjusting the coefficients ω_i described in Section 9.8. As can be seen from the figure, an increase in the coefficient ω_2 for the set of links U_2, results in solutions with better objective function, but the number of the total feasible solutions found by the randomized algorithm is decreased.

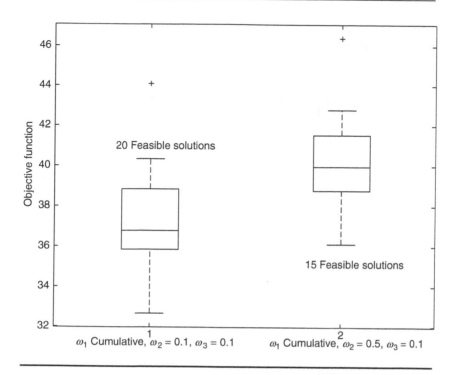

Figure 9.8 Feasibility vs. optimality of the solution depending on the selection of ω factors.

9.11 OPEN RESEARCH PROBLEMS

9.11.1 Scheduling Under Uncertainty

The model formulations that we have encounterd so far in the chapter, together with the related work that have been reviewed, are deterministic in nature. This fundamental assumption implies that the variability of the inherently uncertain parameters such as the link gains and traffic demands is not taken into account and the focus is on complexity and algorithmic aspects of the scheduling schemes. In reality, link gains and traffic demands are stochastic in nature and mean values may not be representative. Being able to take this randomness into account is critical for scheduling in WMNs and this requirement may need a complete new set of problem formulations and solving techniques. We could for example consider the requirement on the SIR as a probabilistic constraint. By imposing a reliability restriction, η, on the SIR system constraints, the linear program can be transformed into a chance constrained program (CCP) by altering the SIR constraint as

$$\Pr\left\{\frac{\bar{g}_{ij}x_{ij}^t}{\displaystyle\sum_{(k,m)/(i,j)\in U}\bar{g}_{kj}x_{km}^t + N} \geq \gamma\right\} \geq \eta \qquad (9.27)$$

This probabilistic constraint can provide an insight into the trade-offs between the risk of an SIR threshold violation and the total value of the objective function, which directly relates to the aggregate throughput. Keeping in mind that violation of the SIR inequality is synonymous to an erroneous transmission, a study of this trade-off is important, since it allows a more complete picture on the system performance.

To formulate and solve the link scheduling problem under uncertainty, results from stochastic and chance constraint programming [25] or stochastic approximate dynamic programming [26] may need to be deployed.

9.11.2 Distributed Link Scheduling

The mathematical programs discussed in this chapter have two fundamental properties: they are centralized and the interference produced by concurrent transmissions is explicitly taken into account (SIR constraint). A different approach, which has been widely used, is the graph theoretic based STDMA scheduling [27]. Graph theoretic scheduling algorithms are based on the protocol interference model [28], in which actual collisions can only occur due to primary or secondary interference (resembling the indegree, outdegree, and degree constraints). By taking into account only the set of one or two hops apart nodes, the issue of collision-free scheduling can be considered as a coloring problem, i.e., one hop or two hop neighbor nodes must be scheduled to transmit in different timeslots [29]. Due to this assumption, graph theoretic scheduling algorithms are more amenable to distributed implementation as opposed to interference-based algorithms. Despite this fact, previous work has shown that by ignoring the accumulative effect of interference in the wireless network, graph-based schemes can perform poorly [30,31]. Therefore, efficient distributed algorithms based on the interference model for STDMA scheduling are an open research issue that requires further investigations.

9.11.3 Layering as Optimization Decomposition

One of the fundamental open research problems in cross-layer designs for wireless networks is the realization of efficient distributed algorithms that incorporate routing and flow control decisions. The issue of routing and/or congestion control over the Internet perceived as a network utility maximization problem was first studied in the seminal papers of Kelly [32] and Low [33]. Since then there has been a significant research attention to jointly optimize routing and congestion control decisions on wired networks using primal–dual decomposition techniques [34]. The basic assumption of these algorithms is that link capacities are fixed and therefore can not be applied for wireless networks where interference, mobility, and channel impairments create links with time-varying capacity. Only recently the implications of the wireless time-varying channel together with scheduling constraints have been considered to formulate joint scheduling, routing, and congestion control algorithms in a systematic way [35,36]. The interactions of these distributed schemes with different scheduling algorithms together with performance aspects in terms of overhead, stability, and achievable rates is an open-ended research challenge.

9.12 CONCLUSIONS

In this chapter we have discussed architectural aspects of all IP-based WMNs, focusing mainly on the formulation of mixed integer linear programs for link scheduling that utilize cross-layer information. Since these programs are intractable, approximation engines based on linear programming relaxation and randomized rounding, that can provide near-optimal decision making, have been studied. Performance evaluation of these schemes in a rectangular network topology with different number of active links has also been included.

In order to gauge the potential impact of WMNs in the future, a significant number of research challenges need to be met. The most prominent of them are in the area of distributed algorithms for performing joint scheduling, routing, and congestion control. Robust architectural paradigms based on such cross-layer designs are wanting.

ACKNOWLEDGMENTS

This work has been carried out while the first author was holding a British Telecom Research Fellowship. The hospitality, assistance, and encouragement from all members of BT Mobility Research Center during the numerous site visits at Adastral Park is greatly appreciated.

REFERENCES

1. S. Cass, Viva Mesh Vegas: The gambling capital antes up for a new mobile broadband technology, *IEEE Spectrum Magazine*, vol. 53, p. 48, January 2005.
2. Y. Bejerano, Efficient integration of multi-hop wireless and wired networks with QoS constraints, *ACM MOBICOM*, September 2002.
3. P. Gupta, P.R. Kumar, The capacity of wireless networks, *IEEE Transactions on Information Theory*, vol. 46, pp. 388–404, 2000.
4. IEEE 802.16-2001, IEEE Standard for Local and metropolitan area networks. Part 16: Air Interface for Fixed Broadband Wireless Access Systems, April 2002.
5. Ian F. Akyildiz, Xudong Wang, Weilin Wang, Wireless mesh networks: A survey, *Computer Networks*, vol. 47, no. 4, pp. 445–487, March 2005.
6. R. Nelson, L. Kleinrock, Spatial-TDMA: A collision-free multihop channel access protocol, *IEEE Transactions on Communications*, vol. 33, no. 9, pp. 934–944, September 1985.
7. A. Ephremides, T.V. Truong, Scheduling broadcasts in multihop radio networks, *IEEE Transactions on Communications*, vol. 38, pp. 456–461, April 1990.
8. Mung Chiang, To layer or not to layer: Balancing transport and physical layers in wireless multihop networks, *IEEE INFOCOM'04*, 2004.
9. A. Ephremides, B. Hajek, Information theory and communication networks: An unconsummated union, *IEEE Transactions on Information Theory*, vol. 44, no. 6, pp. 2416–2434, 1998.
10. V. Kawadia, P.R. Kumar, A cautionary perspective on cross-layer design, *IEEE Wireless Communication*, February 2005.
11. Arunabha Sen, Mark L. Huson, A new model for scheduling packet radio networks, *Wireless Networks*, vol. 3, pp. 71–81, 1997.
12. Dave Wisely, Philip Eardley, Louise Burness, *IP for 3G: Networking Technologies for Mobile Communications*, John Wiley, 2002.
13. Peter Varbrand, Di Yuan, Patrik Bjorklund, Resource optimization of spatial TDMA in ad hoc radio networks: A column generation approach, *IEEE INFOCOM'03*, 2003.
14. A.M. Chou, V.O. Li, Slot allocation strategies for TDMA protocols in multihop packet radio networks, *IEEE INFOCOM'92*, 1992.
15. B. Hajek, G. Sasaki. Link scheduling in polynomial time, *IEEE Transactions on Information Theory*, vol. 34, no. 5, pp. 910–917, September 1988.
16. J. Gronkvist, Traffic controlled spatial reuse TDMA for multihop radio networks, *IEEE PIMRC*, vol. 3, pp. 1203–1207, 1998.
17. Leonid G. Khaciyan. A polynomial algorithm for linear programming, *Doklady Akademii Nauk SSSR*, vol. 244, pp. 1093–1096, 1979.
18. Narendra Karmarkar, A new polynomial-time algorithm for linear programming, *Combinatorica*, vol. 4, pp. 373–395, 1984.

19. George B. Dantzig, *Linear Programming and Extensions*, Princeton University Press, Princeton, 1963.

20. P. Raghavan, Probabilistic construction of deterministic algorithms: Approximating packing integer programs, *Journal of Computer and System Sciences*, vol. 37, pp. 130–143, 1988.

21. P. Raghavan, C.D. Thompson, Randomized rounding: A technique for provably good algorithms and algorithmic proofs, *Combinatorica*, vol. 7, pp. 365–374, 1987.

22. R.E. Steuer, *Multiple Criteria Optimization: Theory, Computation and Application*, Wiley & Sons, New York, 1986.

23. S. Arora, A. Frieze, H. Kaplan, A new randomized procedure for the assingnment problem with applications to dense graph arrangement problems, *Journal of Mathematical Programming*, vol. 97, no. 1–2, pp. 43–69, July 2003.

24. Seung-Jong Park, Raghupathy Sivakumar, Quantitative analysis of transmission power control in wireless ad-hoc networks, *International Conference on Parallel Processing Workshops (ICPPW'02)*, 2002.

25. A. Ruszcyński, A. Shapiro, Stochastic programming, *Handbooks in Operational Research and Management Science*, vol. 10, Elsevier, 2003.

26. Katerina P. Papadaki, Warren B. Powell, An adaptive dynamic programming algorithm for a stochastic multiproduct batch dispatch problem, *Naval Research Logistics*, vol. 50, no. 7, pp. 742–769, 2003.

27. S. Ramanathan, A unified framework and algorithm for channel assignment in wireless networks, *ACM Wireless Networks*, vol. 5, 1999.

28. P. Gupta and P.R. Kumar, The capacity of wireless networks, *IEEE Transactions Information Theory*, March 2000.

29. Jaehyun Yeo, Heesoo Lee, Sehun Kim, An efficient broadcast scheduling algorithm for TDMA ad hoc networks, *Computers & Operations Research*, vol. 29, pp. 1793–1806, 2002.

30. A. Behzad, I. Rubin, On the performance of graph-based scheduling algorithms for packet radio networks, *IEEE Globecom, San Francisco, CA, USA*, vol. 1–5 pp. 3432–3436, December 2003.

31. J. Grnkvist, A. Hansson, Comparison between graph based and interference-based STDMA scheduling, *In Proc. ACM MOBIHOC*, 2001.

32. F. Kelly, A. Maulloo, D. Tan, Rate control for communication networks: Shadow prices, proportional fairness, and stability, *Journal of Operations Research Society*, vol. 49, no. 3, pp. 237–252, March 1998.

33. S.H. Low, D.E. Lapsley, Optimization flow control, I: Basic algorithm and convergence, *IEEE/ACM Transactions on Networking*, vol. 7, no. 6, pp. 861–874, 1999.

34. Jiantao Wang, Lun Li, Steven H. Low, John C. Doyle, Cross-layer optimization in TCP/IP networks, *IEEE/ACM Transactions on Networking*, vol. 13, no. 3, pp. 582–568, June 2005.

35. Leonidas Georgiadis, Michael J. Neely, Leandros Tassiulas, Resource allocation and cross-layer control in wireless networks, *Foundations and Trends in Networking*, vol. 1, no. 1, pp. 1–144, 2006.

36. L. Chen, S.H. Low, M. Chiang, J.C. Doyle, Cross-layer congestion control, routing and scheduling design in ad hoc wireless networks, *IEEE INFOCOM*, Barcelona, Spain, April 2006.

The last decade has shown a phenomenal growth in wireless communication technologies. At the same time, Internet-like multimedia applications are becoming more and more attractive to mobile users. Wireless mesh networks (WMNs) currently appear to be a promising evolution of traditional wireless communications, and recent research efforts addressed the main challenges posed by WMNs. For instance, new routing, autoconfiguration, and self-healing strategies have been proposed. However, limited bandwidth, scarcity of wireless channels, and multihop connections coupled with a highly dynamic topology pose a severe challenge to the quality and interactivity levels of multimedia communications. This chapter will first analyze the quality of service (QoS) requirements of multimedia communications, and how mesh network environments are challenging from the multimedia delivery perspective. Then recent advances at the transport level will be summarized, focusing on the most important proposals aimed at maximizing the quality experienced by the user. In particular, protocols will be evaluated with respect to the QoS guarantees that they can provide for multimedia applications. Finally, we will analyze the architectural requirements that are necessary to deploy successful innovative multimedia services in wireless mesh scenarios. While analyzing current solutions and proposals, particular attention will be given to open problems and challenges.

10.1 INTRODUCTION

A number of multimedia services are already available over wireline packet networks such as the Internet, ranging from interactive voice communications (voice over Internet Protocol [VoIP]) to live and on demand multimedia streaming. At the same time, rapid diffusion of wireless devices is increasingly shifting the users toward wireless technologies. Significant challenges must be addressed to provide successful real-time multimedia communications in wireless scenarios due to the intrinsic unreliability of wireless channels. Users moving to wireless communication networks, however, expect the same level of performance as their wireline counterparts. Therefore, much research effort has been devoted to investigating how to optimally address the challenges posed by wireless multimedia communications.

WMNs, in particular, require to address a number of additional issues compared to traditional wireless networks, such as the potentially high number of traversed hops, which may negatively affect the performance of real-time multimedia applications. However, peculiar characteristics of mesh networks such as the presence of a potentially large number of densely interconnected nodes might be exploited by new transmission schemes to overcome the limitations caused by the unreliability of wireless channels and the highly dynamic behavior of network nodes.

This chapter first analyzes the typical communication and robustness requisites of real-time multimedia applications to understand the main issues involved in designing successful multimedia applications over WMNs. Then, a review of the current research efforts focused at providing QoS for multimedia applications at the transport level is presented, highlighting open research issues and challenges. Finally, some innovative multimedia applications over WMNs are examined, with particular attention to the architectural requirements and open issues.

10.2 MULTIMEDIA CHARACTERISTICS AND QUALITY OF SERVICE REQUIREMENTS

10.2.1 Communication Requirements

In contrast to generic data communications, multimedia communications have more stringent QoS requirements that must be fulfilled to provide an acceptable service. In particular, multimedia traffic is characterized by strong time sensitivity and inelastic bandwidth requirements. Multimedia packets, in fact, must be available at the decoder before their playback time (deadline) to allow an undistorted media reconstruction. Packets that are not received before their deadline become useless. Excessive end-to-end delays might negatively affect user experience as well, for instance, impairing the ability to effectively interact with other users. Regarding bandwidth requirements, multimedia data are generally encoded at a fixed data rate, as in the voice and audio cases. If the bandwidth required by the compressed data exceeds the channel capacity, packet losses will occur causing distortions in the decoded data.

Figure 10.1 shows the main components of a generic multimedia communication system. In addition to the encoder/decoder elements,

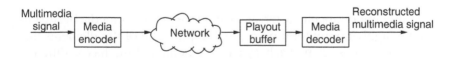

Figure 10.1   Typical multimedia transmission chain.

a playout buffer called dejitter buffer is generally introduced before the media decoder to compensate the unequal packet delays caused by the network. The playout buffer allows to trade off a reduction of excessively delayed packets for an increase of the overall end-to-end delay.

From a communication point of view, multimedia applications can be divided into two classes:

1. Interactive applications, such as voice communications and videoconferencing
2. Streaming applications, such as audio and video on demand (VoD), live streaming, and video surveillance

The first group includes all the applications that require some form of interaction between the end nodes. This group of applications is very demanding in terms of delay requirements. For instance, the typical maximum allowed end-to-end delay to achieve a transparent interaction is ~150 ms [1]. Such a stringent requirement usually implies that the whole system, including sender and receiver nodes, must be designed and optimized to fulfill the maximum delay constraint. If an error control strategy is used, for instance, it should rely on low-delay mechanisms, such as forward error correction (FEC). Conversely, high-delay mechanisms such as end-to-end retransmission requests should be avoided.

For streaming applications an end-to-end delay of a few seconds is usually acceptable; therefore, the effect of packet losses and errors may be mitigated using relatively slow but very efficient automatic repeat request (ARQ) mechanisms, often combined with FEC schemes [2,3]. Moreover, in streaming applications the relaxed delay requirements allow the transmitter to perform complex optimizations. For instance, rate-distortion optimized packet scheduling and retransmission [3] based on the importance of the packets from the multimedia decoding perspective have been shown to optimize network resource utilization as well as maximize performance.

10.2.2 Robustness Issues

Another important characteristic of compressed multimedia bitstreams is the highly nonuniform perceptual importance of the elements that compose the stream. Each syntax element, if lost, has a different perceptual impact on decoded data. This characteristic has to be carefully considered in multimedia communication systems, otherwise

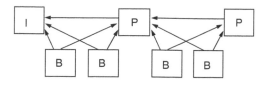

Figure 10.2 Typical data dependency in a compressed MPEG video stream.

it is possible that a few but unlucky errors might cause a significant distortion in the decoded media. As an example, consider the moving picture experts group (MPEG) coded video stream shown in Figure 10.2. If an I-type frame is lost, the number of potentially affected frames is very high, i.e., all the frames until the next I-type frame, while the loss of a B-type frame does not affect any other frame.

A number of error control strategies have been proposed to enhance the error resilience of multimedia communications. The introduction of error-resilient coding modes during the encoding process helps to limit the artifacts caused by errors and packet losses [4]. At the transport level, channel coding techniques based on error-correcting codes can greatly reduce the residual error and packet loss rate. At the same time their use must be carefully tuned because they reduce the bandwidth available for the multimedia encoder.

Application and transport level aspects can also be jointly taken into account. This has led to a number of optimized schemes. For instance, in the widely studied unequal error protection (UEP) schemes important elements are protected more effectively than others to reduce the probability that errors cause a significant distortion in the decoded media data [5,6]. Other schemes take advantage of the different QoS levels offered by the network [7,8]. Finally, some techniques aim to optimize the system as a whole, concurrently exploiting both application level error-resilient coding modes and transport level protection schemes, leading to the so-called joint source-channel coding [9–11].

10.2.3 Perceived Quality Evaluation

Assessment of the quality as perceived by the user is usually carried out to evaluate the performance of a multimedia communication system. Users' satisfaction is, in fact, the ultimate metric to evaluate a multimedia service. Given the nonuniform importance of the various elements in a compressed multimedia bitstream, network-level

with the increasing size and complexity of multihop mesh networks. While data flows (web browsing, email delivery, file transfer) may be almost arbitrarily curtailed and still be useful, multimedia communications are more demanding in terms of QoS. In fact, if delay, bandwidth, or packet loss rate are not within a given range, delivering of voice or video data is of no use, e.g., voice communications may lose intelligibility. However, before considering specific multimedia applications and how QoS may be pursued, it is recommended to study and understand the major technical challenges of mesh networking.

In Section 10.3 and 10.4, we focus on 802.11 networks, which appear to be the most promising networking technology for multimedia services over WMNs [18,19]. In fact, the IEEE 802.11 working group is very active in the standardization of new interoperable 802.11-based standards that, in the near future, will provide some interesting capabilities for multimedia communications such as speeds up to 100 Mb/s (and above), QoS support, fast handoff, and mesh functionalities. Particularly relevant to the mesh networking is the development of the 802.11s standard by the extended service set (ESS) mesh networking task group. Other IEEE 802 working groups are currently involved in the definition of mesh networking extensions to the wireless standards (e.g., 802.15.5, 802.16a, and 802.20). However, with regard to multimedia transmission, and in particular to real-time and interactive services such as videoconferencing systems, very little attention has been devoted to technologies other than 802.11. In fact, wireless sensor networks with low data rates, as well as wireless personal area networks with limited resource devices and short-range communications, severely limit the scenarios for multimedia communications. For example, high data rate 802.15 ultra-wideband (UWB) technology is mainly intended for high-speed wire replacement for consumer electronics. Probably of more interest is the proposal submitted to the 802.16 standard committee for the medium access control (MAC) layer mesh extension of the WiMax point-to-multipoint architecture [20]. However, additional research and standardization work is needed to bring the full benefits of mesh architecture to 802.16/WiMax [21]. Nowadays, solutions based on 802.11 appear more advantageous because they are widely deployed and operate in unlicensed cost-free frequency bands. Nevertheless, it is possible that in the future Wi-Fi and WiMax will be integrated together so that 802.16 wireless links will introduce additional capacity in the mesh and expand the network coverage [22].

10.3.1 Network Capacity

Wireless networks based on the 802.11 standards are widely deployed in homes, enterprises, and public hot spots. Maximum nominal data rates vary from 11 Mbps for 802.11b to 54 Mbps for the 802.11g and 802.11a standards. An additional task group 802.11n is working on higher maximum data rates estimated in a theoretical value of 100–200 Mb/s [23] using multiple transmitter and receiver antennas (multiple-input multiple-output [MIMO] technology). However, the maximum achievable throughput for 802.11 networks is far lower than the nominal data rate due to the nature of the wireless channel. For example, in 802.11b the maximum experimental throughput is ~6.2 Mb/s [24] and decreases as more stations are connected. In particular, when we consider the maximum possible number of VoIP connections in a single 802.11b cell, the protocol inefficiency plays an even more significant role. The fixed overhead for the transmission of 10 ms speech packets limits the number of bidirectional 64 Kb/s G.711 voice calls to six, with an overall throughput of just 768 Kb/s over the nominal 11 Mb/s data rate [25]. In Section 10.4.2, we will see that in this case the packet-sending rate plays a more important role than the transmission rate in determining the network capacity. Thus multimedia communications in wireless networks require adequate study to mitigate the effect of packetization, and admission control policies may be necessary [26].

First-generation wireless network architectures were based on infrastructure access points (APs) or on direct communication between nodes. Nowadays, network nodes are gaining the ability to freely connect among themselves, operating not only as host but also as router. That is, they can forward packets on behalf of other nodes that may not be within the direct wireless transmission range of their destination. As a consequence, in a mesh network a packet destined to a node in the network may hop through multiple nodes to reach its destination. Analysis of the capacity of such networks [27] shows that they suffer from scalability issues, i.e., when the size of the network increases, their capacity degrades significantly, almost proportionally to the number of nodes. Moreover, research demonstrated that, to maximize the network performance in terms of bandwidth, a node should communicate with nearby nodes only [28]. However, in this case the large number of consecutive hops required to deliver the packets may severely limit the QoS experienced by real-time multimedia communications, especially with regard to performance metrics such as end-to-end delay, jitter, and packet loss ratio.

10.3.2 Network Latency

As multimedia applications become a key driver for the deployment of mesh networks, this converged scenario supporting data services as well as voice and video applications will start showing very little tolerance for network latency and jitter. When a packet traverses the network from node to node, processing and transmission delays are naturally introduced; latency of several milliseconds per hop may preclude, after few hops, delay-intolerant applications such as voice and real-time interactive video.

This problem is mainly due to the single-radio channel nature of early-generation mesh networks where each node operates in half-duplex mode and shares the same radio frequency, i.e., all radios on the same channel must remain silent until the packet completes its hops within the same collision domain. Use of more sophisticated, and more expensive, multiradio mesh networks can increase the system scalability with the creation of a wireless "backbone" called backhaul network, which interconnects all nodes and handles traffic between nodes [29]. However, while excessive hop counts can be sometimes minimized through good network architecture design, this is not the case for spontaneous, unstructured, ad hoc mesh networks that require support at the application level to mitigate the effect of excessive delay and jitter.

Experimental studies [18] evidence how packet jitter variations are significant in current 802.11 networks. In the University of California, Santa Barbara (UCSB) MeshNet test bed experiments consisting of video streams and voice traffic evaluated the performance of multimedia data delivery through multihop wireless paths. Results confirmed that as the length of the transmission path increases the multimedia transmission performance degrades, and the latency and loss rate increase. Average packet latency for video traffic that may allow 10 concurrent flows when a single-hop transmission is considered, rapidly drops to only two reliable flows when the number of hops increases to three (with 150 ms as the delay threshold for interactive applications). New solutions are then needed to dampen the overall delay, and delay variation for real-time traffic delivery over WMNs.

10.3.3 Handoff

As shown in Figure 10.3, one of the main characteristics of mesh networks is that they have only a few wireless gateways (WGs) connected to a wired network while the wireless routers (WR) provide network access to mobile clients (i.e., they act as APs to the clients). Within the range of a given WR the client may move freely, but as it moves away from the WR and closer to another, it should hand off all its open connections to the new one to preserve network connectivity. Ideally the handoff should be completely transparent to mobile clients with no interruption, loss of connectivity, transmission "hiccups," or degradation of voice quality if VoIP communications are involved.

In cellular data and voice systems the handoff problem is typically coordinated by the network itself using signaling embedded in the low-level protocols that are able to leverage considerable information about the network topology and client proximity. In contrast, 802.11 networks lack efficient and transparent handoff solutions. Consequently, as a mobile 802.11 client reaches the limits of its current coverage region inside the mesh, it must abandon its current WR, it must actively probe the network to discover alternatives, and only then it can reconnect to the current best WR. For some applications (e.g., file transfer, email delivery, and web browsing) this delay is acceptable; however, it may be too long for real-time traffic such as voice-over IP and videoconferencing.

To become widely deployed, wireless mesh telephony services need clients who are able to rapidly disassociate from one WR and connect to another. In this particular case, the maximum allowed delay that can occur during handoff cannot exceed 50 ms, which is the interval detectable by the human ear. The 802.11r working group [30] of the IEEE is drafting a protocol that will facilitate the deployment of IP-based telephony over 802.11-enabled phones by speeding up handoffs between APs (previously called WRs in the WMN context). Under 802.11r, clients will be able to use the current AP as a conduit to other APs. APs will continuously monitor the connectivity quality of any client in their vicinity. This information is then efficiently shared with other APs in the vicinity of that client to coordinate those that should serve the client.

The major factors affecting latency associated with layer 2 roaming are scanning, reassociation, and re-authentication. Since 802.11 does not provide a shared control channel, the client must explicitly scan each channel for potential APs. Passive scanning is not feasible because in this case the client should try to listen to periodic beacon frames that are typically sent every 100 ms leading to a latency well over 1 second. Active scanning is generally used for voice-over WLAN where the client actively broadcasts a probe packet on each channel to force an AP to respond immediately. Even in this case, however, scanning all the 802.11b/g channels could exceed the requirements of real-time communications, e.g., taking ∼20–300 ms. To overcome this problem, preemptive scanning is preferred [31], where the station prescans at regular intervals the existence of an AP on a single channel while it is still associated with the current AP.

Nevertheless, in the case of voice communications, where packets are sent every 20 ms or less, a significant increase in latency and jitter is introduced every time the client probes a new channel. As a consequence, to seamlessly handle a handoff between two WR, it is important that the client dejitter buffer is capable of rapidly increasing the delay without audible distortion as described in Section 10.4.2.

Another problem with current 802.11 wireless mesh implementations is that a mobile device cannot know if necessary QoS resources are available at a new AP until after the handoff. Thus, it is not possible to know whether a transition will lead to a satisfactory application performance. As mentioned earlier, while data flows may be curtailed and still be useful, multimedia flows need to be given the full resources they desire. In fact, placing an additional call that exceeds the capacity of the wireless network will result in unacceptable quality for all ongoing multimedia calls. Thus, taking into account the low number of possible interactive multimedia connections, the need for admission control is apparent. Specifically, the 802.11e standard [32] deals with QoS for wireless networks and it introduces the support for admission control by means of a component called Wi-Fi multimedia (WMM).

10.3.4 Network Routing

Different approaches to wireless mesh networking are possible, but it is intrinsic to all mesh networks that user traffic must travel through several nodes before exiting the network (e.g., to reach the wired local area networking [LAN]). The number of hops that user traffic must traverse to reach its destination will depend on network design, length of the links, technology used, and routing protocol. Two common approaches to construct mesh networks are the structured and the unstructured approach. In the structured approach, the multihop network is carefully deployed with nodes in chosen locations and directional antennas might be aimed to provide good connectivity, high-quality radio links, high throughput, and high performance, but this requires well-coordinated groups with technical expertise. In the unstructured approach, mesh networks are completely spontaneous and aim at operating without central planning or management, but still providing wide coverage and acceptable performance. Based on omnidirectional antennas and multihop routing to improve coverage and performance despite the lack of a centralized configuration and planning, they should rely on optimized routing for throughput maximization to route data through whichever neighbors a node can transmit to.

Routing protocol is in charge of maintaining information on the topology of the network to calculate routes for packet forwarding. These algorithms may be classified in proactive, e.g., optimized link state routing (OLSR) [33] and topology broadcast–based reverse-path forwarding (TBRPF) [34], and reactive, e.g., ad hoc on demand distance vector routing (AODV) [35] and dynamic source routing (DSR) [36]. Proactive algorithms find routes to all neighbors ahead of time and have them ready when needed. In contrast, reactive routing mechanisms only seek to find routes to destinations if they are needed. On the one hand, the former approach definitely introduces more overhead and it can cause a waste of resources when a path to a certain node is calculated but never used. On the other hand, proactive protocols provide more responsive forwarding of multimedia packets since a packet to be forwarded can be sent immediately because a path is always available. This cannot be guaranteed with reactive protocols. With a reactive approach, in fact, a packet cannot be forwarded until a path is found from the sender to the receiver [37].

Considerable research has addressed the problem of routing in wireless multihop networks [38]. The routing mechanism may choose to use information about the underlying topology of the network to collect count of hops or distances of each node to all the other nodes, or to determine where nodes are connected to each other. Some proposals utilize the shortest hop count metric as the path selection metric. This metric has been shown to result in poor network throughput because it favors long, low-bandwidth links over short, high-bandwidth links [39]. Recently, proposals have aimed to improve routing performance by utilizing route selection metrics [40], which considers not only the throughput, but also the contribution of both bandwidth and delay. These advanced metrics, such as estimated transmission count (ETX [41]) and estimated transmission time (ETT [42]), may successfully be applied for real-time voice and video services. For instance, ETT predicts the total amount of time it would take to send a data packet along a route, taking into account each link highest-throughput transmit bit rate and its delivery probability at that bit rate. ETX and ETT metrics achieve very good performance especially for stationary nodes. The minimum hop count method, however, has been shown to perform well in scenarios where nodes are highly mobile. The reason is that, as the sender and receiver move, link quality metrics cannot quickly track the change in the link quality. However, better performance metrics needs to be discovered and utilized to improve the performance, and routing protocols integrating multiple performance metrics are necessary for WMNs.

When considering real-time multimedia services, strict requirements in terms of latency, jitter, and packet loss rate need to be imposed on the underlying network. To meet these demands QoS extensions should be introduced to improve the performance of the existing routing protocols [43]. As the ability to provide QoS is heavily dependent on the resource management at the MAC layer, for each flow the QoS routing protocol must interact with it to find both the route and the resources for each link on the route. In situations with high load, real-time packets risk becoming obsolete, so priority queuing should be employed to make sure that real-time packets are transmitted ahead of other packets. Still, real-time packets might have to wait too long in the high-priority queue. Thus, to prevent the transmission of outdated packets, hop-constrained queue timeouts may drop obsolete real-time packets and save bandwidth. In addition, neighbor-aware rate control policies may be used to limit the amount of low-priority traffic. If some nodes have high-priority traffic to be sent, other nodes that are sending low-priority traffic may be temporarily excluded from the occupation of the communication channel. Comparisons of different routing mechanisms and their effect on QoS have been discussed further [44]. At present, there is no perfect protocol for QoS, but understanding how each protocol affects quality in a WMN is important to design a reliable and robust QoS framework.

The mesh topology of ad hoc networks allows the existence of multiple routes between two endpoints. Such routes may be utilized (together with the previously proposed solutions) by multipath routing techniques to increase the user's perceived quality in multimedia transmission. These techniques enable a source to discover several alternate paths to the destination, all of which are capable of providing the required QoS. If the current path becomes unusable, the traffic flow can then quickly switch to one of the alternate paths without loss of performance. Thus, without waiting for setting up a new routing path, the end-to-end delay, jitter, throughput, and fault tolerance can be improved. Multipath multimedia streaming has several advantages. First, it can potentially provide higher aggregate bandwidth to real-time multimedia applications (given that the bottleneck is not shared by the paths). Second, data partitioning over multiple paths can reduce the short-term correlation in real-time traffic thus improving the performance of multimedia streaming applications. Third, the existence of multiple paths can help to reduce the chance of interrupting the service due to node mobility [45]. Clearly, monitoring of each alternate path is required to ensure that each one can provide the required QoS [46].

Multihop ad hoc networking, high traffic load, lack of coordination among nodes, no facility for route reservation or clustering, and other issues all contribute to a challenging scenario for real-time multimedia communications. Nevertheless, future advances in the aforementioned areas and improved techniques at the application level (as discussed in the following sections) will surely make mesh networks a viable approach to handling real-time traffic in wireless environments.

10.4 INNOVATIVE MULTIMEDIA APPLICATIONS

The interest in WMNs is rapidly growing. One of the reasons is that they can provide connectivity in scenarios and conditions that cannot be easily supported by other architectures. In future, the new functionalities offered by WMNs will be exploited by a number of advanced applications ranging from traditional multimedia applications to innovative ones, such as intervehicle multimedia communications and multiplayer gaming.

However, typical users expect to be able to use multimedia services regardless of the technological infrastructure supporting the communication. Therefore, a great deal of effort has been devoted to provide a satisfactory QoS level in WMNs despite the harder constraints posed by that architecture. In this regard, peculiar features of the mesh scenario such as the potential presence of multiple paths between source and destination are useful to achieve an acceptable performance for multimedia services and to provide a graceful degradation in case of unreliable network conditions.

The flexibility offered by the WMN architecture is useful to design innovative multimedia applications. For instance, intervehicle communications and multiplayer wireless gaming are two promising applications. A WMN architecture, in fact, can be employed opportunistically to create a network between vehicles, as well as to provide vehicles with Internet connection using roadside APs. A WMN architecture can also be useful for multiplayer gaming applications. However, in this scenario QoS issues play a critical role in users' experience; hence, a careful design is needed to ensure a successful deployment of multiplayer gaming applications over WMNs.

In Section 10.4.1 the main architectural requirements concerning the support of streaming applications in WMNs are analyzed. Then, VoIP applications support in WMNs is discussed, emphasizing the most suitable communication schemes. Finally, the main issues posed by the previously mentioned innovative applications, namely intervehicle communications and multiplayer gaming, are investigated from the point of view of WMNs while pointing out open research problems.

10.4.1 Streaming Services

The number of users of multimedia streaming services has grown dramatically in recent years. Typical users expect to be able to use streaming services regardless of the technological infrastructure supporting the communication. Previous sections highlighted the most significant challenges posed by the WMN architecture to real-time multimedia transmissions. However, peculiar characteristics of WMNs such as the presence of a potentially large number of densely interconnected nodes might be exploited by new transmission schemes to overcome the limitations caused by the unreliability of wireless channels.

Routing operations in WMNs might be challenging due to various factors, including node mobility and variability of wireless channels. Thus a number of routing protocols have been developed for wireless mesh scenarios. Interestingly, some of them [47] can maintain several partially independent routes. Multimedia streaming services can exploit the presence of multiple transmission paths between nodes in the network to improve their performance in terms of reliability and quality.

One of the main issues when multiple paths are involved is how to optimally exploit them. The application level coding technique known as multiple description coding (MDC) [48] has been developed to take advantage of the availability of multiple paths. An MDC scheme encodes a multimedia source into a number of independent descriptions, with the property that a reduced quality version of the multimedia content can be decoded even from a single description. Combining more than one description could increase the quality of the decoded data, until reaching the full quality if all descriptions are used.

When an MDC scheme is employed for multimedia communications over packet networks, each description is sent on a different

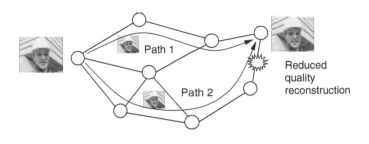

Figure 10.4 Multiple description coding communication in a wireless mesh network (WMN).

path (Figure 10.4). Better the independence between different paths, better will be the performance because the probability to lose all the description of a given element is minimized. Moreover, MDC schemes are well suited for networks in which paths are frequently subject to change due to the dynamic behavior of network nodes. Figure 10.4 shows an example in which path 2 is suddenly interrupted. Using an MDC coding scheme, even if some paths are interrupted and routes need to be recomputed—a potentially time-consuming task—the receiver can continue to decode the data coming from the other paths, providing a reduced yet acceptable quality.

In the context of MDC, the work of Vetterli et al. [49] considers all combinations of single and MDC schemes with single or multipath routing approaches. The results show that the most sophisticated scheme, i.e., MDC and multipath routing, performs significantly better than the usual single-path and single-description scheme. The performance improvement is higher in case of low rates and with stringent maximum delay constraints.

Coding techniques different from MDC, such as layered coding, can also take advantage of multiple paths in WMNs to combat transmission errors. For instance, the work of Mao et al. [50] compares the performance of various transmission schemes in the context of multipath video transmission. The results show that great improvements in video quality can be achieved over the standard schemes with limited additional costs. The work identifies the most suitable coding and transport scheme in each tested environment depending on the end-to-end delay constraints, the error characteristics of each path, and the availability of a feedback channel.

The work of Setton et al. [51] investigates how to optimally subdivide multimedia traffic over different paths, while taking into account the different degrees of reliability and average delay that the various paths can offer. A framework to optimize low-latency video streaming over networks that offer multiple paths is introduced. Moreover, the work of Setton et al. [52] shows that taking into account the overall impact of the actions of each node in ad hoc or mesh networks is important to achieve good communication performance.

10.4.2 Interactive Voice Services

The expanding interest in WMNs is creating a great deal of effort to support interactive voice services. The challenges in deploying VoIP over WMNs stem mainly from issues related to network congestion, delay, and link quality. When several VoIP users are connected to the same network, delay, jitter, and packet loss can be very significant, especially because of the large number of data packets VoIP communications send through the network, i.e., about 100 packets/s. Thus, the efficiency of the system quickly deteriorates when the number of calls increases.

A natural question to be answered is how many voice calls can be supported in a wireless 802.11 environment. As shown in the literature this question does not have a unique answer. Rather, the capacity is strictly related to the channel bandwidth, voice codec, packetization interval, and data traffic in the system [53]. Considering a typical communication with G.711 speech coded at 64 Kb/s and packetized in frames of 10 ms, a huge amount of overhead is introduced by the network protocol. Experiments [25] as well as theoretical calculations [54] show that in an 802.11b network the maximum throughput when using only VoIP traffic can be as low as 800 Kb/s using the G.711 VoIP codec. Consequently, the maximum number of simultaneous VoIP calls that can be placed in a WMN cell is ~6. High-bandwidth systems such as 802.11a and 802.11g can offer a capacity that is approximately four times that of 802.11b. Interestingly, using a higher compression speech codec does not significantly increase the number of channels that could be handled. For example, although the output bit rate of a G.711 encoder is eight times that of G.729 encoder, the reduction in capacity when G.711 is used is less than 50% [54]. The reason is that node congestion depends more on the number of packets that need to be processed than on the actual bandwidth. Voice packets are small and are sent very frequently, which explains the low throughput. Because of this limitation it is useful to put more than one voice frame into the same packet. This technique reduces the number of packets and hence increases the throughput, but, as a result, the delay increases. While the channel efficiency improves as more frames are included in the same packet, it tends to saturate as the packetization interval is increased. Moreover, increasing the packetization interval beyond 80–100 ms would drastically interfere with the end-to-end delay budget of 150 ms [55]. Figure 10.5 plots the optimal choice of coding rate and the number of packets per frame that optimize the conversational quality with respect to the available channel bandwidth. Experiments refer to an IEEE 802.11b WLAN link.

So, even if it may seem that packets should contain as much voice data as possible for improved channel capacity, other call quality requirements such as delay constraints and sensitivity to channel errors must be considered. In fact, long packets are more susceptible to errors than short packets, the packetization delay is higher, and the loss of such packets is also harder to be concealed than smaller packets [56].

It may be observed that the number of hops traversed in a VoIP transmission influences the communication performance. Experiments

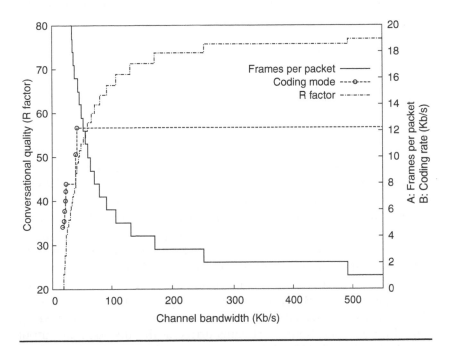

Figure 10.5 Optimal coding rate and packetization for voice transmission on an IEEE 802.11b wireless link. (From Mahlo, C., Hoene, C., Rostami, A., and Wolisz, A., "Adaptive coding and packet rates for TCP-friendly VoIP flows," in *Proceedings of 3rd International Symposium on Telecommunications (IST 2005)*, Shiraz, Iran, September 2005. With permission.)

in the UCSB WMN show that, as the length of the transmission path increases, the performance degrades and the latency and loss rate increase [18]. The network capacity is clearly constrained by the number of hops, i.e., the acceptable number of voice flows drops from around ten (with a single hop) to one when four hops are traversed. Additionally, traffic delay variations, that with two hops and four voice flows range from 5 to 200 ms, increase with more concurrent flows and severely impact received video/voice quality. Clearly new solutions are needed to dampen the delay variation of real-time traffic delivery. For example, results show that the maximum number of voice transmissions can be increased when medium access time is reduced by means of a service differentiation mechanism applied to the MAC layer [57]. Implementing QoS mechanisms is, in fact, a powerful approach to improve the capacity of WMNs for VoIP traffic. The goal is to introduce additions to the current 802.11 MAC layer that take into account the different requirements of regular data traffic and time-sensitive voice traffic. Prioritization of voice (and video) packets is the first level of QoS that is typically introduced [57]. However, the results of recent case studies suggest that prioritization is not enough since only a few VoIP calls could be carried even if no data traffic were present. Therefore, to be successful, WMN protocols must compliment QoS work with changes to the 802.11 standards that will reduce the jitter and delay by changing the transmission scheme [32].

To exploit the availability of multiple paths in a WMN, many multipath routing protocols have been proposed for real-time communications [58]. In this scenario, characterized by severe network conditions and multiple connections between nodes, MDC appears to be a very promising approach. Although in the literature relatively little attention has been given to multiple description speech coding [59,60], multiple paths in WMNs may also be exploited to increase voice communication quality. Using a multiple description codec for voice communication over MANETS was first suggested by Dong et al. [61], where the authors proposed to use MDC during severe channel conditions. The encoder splits the bitstream of the adaptive multirate wideband (AMR-WB) encoder into two redundant substreams by directly selecting overlapping subsets of encoded data generated for each frame. When both substreams arrive at the decoder, an output identical to that of AMR-WB is recovered. If only one substream arrives at the decoder, degraded but still acceptable speech quality is obtained. For further investigations on this approach refer to [62].

When roaming around in a wireless network, the link quality and the transmission delay vary rapidly as a result of the movement [63].

Additionally, at some point the mobile device will have to switch the AP it is associated with, otherwise mobility can cause a change in the routing path to the destination. In a WMN it is common that such a change introduces tens of milliseconds of delay, which have a very audible impact on the call quality. This is a severe drawback, since a WMN is often introduced to provide increased mobility and a wireless VoIP user can be expected to roam around the coverage area. The result of mobility issues and bad link quality is that packets are lost or arrive too late to be useful. To seamlessly handle late packets, it is important that the receiver dejitter buffer is capable of accommodating for sudden changes in the end-to-end delay. Therefore, dejitter buffers with dynamic size allocation, the so-called adaptive dejitter buffers, are becoming very common [64].

A new algorithm has recently been introduced that combines an advanced adaptive dejitter buffer control with audio time-scaling [65]. As shown in Figure 10.6, this approach allows to optimize the trade-off between short buffering delay and robustness to late packets by adaptively adjusting the dejitter buffer size to the network conditions so that the end-to-end delay is always kept as small as possible. As a consequence, dynamic adaptation of the dejitter buffer size requires changes to the playout scheduling of audio frames. The continuous output of high-quality audio is then assured by expanding or

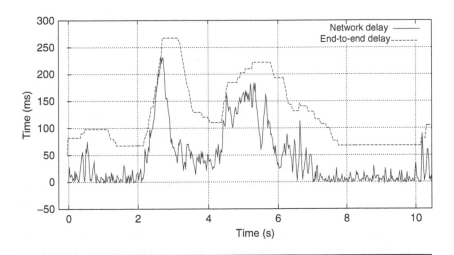

Figure 10.6 Dejitter buffer control algorithm adapting the buffering delay (the gap between the two lines) to the network delay. Frames duration is scaled accordingly using an audio time-scaling algorithm.

compressing the audio frames with the waveform similarity overlap add (WSOLA) timescale modification technique that can change the frame duration without altering the audio frequency content [66]. As a consequence, the media is played slower than in real-time when the buffer occupancy is below a desired level and faster when the channel conditions are good to eliminate excessive latency accumulated during bad channel periods. In a typical mesh scenario providing uninterrupted communications requires maintaining client-side buffer of proper size with enough audio content to play during the handoff process until the mobile node is reconnected to the new wireless mesh router. Handoff prediction [67] together with audio time-scaling can enable the adaptive management of client buffers by increasing the buffer size (of the amount expectedly needed), and consequently slowing down the playout, in anticipation of client handoffs. When the handoff is completed, the delay may be quickly reduced again for best interactivity and reduced latency.

10.4.3 Intervehicle Communications

The presence of wireless devices on public and private vehicles is expected to rapidly increase in future. For instance, some airplanes and trains already allow Internet access using the 802.11 protocol by onboard passengers, and several cars already include an intravehicle wireless platform that allows easy integration of various devices, such as mobile phones, with the onboard systems. For the specific case of the automotive industry, intervehicle wireless communications are also expected to gain popularity in future, as shown by the numerous research projects that are currently under development [68–70] and by standardization efforts such as 802.11p [71]. Potential applications of intervehicle communications include multivehicle-based visual processing of road information, and multivehicle radar systems for obstacle avoidance and automatic driving. Intervehicle communication networks will also make a new class of applications possible, e.g., "swarm" communications among cars traveling along the same road, network gaming among passengers of adjacent cars, and virtual meetings among coworkers traveling in different vehicles.

Automotive networks will heavily rely on mobile nodes (vehicles) to perform message routing and network management functions. Figure 10.7 shows an example of intervehicle communications scenarios. Intervehicle networks, however, present a fundamentally different behavior when compared to generic MANET networks. A number

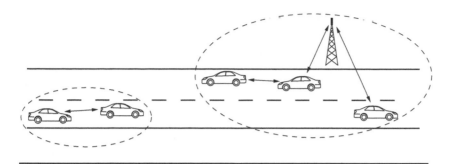

Figure 10.7   Example of intervehicle communications scenarios.

of factors, such as driver behavior, mobility constraints, and high speeds impact on the characteristics of intervehicle communication networks. With respect to the models currently used for MANETS, intervehicle networks present peculiar characteristics such as rapid but somewhat predictable changes in topology.

An important issue in intervehicle networks is the routing layer, because it can severely affect the performance of communications. Results of accurate simulations taking into account both vehicle mobility and wireless network behavior [72] show that rapid changes in intervehicle networks are difficult to manage, that network fragmentation is frequent, and that the effective network diameter is small. Moreover, the presence of redundant paths is often limited. Other results show that for an intervehicle network the number of nodes within the communication range grows linearly with the increase in the radio range, compared to a quadratic increase with a completely random mobility model.

Moreover, fixed nodes such as roadside APs can participate in network formation and communication (Figure 10.7). The work of Ott and Kutscher [73] investigates the main challenges posed by such scenarios. A number of experiments demonstrate that it is possible to communicate even at high relative speeds but the time interval in which the link is available is limited to a few seconds. It is, however, possible to take advantage of multiple connection points to increase the time interval in which roadside communication is possible, but devices as well as application protocols must be prepared to deal with such situations, e.g., performing fast handoffs. Other studies have assessed the performance of 802.11 wireless intervehicle communications in terms of throughput and signal-to-noise ratio (SNR) as a function of various parameters, such as the relative and absolute velocities and distances [74]. These results can constitute the basis for the design of several innovative applications.

As described earlier, providing multimedia services over intervehicle networks is a very challenging task. A cross-layer design should be employed to optimize all the parameters involved in the multimedia transmission system. Coding parameters, for instance, should be dynamically adapted to cope with the varying characteristics of the driving scenarios [75]. Bucciol et al. evaluate the characteristics of an intervehicle 802.11 wireless channel, in terms of link availability and SNR, in two scenarios: urban and highway. The results show that those parameters strongly depend on the scenario. For instance, the wireless devices can communicate for over 97% of the time in the urban scenario, while in the highway scenario the link is available for less than half of the time. In the highway scenario, however, the average SNR of the signal is higher compared with the urban one. On the basis of these results, an adaptive coding and packetization algorithm has been designed with the aim of creating small multimedia packets while driving in the urban scenario—which present a higher bit error probability due to the generally low SNR—and conversely larger packets in the highway scenario.

However, the intervehicle communications technology is in its infancy, and a lot of work is still needed to explore the full potential of this promising scenario. Currently, high-speed mobility of network nodes and the need to extend the network coverage by means of multihop communications appear to be the two main issues that currently need to be addressed by the research community.

10.4.4 Real-Time Multiplayer Games

Popularity of multiplayer games is rapidly increasing since the support of real-time online gaming has been introduced. Although a big business in today's Internet, it is expected to increase its revenue even more in future. Real-time multiplayer games have some requirements in common with multimedia applications. For example, critical information, such as the movements of the game characters, has to be delivered as fast as possible to the counterparts. High latency in delivering this data can clearly make the game not playable any more. In particular, some kind of games such as person shooters, real-time strategy games, or sport games, have the highest demands on QoS requirements and they can only tolerate round-trip delays up to 150 ms, with minimum jitter and low packet loss rate (e.g., below 5%) [76]. Bandwidth requirements are not so important and are generally easily satisfied.

Evolution of mobile devices with integrated wireless networking support is pushing the market toward new scenarios. Portable devices will no more be limited to providing multimedia communications services, and real-time multiplayer games will also be possible. Then spontaneous and self-organized wireless multihop mesh networks, where mobile game consoles can freely connect to each other, will become the preferred multiplayer game environment. But, differently from the wired Internet, wireless networks QoS mechanism are not yet ready to meet the demands of these applications.

An analysis on the effects of jitter reveals that players' perception of jitter is game-dependent [77]. When the jitter is high, packets tend to arrive too late to be useful, so prediction mechanisms such as dead-reckoning must be used to estimate players' future position as if speed, time, and course remain fixed. This is not always adequate, so the quality of these prediction mechanisms may fail for some games and be useful for others. Since prediction mechanisms can fail, the players' experience degrades if jitter levels are not kept as low as possible.

Design of routing protocols and QoS techniques targeted to real-time requirements of network games is very challenging and has been an open field of research in the past couple of years. Common QoS extensions to 802.11 wireless communications were analyzed and evaluated focusing specifically on the requirements of real-time multiplayer computer games [78]. The AODV routing protocol was selected against other techniques such as DSR, destination sequenced distance vector, and OLSR because it showed the best overall performance for real-time applications [43]. However, additional mechanisms need to be integrated in this protocol to meet the stringent demands of multiplayer games in terms of end-to-end communication delay and jitter. To overcome the issues related to mobility and the time-varying properties of the wireless channels, QoS provisioning has to be managed on several layers of the protocol stack.

A cross-layer design has been proposed [78] to introduce various enhancements at different levels of wireless mesh infrastructure: routing, traffic management, and MAC. At the routing level, the AODV protocol is modified, disabling its local repair property and implementing a backup route mechanism. A local repair mechanism allows intermediate nodes that have detected a link failure to temporarily queue packets and repair the route. As a consequence, it increases the number of packets that arrive too late for being useful because the repair process takes too much time for delivering real-time packets and because repair routes tend to be longer than the original routes.

The backup route mechanism allows, instead, to repair broken links transparently and without any delay since a backup route—a path with the same hop count as the default path, but with a different next hop—is always available. At the traffic management level, mechanisms such as priority queuing, timeouts, and real-time neighbor-aware rate control have been introduced. Hop-constrained queue timeouts prevent the transmission of outdated packets, thus saving bandwidth for more recent packets. Obsolete real-time packets are dropped, applying a gradually decreasing timeout timer in each routing node. At the MAC layer, advanced mechanisms like broken link detection, signal strength monitoring, neighbor detection, and request-to-send/ clear-to-send (RTS/CTS) adaptation may increase the speed of finding a new route or handing over a new AP.

Simulations prove that these QoS techniques can significantly reduce end-to-end communication delay, jitter, and loss rate. For connections up to three hops, AODV with the previously mentioned QoS mechanisms can meet the demands of real-time multiplayer games [78]. Player nodes were assumed to move only slightly, within a distance of 0–15 m, because it is rather difficult to move and play at the same time with today's available mobile gaming devices. Typical multiplayer game traffic between the player nodes was simulated as high-priority bidirectional user datagram protocol (UDP) data flows, with a constant bit rate (CBR) traffic of 20 packets/s and a packet size of 64 bytes [79].

10.5 CONCLUSIONS

This chapter presented a survey of current research efforts for multimedia communications support over WMNs. WMNs are particularly challenging because of their multihop structure, the mobility of network nodes, and the unreliable nature of the wireless channel. While most of the studies (routing protocols, priority mechanisms, and rate control techniques) have focused on throughput and packet loss performance of WMNs, in general, further investigation is required to provide QoS for real-time multimedia applications. At the network and transport levels new techniques and protocols should be implemented to control, above all, latency and jitter. At the application level, reliable multimedia delivery should be supported by MDC and multipath transmission, adaptive playout techniques, and perceptually aware cross-layer schemes. The deployment and success of multimedia services in future WMNs will depend not only on the increasing speed of the physical transmission rate, but also surely on the availability of techniques that can provide the desired level of QoS in terms of latency, jitter, rate, and packet loss.

REFERENCES

1. ITU-T, "One way transmission time," *ITU-T Rec. G.114*, May 2003.
2. A. Majumdar, D.G. Sachs, I.V. Kozintsev, K. Ramchandran, and M.M. Yeung, "Multicast and unicast real-time video streaming over wireless LANs," *IEEE Journal on Selected Areas in Communications*, vol. 12, pp. 524–534, June 2002.
3. P.A. Chou and Z. Miao, "Rate-distortion optimized streaming of packetized media," *IEEE Transactions on Multimedia*, vol. 8, pp. 390–404, April 2006.
4. Y. Wang, S. Wengerm, J. Wen, and A.K. Katsaggelos, "Error resilient video coding techniques," *IEEE Signal Processing Magazine*, vol. 17, pp. 61–82, July 2000.
5. M. Gallant and F. Kossentini, "Rate-distortion optimized layered coding with unequal error protection for robust internet video," *IEEE Transactions on Circuits and Systems for Video Technology*, vol. 11, pp. 357–372, March 2001.
6. Y. Shan and A. Zakhor, "Cross layer techniques for adaptive video streaming over wireless networks," in *Proceedings of the IEEE International Conference on Multimedia and Expo*, vol. 1, pp. 277–280, August 2002.
7. J. Shin, J. Kim, and C.-C.J. Kuo, "Quality-of-service mapping mechanism for packet video in differentiated services network," *IEEE Transactions on Multimedia*, vol. 3, pp. 219–231, June 2001.
8. A. Ksentini, M. Naimi, and A. Gueroui, "Toward an improvement of H.264 video transmission over IEEE 802.11e through a cross-layer architecture," in *IEEE Communications Magazine*, vol. 44, pp. 107–114, June 2006.
9. P. Frossard and O. Verscheure, "Joint source/FEC rate selection for quality-optimal MPEG-2 video delivery," *IEEE Transactions on Image Processing*, vol. 10, pp. 1815–1825, December 2001.
10. C. Luna, Y. Eisenberg, R. Berry, T. Pappas, and A. Katsaggelos, "Joint source coding and packet marking for video transmission over DiffServ networks," in *Tyrrhenian International Workshop on Digital Communications (IWDC)*, Capri, Italy, September 2002.
11. F. Zhai, Y. Eisenberg, T.N. Pappas, R. Berry, and A.K. Katsaggelos, "Rate-distortion optimized hybrid error control for real-time packetized video transmission," *IEEE Transactions on Image Processing*, vol. 15, pp. 40–53, January 2006.
12. ITU-T, "Methods for subjective determination of transmission quality," *ITU-T Rec. P.800*, August 1996.
13. ITU-R, "Methodology for the subjective assesment of the quality of television pictures," *ITU-R Rec. BT.500-10*, March 2000.
14. ITU-T, "Subjective video quality assessment methods for multimedia applications," *ITU-T Rec. P.910*, September 1999.

15. ITU-T, "Perceptual evaluation of speech quality (PESQ), an objective method for end-to-end speech quality assessment of narrowband telephone networks and speech codecs," *ITU-T Rec. P.862*, February 2001.

16. ITU-T, "Method for objective measurements of perceived audio quality (PEAQ)," *ITU-T Rec. BS.1387*, November 2001.

17. E. Ong, X. Yang, W. Lin, Z. Lu, and S. Yao, "Video quality metric for low bitrate compressed videos," in *Proceedings of the IEEE International Conference on Image Processing*, vol. 5, Genova, Italy, pp. 3531–3534, October 2004.

18. Y. Sun, I. Sheriff, E. Belding-Royer, and K. Almeroth, "An experimental study of multimedia traffic performance in mesh networks," in *Proceedings of the International Workshop on Wireless Traffic Measurements and Modeling (WitMeMo)*, Seattle, WA, USA, June 2005.

19. C. Chou and A. Misra, "Low latency multimedia broadcast in multi-rate wireless meshes," in *Proceedings of the First IEEE Workshop on Wireless Mesh Networks*, September 2005.

20. D. Beyer, N. Waes, and K. Eklund, "Tutorial: 802.16 MAC-layer mesh extensions," *IEEE 802.16 Standard Group Discussions*, February 2002.

21. V. Gunasekaran and F. Harmantzis, "Affordable infrastructure for deploying wimax systems: Mesh v. non mesh," in *Proceedings of the IEEE 61st Vehicular Technology Conference (VTC)*, vol. 5, pp. 2979–2983, May–June 2005.

22. R. Bruno, M. Conti, and E. Gregori, "Mesh networks: commodity multihop ad hoc networks," *IEEE Communications Magazine*, vol. 43, pp. 123–131, March 2005.

23. IEEE 802.11-TG N, "Status of project IEEE 802.11n," URL: http://grouper. ieee.org/groups/802/11/Reports/tgn_update.htm, January 2006.

24. A. Vasan and A. Shanker, "An empirical characterization of instantaneous throughput in 802.11b WLANs," in *Technical Report*, CS-TR-4389, University of Maryland, 2002.

25. S. Garg and M. Kappes, "An experimental study of throughput for UDP and VoIP traffic in 802.11b networks," in *Proceedings of the IEEE Wireless Communications and Networking Conference*, vol. 3, pp. 1748–1753, March 2003.

26. S. Garg and M. Kappes, "Admission control for VoIP traffic in IEEE 802.11 networks," in *Proceedings of the IEEE Global Telecommunications Conference (GLOBECOM)*, vol. 6, pp. 3514–3518, December 2003.

27. P. Gupta and P. Kumar, "The capacity of wireless networks," *IEEE Transactions on Information Theory*, vol. 46, pp. 377–404, March 2000.

28. J. Li, C. Blake, D.D. Couto, H. Lee, and R. Morris, "Capacity of ad hoc wireless networks," in *Proceedings of the 7th International Conference on Mobile Computing and Networking*, Rome, Italy, pp. 61–69, 2001.

29. I. Akyildiz, X. Wang, and W. Wang, "Wireless mesh networks: a survey," *Elsevier Computer Networks*, vol. 47, pp. 445–487, 2005.

30. IEEE 802.11-TG R, "Status of project IEEE 802.11r," URL: http://grouper. ieee.org/groups/802/11/Reports/tgr_update.htm, January 2006.

31. I. Ramani and S. Savage, "Syncscan: practical fast handoff for 802.11 infrastructure networks," in *Proceedings of the IEEE 24th Annual Joint Conference of the Computer and Communications Societies*, vol. 1, pp. 675–684, March 2005.

32. IEEE 802.11 Committee, "IEEE standard for information technology—telecommunications and information exchange between systems—local and

metropolitan area networks—specific requirements part 11: wireless LAN medium access control (MAC) and physical layer (PHY) specifications amendment 8: medium access control (MAC) quality of service enhancements," *IEEE Std 802.11e*, November 2005.

33. T. Clausen and P. Jacquet, "Optimized link state routing protocol (OLSR)," *RFC 3626*, October 2003.
34. R. Ogier, F. Templin, and M. Lewis, "Topology dissemination based on reverse-path forwarding (TBRPF)," *RFC 3684*, February 2004.
35. C. Perkins, E. Belding-Royer, and I. Chakeres, "Ad hoc on demand distance vector (AODV) routing," *RFC 3561*, July 2003.
36. D. Johnson, D. Maltz, and Y.-C. Hu, "The dynamic source routing protocol for mobile ad hoc networks (DSR)," Internet Draft: draft-ietf-manet-dsr-09.txt, April 2003.
37. S. Armenia, L. Galluccio, A. Leonardi, and S. Palazzo, "Transmission of VoIP traffic in multihop ad hoc IEEE 802.11b networks: experimental results," in *Proceedings of the First International Conference on Wireless Internet*, Budapest, Hungary, pp. 148–155, July 2005.
38. S. Gray, D. Kotz, C. Newport, N. Dubrovsky, A. Fiske, J. Liu, C. Masone, S. McGrath, and Y. Yuan, "Outdoor experimental comparison of four ad hoc routing algorithms," in *Proceedings of the 7th ACM International Symposium on Modeling, Analysis and Simulation of Wireless and Mobile Systems (MSWiM)*, Venezia, Italy, pp. 220–229, December 2004.
39. D.D. Couto, S. Aguayo, B. Chambers, and R. Morris, "Performance of multihop wireless networks: shortest path is not enough," *ACM SIGCOMM Computer Communication Review*, vol. 33, pp. 83–88, January 2003.
40. R. Draves, J. Padhye, and B. Zill, "Comparison of routing metrics for static multi-hop wireless networks," in *Proceedings of the ACM Annual Conference of the Special Interest Group on Data Communication (SIGCOMM)*, pp. 133–144, August 2004.
41. D.D. Couto, D. Aguayo, J. Bicket, and R. Morris, "A high-throughput path metric for multi-hop wireless routing," in *Proceedings of the ACM Annual MobiCom*, San Diego, CA, pp. 134–146, 2003.
42. R. Draves, J. Padhye, and B. Zill, "Routing in multi-radio, multi-hop wireless mesh networks," in *Proc. ACM Annual MobiCom*, Philadelphia, PA, September 2004.
43. K. Farkas, D. Budke, O. Wellnitz, B. Plattner, and L. Wolf, "QoS extensions to mobile ad hoc routing supporting real-time applications," *Proceedings of the 4th ACS/IEEE International Conference on Computer Systems and Applications (AICCSA)*, March 2006.
44. J. Novatnack, L. Greenwald, and H. Arora, "Evaluating ad hoc routing protocols with respect to quality of service," in *Technical Report DU-CS-04-05*, Department of Computer Science, Drexel University, October 2004.
45. K. Rojviboonchai, F. Yang, Q. Zhang, H. Aida, and W. Zhu, "AMTP: a multipath multimedia streaming protocol for mobile ad hoc networks," in *Proceedings of the IEEE International Conference on Communications*, vol. 2, pp. 1246–1250, May 2005.
46. E. Belding-Royer and A. Lindgren, "Multi-path admission control for mobile ad hoc networks," in *Proceedings of the 2nd Annual International*

Conference on Mobile and Ubiquitous Systems Networking and Services (*MobiQuitous*), pp. 407–417, July 2005.

47. M.K. Marina and S.R. Das, "On-demand multipath distance vector routing in ad hoc networks," in *Proceedings of the IEEE International Conference on Network Protocols*, Riverside, CA, pp. 14–23, November 2001.

48. V.K. Goyal, "Multiple description coding: compression meets the network," *IEEE Signal Processing Magazine*, vol. 18, pp. 74–93, September 2001.

49. M. Vetterli, P.L. Dragotti, A. Verma, G. Barrenechea, and B. Beferull-Lozano, "Multiple description source coding and diversity routing: a joint source channel coding approach to real-time services over dense networks," in *Proceedings of the Packet Video Workshop*, Nantes, France, April 2003.

50. S. Mao, S. Lin, S.S. Panwar, Y. Wang, and E. Celebi, "Video transport over ad hoc networks: multistream coding with multipath transport," *IEEE Journal on Selected Areas in Communications*, vol. 21, pp. 1721–1737, December 2003.

51. E. Setton, X. Zhu, and B. Girod, "Minimizing distortion for multi-path video streaming over ad hoc networks," in *Proceedings of the IEEE International Conference on Image Processing*, vol. 3, Genoa, Italy, pp. 1751–1754, October 2004.

52. E. Setton, X. Zhu, and B. Girod, "Congestion-optimized multi-path streaming of video over ad hoc wireless networks," in *Proceedings of the IEEE International Conference on Multimedia and Expo*, vol. 3, Taipei, Taiwan, pp. 1619–1622, June 2004.

53. K. Madepalli, P. Gopalakrishnan, D. Famolari, and T. Kodama, "Voice capacity of IEEE 802.11b, 802.11a and 802.11g wireless LANs," in *Proceedings of the IEEE Global Telecommunications Conference* (*GLOBECOM*), vol. 3, pp. 1549–1553, December 2004.

54. S. Garg and M. Kappes, "Can I add a VoIP call?," in *Proceedings of the IEEE International Conference on Communications* (*ICC*), vol. 2, pp. 779–783, May 2003.

55. C. Hoene, H. Karl, and A. Wolisz, "A perceptual quality model intended for adaptive VoIP applications," *Wiley International Journal of Communication Systems*, vol. 19, no. 3, pp. 299–316, 2006.

56. C. Perkins, O. Hodson, and V. Hardman, "A survey of packet loss recovery techniques for streaming audio," *IEEE Network*, vol. 12, pp. 40–48, October 1998.

57. P. Velloso, M. Rubinstein, and O. Duarte, "Analyzing voice transmission capacity on ad hoc networks," in *Proceedings of the International Conference on Communication Technology* (*ICCT*), vol. 2, pp. 1254–1257, April 2003.

58. M. Mosko and J. Garcia-Luna-Aceves, "Multipath routing in wireless mesh networks," in *Proceedings of the First IEEE Workshop on Wireless Mesh Networks*, Santa Clara, CA, September 2005.

59. N. Jayant, "Subsampling of a DPCM speech channel to provide two self-contained, half-rate channels," *Bell Systems Technical Journal*, vol. 60, pp. 501–509, 1981.

60. D. Lin and B. Wah, "LSP-based multiple-description coding for real-time low bit-rate voice transmissions," in *Proceedings of the IEEE International Conference on Multimedia and Expo*, vol. 2, pp. 597–600, 2002.

61. H. Dong, A. Gersho, J. Gibson, and V. Cuperman, "A multiple description speech coder based on AMR-WB for mobile ad hoc networks," in *Proceedings of the IEEE International Conference on Acoustic, Speech, and Signal Processing*, vol. 1, pp. 277–280, 2004.

62. J. Balam and J. Gibson, "Path diversity and multiple descriptions with rate dependent packet losses," in *Proceedings of the Information Theory and Applications Workshop*, February 2006.

63. C. Hoene, A. Guenther, and A. Wolisz, "Measuring the impact of slow user motion on packet loss and delay over IEEE 802.11b wireless links," in *Proceedings of Workshop on Wireless Local Networks (WLN)*, Bonn, Germany, October 2003.

64. R. Ramjee, J. Kurose, D. Towsley, and H. Schulzrinne, "Adaptive playout mechanisms for packetized audio applications in wide-area networks," in *Proceedings of IEEE INFOCOM*, vol. 2, pp. 680–688, June 1994.

65. Y. Liang, N. Färber, and B. Girod, "Adaptive playout scheduling and loss concealment for voice communication over IP networks," *IEEE Transactions on Multimedia*, vol. 5, pp. 532–543, December 2003.

66. W. Verhelst, "Overlap-add methods for time-scaling of speech," *Speech Communication*, vol. 30, no. 4, pp. 207–221, 2000.

67. P. Bellavista, A. Corradi, and C. Giannelli, "Adaptive buffering based on handoff prediction for wireless Internet continuous services," in *Proceedings of the International Conference on High Performance Computing and Communications (HPCC)*, Sorrento, Italy, pp. 1021–1032, September 2005.

68. "Self–Organizing Traffic Information System Home Page," http://www.et2.tu-harburg.de/Mitarbeiter/Wischhof/sotis/sotis.htm.

69. "Car 2 Car Communication Consortium," http://www.car-2-car.org.

70. "Spectrum Efficient Uni- and Multicast Services over Dynamic Multi-Radio Networks in Vehicular Environments," http://www.ist-overdrive.org.

71. IEEE 802 Committee, "Draft amendment to standard for information technology—telecommunications and information exchange between systems—local and metropolitan networks—specific requirements—part 11: wireless medium access control (MAC) and physical layer (PHY) specifications: wireless access in vehicular environments (WAVE)," *IEEE Std 802.11p Draft*, November 2005.

72. J.J. Blum, A. Eskandarian, and L.J. Hoffman, "Challenges of intervehicle ad hoc networks," *IEEE Transactions on Intelligent Transportation Systems*, vol. 5, pp. 347–351, December 2004.

73. J. Ott and D. Kutscher, "Drive-thru Internet: IEEE 802.11b for automobile users," in *Proceedings of IEEE Infocom*, Hong Kong, pp. 373–376, March 2004.

74. J.P. Singh, N. Bambos, B. Srinivasan, and D. Clawin, "Wireless LAN performance under varied stress conditions in vehicular traffic scenarios," in *Proceedings of the IEEE 56th Vehicular Technology Conference*, pp. 743–747, September 2002.

75. P. Bucciol, E. Masala, and J.C. De Martin, "Dynamic packet size selection for 802.11 inter-vehicular video communications," in *Proceedings of Vehicle to Vehicle Communications Workshop (V2VCOM)*, San Diego, CA, July 2005.

76. T. Beigbeder, R. Coughlan, C. Lusher, J. Plunkett, E. Agu, and M. Claypool, "The effects of loss and latency on user performance in Unreal tournament

2003," in *Proceedings of Workshop on Network and System Support for Games*, pp. 144–151, August 2004.

77. M. Dick, O. Wellnitz, and L. Wolf, "Analysis of factors affecting players' performance and perception in multiplayer games," in *Proceedings of Workshop on Network and System Support for Games*, Hawthorne, USA, October 2005.

78. D. Budke, K. Farkas, O. Wellnitz, B. Plattner, and L. Wolf, "Real-time multi-player game support using QoS mechanisms in mobile ad hoc networks," in *Proceedings of the Third IEEE/IFIP Annual Conference on Wireless On demand Network Systems and Services (WONS)*, Les Menuires, France, January 2006.

79. J. Färber, "Network game traffic modelling," in *Proceedings of Workshop on Network and System Support for Games*, pp. 53–57, April 2002.

80. C. Mahlo, C. Hoene, A. Rostami, and A. Wolisz, "Adaptive coding and packet rates for TCP-friendly VoIP flows," in *Proceedings of 3rd International Symposium on Telecommunications (IST2005)*, Shiraz, Iran, September 2005.

11

MULTIPLE ANTENNA TECHNIQUES FOR WIRELESS MESH NETWORKS

Honglin Hu, Jijun Luo, and Xiaodong Zhang

CONTENTS

11.1   INTRODUCTION

Wireless mesh networks (WMNs) have emerged as important architectures for the future wireless communications. WMNs consist of mesh routers and clients, and could be independently implemented or integrated with other communication systems such as the conventional cellular systems [1].

Although the current research on the WMNs mainly focuses on the media access control (MAC) protocol and network layer, the physical layer plays a more fundamental role in the performance of the WMNs. However, the techniques of the physical layer of the conventional cellular systems could not be used directly in the WMNs due to the architecture differences.

Moreover, the performances of the physical layer techniques in the ad hoc networks is different from those in the WMNs. So, the investigation of the physical layer techniques for the WMNs is highly demanding [2].

Among the various physical layer techniques, multiple antenna techniques are most important. By using multiple antenna techniques, the capacity and throughput of the WMNs could be enlarged, and the routing performance be improved. In addition, the WMNs benefit from the multiple antenna techniques in the aspects such as increased capacity and throughput, improved routing performance, increased energy efficiency, better quality of service (QoS), and improved location management. Section 11.2 and Section 11.3 give brief surveys of the multiple antenna techniques and the WMNs. Then, we review the multiple antenna techniques for the WMNs.

11.2 SURVEY OF MULTIPLE ANTENNA TECHNIQUES

The rapid growth in mobile communications leads to an increasing demand for wideband high data rate communication services. In recent years, multiple antennas are equipped at the base station (BS) in cellular systems, and hence, smart antennas techniques have been identified as an enabling technique for high-rate multimedia transmissions over wireless channels [3]. Smart antennas techniques have been widely used in mobile communication systems to overcome the problem of limited channel bandwidth, satisfying a growing demand for a larger number of mobiles on communication channels. Smart antennas, when used appropriately, help in improving the system performance by increasing channel capacity and spectrum efficiency, extending range coverage, and compensating electronically for aperture distortion.

With the development of the hardware, e.g., high speed digital signal processors and the powerful small-sized RF headers, multiple antennas could also be implemented in the mobile station (MS) side.

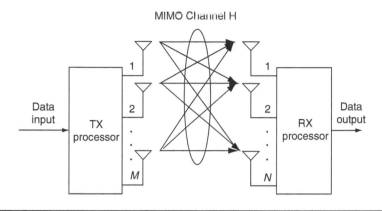

Figure 11.1   Schematic diagram of a MIMO wireless system.

Thus, multiple-input multiple-output (MIMO) channels are built between the transmitters and the receivers, as illustrated in Figure 11.1. On the other hand, the breakthrough in the information theory has proved that higher capacity could be achieved by using the MIMO techniques [4]. The MIMO systems can be viewed as an extension of, but much more than, the conventional smart antennas techniques, which have been investigated for decades.

The conventional smart antennas techniques commonly concern the case that multiple antennas are implemented at the BS side while only one antenna element is implemented at the MS side, i.e., multiple-input-single-output (MISO) case. With multiple antennas at the BS side, the key concept in smart antennas, i.e., beamforming, could be realized. By forming the beams toward the direction of arrival (DOA) of the desired signal, the average signal-to-noise ratio (SNR) at the receiver could be remarkably increased. On the other hand, the antenna arrays at the BS side could also form the nulls toward the DOAs of the interference signal, i.e., antenna nulling, to reduce the received power of the interference. By pointing the beams to the desired signals and the nulls toward those interfering signals, the performance of the wireless communication systems such as capacity, consumed power reduction, and spectrum efficiency could be greatly improved.

When multiple antennas are implemented at the MS side as well to form a MIMO link, most benefits of smart antennas are retained since the optimization of the transmitting and receiving antenna elements can be carried out in a larger space. The most interesting aspect of the MIMO systems lies in that multipath propagation, which is harmful to the conventional wireless communications, can be well utilized and become an important factor to increase the capacity of the system.

MIMO system can provide two types of gains: diversity gain and spatial multiplexing gain. It is well known that diversity could be used to combat fading. The principle of both the transmit diversity and the receive diversity is to send or receive signals that carry the same information through different paths. In this way, multiple independently faded replicas of the data symbol can be obtained at the receiver and thus more reliable reception could be achieved. Comparably, the multiplexing gain is achieved by transmitting independent information streams in parallel through the spatial channels. As a result, the data rate could be increased. For example, the Bell Labs space–time architecture (BLAST) is a kind of MIMO system which explores the multiplexing gain [5]. Here, we should note that the diversity gain could also be achieved by MISO or single-input-multiple-output (SIMO) systems. However, the multiplexing gain could only be achieved by the MIMO system. It is shown [6] that to achieve a higher diversity gain, we have to sacrifice some multiplexing gain. Therefore, in a practical wireless communications system with MIMO channels, it is important to keep the fundamental trade-off between the diversity gain and the multiplexing gain.

Generally, when multiple antennas are implemented at one or both sides of the transmitter and the receiver, many different kinds of related techniques could be used to improve the performance of the system. Section 11.2.1 through Section 11.2.3 introduce several typical multiple antenna techniques.

11.2.1 Beamforming and Antenna Nulling

The beamforming and antenna nulling techniques have been used in the wireless system equipped with multiple antennas for a long time. The adaptive antenna arrays concept originally used in the radar system could also be classified into the beamforming and antenna nulling techniques. By multiplying a group of weighting coefficient on the received/transmitted signal from each antenna element, a beam could be formed toward the target signal, which is named as beamforming technique. The beamforming process will increase the power of the desired signal at the receiver side. On the other hand, the antenna nulling could reduce the received interfering signal, thus increase the signal-to-interference ratio (SIR) at the receiver side. The number of beams and nulls that could be simultaneously formed by the antenna array is limited by the number of elements of the array.

The beam-width and the antenna gain also depend on the number of the elements of the antenna array.

For the wireless communication systems, e.g., cellular system, only the BS side has been equipped with multiple antennas, due to the size limitation at the MS side. In such a MISO scenario, typically there are two different types of beamforming techniques: adaptive array and switched-beam [7]. The adaptive array scheme tracks each user in a given cell with an individual adaptive beam pattern, while the switched-beam scheme selects one beam pattern for each user out of a number of preset fixed beam patterns, depending on the location of the user. In comparison with the switched-beam scheme, the adaptive array scheme has better performance in most cases. However, adaptive array technique is much more complex to be implemented, and has to rely on the accuracy of channel estimation and the convergent speed of beamforming algorithms [8].

From multiuser diversity point of view, another important beamforming technique, opportunistic or random beamforming was developed. The basic idea of opportunistic or random beamforming is that the BS forms random beams which are changed for each transmission duration, thus induces random fading to the mobile radio channel in order to exploit multiuser diversity [9]. Training sequences are transmitted to allow the MS to estimate the real-time signal-to-interference-plus-noise-ratio (SINR) values which are fed back to the BS. Hence, based on these SINR values the BS can schedule the specific user traffic to the time slots where the most suitable beam for the regarded MS is used. This technique is most beneficial, if many active MSs be supported and are uniformly distributed in the cell. However, if there are only a few users in the cell, the performance of the opportunistic beamforming technique is not as good as other multiple antenna techniques such as conventional coherent beamforming or space-time block coding (STBC). This is because the multiuser diversity will be remarkable only with a large amount of MSs. Moreover, the opportunistic beamforming technique relies on the training sequences to estimate the real-time SINR of each activated beam (or weighting vector) for each MS.

A new beamforming technique, termed as organized beam-hopping (OBH), was proposed by Hu et al. [10]. In the OBH, the beamforming vectors are not randomly selected as in the opportunistic or random beamforming techniques; instead, the beamforming vectors are hopped in an organized pattern. In addition, the organized pattern is changeable according to the distribution of the MSs in the cell. The OBH technique is different from the beam-hopping concept [11], where the received signals from the hopped beams are combined to exploit spatial diversity and mitigate strong directional interference. The OBH technique remains a good performer for variable number of MSs (e.g., from one to several hundreds), together with a mixture of MIMO and MISO cases. In addition, the OBH technique does not need training sequences to estimate the effective SINR of current activated beam for each MS, thus keeps high spectrum efficiency due to the reduction of signaling overhead.

The antenna nulling techniques are mainly used in the military communications instead of the cellular system. This is because the strong directional interferences normally happen in the military communications which are unusual in the cellular systems. Moreover, in the cellular system the density of the cochannel users may be high. Thus, the performance of the antenna nulling techniques is limited.

11.2.2 Diversity and Space–Time Coding

Diversity techniques have been widely used in a single-input-single-output (SISO) scenario, in the form of time or frequency diversity, such as the RAKE receiver in the code division multiple access (CDMA) systems. The principle of diversity is to exploit the multiple fading paths between the transmitter and the receiver. As the number of the involved paths increases, the reliability of the wireless link also increases. When multiple antennas are implemented, it is possible to further exploit spatial diversity. The spatial diversity does not bring about the penalty in data rate, as happens in time or frequency diversity due to the utilization of time or bandwidth to introduce redundancy. Generally, spatial diversity is classified into transmit diversity and receive diversity, where the diversity process mainly happens at the transmitter and the receiver side, respectively. Moreover, spatial diversity could be combined with beamforming. For example, the beam-hopping concept [11] is to combine the signals received from each hopping beam, to mitigate the strong directional interferences.

The receive diversity is similar to the conventional RAKE technique. The desired signals received from different antenna elements are combined at the receiver side. The combining scheme could be selection combining (SC) that chooses the path with the highest SNR, maximal ratio combining (MRC) that optimally (with optimal weighting coefficient multiplied on each path) combines the signals from each antenna element, and equal gain combining (EGC) that simply adds the signals from each antenna element after cophased.

The diversity in a system is characterized by the number of independently fading diversity branches, also known as the diversity order. Moreover, the correlation between the antenna elements will strongly affect the performance of the receive diversity. Higher correlation will result in poorer performance at the receiver, and the diversity order cannot exceed the rank of the spatial correlation matrix of the channel.

On the other hand, the spatial diversity could also be implemented at the transmitter side, namely transmit diversity, by transmitting the same data information from different antenna elements. However, the transmit diversity should be more delicately designed than the receive diversity, because of the mutual interference of the simultaneously transmitting antennas from the same node. Space–time coding is a powerful transmit diversity technique that relies on coding across space (transmit antennas) and time to extract diversity. Generally, the codes used for transmit diversity could be divided into space-time trellis codes (STTC) and STBCs. STBCs are generalizations of the Alamouti code when more than two transmit antennas are used, and we could expect that the linear STBCs are simple to encode and decode. Moreover, it is shown that the linear STBCs with the best error performance do not always demonstrate the best capacity performance. Comparably, STTCs pioneered by Tarokh et al. [12] combine signal processing at the receiver with intelligent coding techniques appropriately mapping the information to be transmitted by multiple antennas. The STTCs provide not only a full diversity order but also coding gain, thus are of particular interests.

Antenna selection could be an enhanced technique for the spatial diversity in the multiple antennas system [13]. It could be used in both the transmit diversity and the receive diversity. By using the antenna selection technique, the cost and the complexity of the multiple antenna systems could be remarkably reduced. For the receive diversity, we could choose L_r branches out of the total N receiving antenna elements that have the highest SNRs for the MRC or EGC process. Thus, beside the reduced cost and complexity, we could also avoid combining poor branches that have very low SNR. The low SNR branches might be harmful to the output SNR after the combining process, especially EGC. For the transmit diversity, the principle of antenna selection is the same, i.e., to choose the L_t branches out of the total M transmitting antennas that provide the highest SNRs at the receiver side. However, unlike that for the receive diversity, the antenna selection for the transmit diversity needs feedback from the receiver.

11.2.3 Spatial Multiplexing

Diversity techniques are used to combat the fading. Comparably, the spatial multiplexing techniques utilize the parallel channels to transmit different data streams. Thus, the spatial multiplexing actually turns the multipath fading effect to be beneficial for the wireless communication systems. In addition, we could easily expect that the spatial multiplexing could achieve a higher data rate than that of diversity schemes in the high SNR regime. The principle behind the spatial multiplexing is that the receiver could possibly descramble signals that transmitted simultaneously from multiple antennas. The spatial multiplexing could only be implemented in a MIMO system, otherwise the parallel data streams could not be transmitted or received correctly [6].

The well-known diagonal BLAST (D-BLAST) and vertical BLAST (V-BLAST) are typical spatial multiplexing systems. D-BLAST utilizes an elegant diagonally layered coding structure in which code blocks are dispersed across diagonals in space–time. However, the diagonal approach suffers from certain implementation complexities which make it inappropriate for initial implementation. Comparably, the simplified version, V-BLAST, is more practical and could be implemented in real-time environment.

In an environment with sufficient rich scattering and the wireless channels having sufficient degrees of freedom, the parallel data streams from the multiple transmitting antennas could be separated by using the multiple antennas at the receiver. For the MIMO system, the capacity of a deterministic and a random MIMO channel H could be expressed as

$$C = \log_2 \left[\det \left(I_N + \frac{\gamma}{M} HH^{11} \right) \right] \text{bits/s/Hz} \qquad (11.1)$$

and

$$C = E_{11} \log_2 \left[\det \left(I_N + \frac{\gamma}{M} HH^{11} \right) \right] \text{bits/s/Hz} \qquad (11.2)$$

respectively, where γ represents the SNR at any receive antenna and E_{11} denotes the expectation with respect to the random channel H. It has been proved that the capacity of the MIMO system grows linearly with $min(M,N)$, provided that the receiver has the channel state information (CSI), while the transmitter employs either the CSI or its distribution information.

Antenna selection techniques could also be used for the spatial multiplexing, by choosing the most suitable antennas from both sides of the transmitter and the receiver. For the receiver side, norm-based or successive selection schemes could be implemented, where the former approach is more suitable for the low SNR regime and the latter fits well for the high SNR regime. However, like that for the transmit diversity, the antenna selection at the transmitter side for spatial multiplexing needs feedback from the receiver side. Fortunately, the feedback required by antenna selection is only a small part of the full CSI.

11.3 OVERVIEW OF WIRELESS MESH NETWORKS

Broadband wireless systems are envisioned to provide ubiquitous access to end users, providing high-quality voice services, video, and other multimedia contents. As various next-generation wireless networks evolve around broadband services, the WMN attract great interests. The WMNs offer huge possibilities for community networks to provide access to homes and businesses, and could be used to connect public access sites throughout a community [2]. The basic architecture of WMNs could be classified into three groups: infrastructure/backbone WMNs, client WMNs, and hybrid WMNs. For the infrastructure/backbone WMN, mesh routers are included to support the functionality of the network. The mesh routers often have multiple interfaces to connect the other networks, e.g., Internet, cellular, Wi-Fi, sensor networks, WiMAX, and so on. In addition, the mesh routers also act as gateway or bridge for the mesh client. The mesh clients usually have only one interface and could not act as gateway or bridge. Comparably, for the client WMNs, there exist only mesh clients and the mesh routers are excluded. Therefore, the client WMNs are more or less similar to the conventional ad hoc wireless network, where a small group of users communicate with each other by multihop techniques. In some scenarios, the architecture named hybrid WMNs, which combines the infrastructure/backbone and client mesh networks is needed. In the hybrid WMNs, the mesh clients could directly communicate with each other and access the network through the mesh router [1].

11.3.1 Characteristic Aspects of Wireless Mesh Networks

The WMNs are mainly developed from the ad hoc networks. So they have the common characteristics with the ad hoc systems.

The communications between different mesh clients, or mesh client and mesh routers, mainly rely on the multihop-related techniques. However, the WMNs are much more than the ad hoc networks, because they have more flexible architectures, and most importantly, they are equipped with mesh routers. Some characteristics of the WMNs are illustrated below.

1. The WMNs are self-organizing, self-roaming, and self-healing networks. Once a mesh router or client detects the availability of a WMN, it could join the WMN by using certain protocols. In addition, when some clients or routers are leaving from a WMN, the network could reorganize the left ones to maintain the communications between them. Without fixed infrastructure, a group of mesh clients and routers can also self-organize to form a WMN. Normally, the mesh routers in the WMNs have less mobility than the mesh clients, and the routers are commonly selected as the head for a cluster of mesh clients. When a mesh router is not available for a group of mesh client, the head of the cluster also is chosen among the clients.

2. The mesh routers normally have more powerful energy supplies than those of mesh clients. Since the infrastructure/backbone and hybrid WMNs are equipped with mesh routers, the more complex but effective algorithms could be implemented in the WMNs. These algorithms improve the performance of the system at different layers and from different aspects. The energy consumption is highly reduced in the WMNs or the ad hoc networks due to the application of multihop-related techniques. However, the limited energy problem is still considered as a bottleneck for the performance of the ad hoc or the sensor networks, because powerful digital signal processing (DSP) could not be conducted in the nodes.

3. The WMNs have the potential to provide more reliable communications than that of the cellular networks. This is because the WMNs use multihop techniques for the transmission. The routing in the WMNs is more flexible than that in the cellular networks, and the WMNs do not necessarily rely on the infrastructure or backbone. In addition, the communications in the WMNs are more reliable than those in the ad hoc network, because the WMNs are commonly equipped with mesh routers serving as wireless infrastructure. The coverage, connectivity, and reliability of mesh routers are much larger than that of the nodes/clients in the ad hoc networks. The mesh routers could be equipped with multiple radios to further increase their

capabilities such as throughput and number of the simultaneously supported services.

4. The WMNs have open wireless architecture, which could be easily integrated with other networks such as cellular, Wi-Fi, WiMAX, and so on. The open wireless architecture is considered as a key feature for the next-generation wireless communication systems, where different industrial standards and various wireless networks could be supported or integrated. The integration of the WMNs with other existing networks is through the mesh routers, where multiple interfaces are equipped. The mesh routers act as gateways or bridges, so that the services in the other networks could be provided for the mesh clients, and the services available in the WMNs could also be offered to the terminals in other networks through the gateways or bridges. Comparably, the conventional ad hoc networks are normally independent and not compatible with other types of networks.

11.3.2 Challenges in Wireless Mesh Networks and Possible Solutions

Although the WMNs are promising for the future wireless communications, they face many challenges in the different layers of the system. This chapter mainly focuses on the performance improvement which could be brought by the multiple antenna techniques for the WMNs. Therefore, in the following, we mainly concern the challenges in the physical layer and illustrate the possible solutions by the application of multiple antenna techniques.

1. In order to be compatible with other existing networks, the open wireless architecture of the WMNs needs both advanced hardware and highly efficient protocols. The mesh routers act as gateway or bridge, so they should be equipped with multiple interfaces and needs much more energy than that of mesh clients. On the other hand, the mesh routers need to be more compact, i.e., easy to be deployed and have low cost. Multiple radios could be a possible solution to integrate multiple interfaces, but the cost and size will inevitably increase as well. In addition, the software-defined radio (SDR) is just one of the implemental modules of the open wireless architecture system, and the related techniques need further evaluation and research. Comparably, multiple antenna techniques could resolve different paths and support parallel data streams, thus

suitable for the mesh routers to act as interfaces. Furthermore, the multiple antenna techniques could at least partly replace the functions of multiple radios.

2. The capacity analysis is complex for the WMNs, because the capacity relies on the allowed number of hops from the source to the destination. In addition, the result for the capacity of the ad hoc networks could not be readily applied in the WMNs, since the mesh routers have much less mobility but higher capabilities than those of common mesh clients. It is shown that the throughput capacity could be increased by deploying relaying nodes and/or grouping the nodes into clusters. However, deploying relaying nodes will inevitably increase the cost/complexity of the WMNs, and the nodes-clustering is difficult for the distributed systems. Multiple antenna techniques have been used in cellular networks to increase capacity, and they are proved to be mature techniques. The multiple antennas could be practically equipped at least at the mesh routers. Therefore, many kinds of multiple antenna techniques could be exploited to improve the capacity of the WMNs. It is shown that the system capacity could be potentially increased three times or even more by the MIMO techniques. Consequently, when the multiple antennas are also installed at the mesh clients, we expect that the capacity of the WMNs will be further increased by the MIMO techniques. A survey of directional antennas technique for the WMNs could be found in Li et al. [14].

3. The WMN requires the routine integration with the user plane traffic. This behavior is analog to *Tunneling*, which requires the packets to be encapsulated, thus increasing the per-packet overhead. This overhead reduces the portion of bandwidth available for application data. For example, real-time traffic with a packet size of 100 bytes would experience a 40% overhead. On a high-cost satellite link such a magnitude of overhead would not be economically viable.

4. The energy consumption is a critical problem for the WMNs, especially for the mesh clients. Multihop techniques on the one hand reduce the transmitted power of the mesh clients. On the other hand, however, they also increase the duration and the frequency of the data transmission at the mesh clients. For example, a mesh client in the WMNs should always be ready to relay for others. Comparably, when a mobile terminal in the cellular network is in the idle mode, it does not relay data for the other users, thus the life of the battery is extended. The employment of multiple antennas at the mesh routers or clients will inevitably require more complex signal process,

thus cost additional energy. However, with multiple antenna techniques, the transmission power could be highly reduced, because the desired signal power could be increased by diversity or beamforming techniques, and the interfering signal could be significantly mitigated. As a result, the total consumed energy in the WMNs is reduced when multiple antenna techniques are employed.

5. The cross-layer design and optimization is crucial for the application of the multiple antenna techniques in the WMNs. Although the multiple antenna techniques could theoretically improve the performance of the WMNs from many different aspects, the compatible higher-layer protocols are intensively needed. For example, the harmful exposed and/or hidden nodes exist in the WMNs and the directional antennas technique could reduce the number of the exposed nodes. However, more hidden nodes will be possibly generated in this case because of the directional transmission. Therefore, advanced MAC protocols should be developed for the directional antennas. In addition, when other new techniques, e.g., orthogonal frequency division multiplexing (OFDM) and ultrawide band (UWB), are applied in the WMNs, the design and selection of multiple antenna techniques should be carefully considered for the WMNs.

11.4 MULTIPLE ANTENNA TECHNIQUES FOR WIRELESS MESH NETWORKS

As seen from Section 11.3, the employment of multiple antennas could improve the performance of the WMNs from different aspects. Generally, in order to meet the challenges in the WMNs, the functions of the multiple antenna techniques could be classified into many different parts: increase the capacity and throughput, improve the routing performance, increase energy efficiency, and many other performance improvements. Section 11.4.1 through Section 11.4.4 illustrates the above-mentioned functions of multiple antenna techniques for the WMNs. However, the multiple antenna techniques could also be exploited in the areas such as broadcasting, antijamming, and so on.

11.4.1 Increase Capacity and Throughput

It has been proved that multiple antenna techniques can be used to improve the capacity of the wireless communication systems. The capacity of the WMNs could also be increased by the multiple antenna techniques. However, due to the multihop and self-organizing features of the WMNs, the exact improvement of capacity for the WMNs is difficult to evaluate and thus still an open topic. Therefore, we have to resort to some results on the capacity of the conventional ad hoc networks. Fortunately, most of the results for the ad hoc networks could be readily used for the WMNs.

Due to the multihop aspect of the WMNs and the ad hoc networks, the directional antennas technique is one of the most suitable multiple antenna techniques. It is proved [15] that capacity gains of $\sqrt{2\pi/\alpha}$, $\sqrt{2\pi/\beta}$, and $2\pi/\sqrt{\alpha\beta}$ could be achieved by using directional transmission and omni-reception, omni-transmission and directional reception, and directional transmission and directional reception, respectively. Here, α and β are the transmitter and receiver antenna beamwidths, respectively, and the above results are valid for arbitrary networks. In addition, capacity for the ad hoc networks consisting of nodes equipped with the directional antennas is studied [16]. The capacity bounds for an abstract and real-world linear array directional antenna models, and how these bounds are affected by important antenna parameters like gain and beamwidth are provided. It is shown that interference from concurrent transmissions limits the maximum achievable capacity. Furthermore, by combining beam steering for transmitting and adaptive beamforming for receiving, the capacity of the ad hoc networks with spatial time division multiple access (STDMA) could increase up to 980%, when eight antenna elements are used for each node [17]. Due to the complexity and practical consideration, the exact capacities of the other more complicated multiple antenna techniques for the ad hoc networks or the WMNs are still under research. The authors propose average rate region and outage capacity region concepts to study the capacity of the ad hoc networks with MIMO [18]. It is shown that the average rate region is a realistic upper bound on the performance of many existing ad hoc routing protocols, and it gives the average system performance over fading or random node positions. In addition, the study shows that the gain from MIMO for the ad hoc networks is similar to that from point-to-point communications.

Throughput is relevant to the capacity, and is thought as a practical realization of it. Generally, capacity bound is calculated by theoretical analysis, while the throughput is mainly by simulation or experimental test. Many researches have proved that the throughputs of the WMNs and the ad hoc networks could be enlarged by the multiple antenna techniques. The overall network throughput is studied for the ad hoc networks with adaptive antennas [19]. The simulation results show that radiation patterns with smaller beamwidths and/or lower side-lobes result in higher network throughput, and adaptive radiation patterns usually outperform the patterns with lower sidelobes but no nulls toward the interfering signals. The above results have also been proved by Thomson [20]. In addition, Bellofiore et al. [19] show that the training periods that occupy more than 20% of the time resource will reduce the throughput considerably, so the fast beamforming algorithms are critical for the high network throughput. The performance of adaptive arrays and switched beam smart antennas techniques for the ad hoc networks are also compared [21]. It is proved by simulation that the throughput of adaptive array is higher than that of switched-beam smart antennas for the ad hoc networks.

Furthermore, experimental results are studied [22] for a mesh network test bed using low-cost analog directional antennas. In the experiment, two communication links are closely placed in a square topology and the network could achieve 90% throughput improvement by using the directional antennas. The results imply that the two communication pairs are almost communicating simultaneously with the directional antennas.

In a typical multihop scenario, the data is routed from an access point over one or more relays to the MS. In order to avoid the throughput reduction of the relaying in the time domain, spatial multiplexing technique could be applied. The throughput of spatial multiplexing technique for the WMNs or the ad hoc networks is simulated [23]. In addition, different levels of abstraction are used to evaluate the throughput gains and a suboptimal approach for relaying topology is proposed. Comparably, the authors [24] utilize more flexible multiple antenna techniques to improve the throughput and fairness for the MIMO ad hoc networks. The proposed protocol employs spatial multiplexing with antenna subset selection for data packet transmission, while using the Alamouti STC for control packet transmission. Furthermore, limiting and saturation throughput of the MIMO ad hoc networks are also studied [25,26].

11.4.2 Improve Routing Performance

Routing is an important issue in the WMNs and the ad hoc networks. Since the increased number of hops will inevitably decrease the network capacity and increase the delay of transmission, many protocols and techniques are proposed to improve the routing performance. By exploiting the multiple antenna techniques, e.g., the beamforming or the diversity, the transmission range of the desired

source mesh routers/client and decrease the cochannel interferences to the other nodes. Since the cochannel interferences are decreased, the other nodes could also reduce their transmitting power to guarantee the same SNR at the destination nodes. However, the reduced transmitting power will result in more necessary hops from the source to the destination nodes, which means more consumed energy in the relays. Therefore, the overall consumed energy could not be saved by simply reducing the transmitting power.

The energy could be more efficiently gathered or utilized in the spatial domain by using multiple antenna techniques such as the beamforming and the diversity. Among the various multiple antenna techniques, the best way to increase the energy efficiency is to use the directional antennas. Nasipuri et al. [33] study the power consumption in the ad hoc networks, when the directional antennas are applied. A power control scheme, where the transmission power is reduced by an additional factor that is based on the minimum SINR required at the destination, is proposed in order to maximally utilize the savings in the average power consumption in the network. The authors assume that the power required for a given transmission distance is proportional to the beamwidth of the directional antennas at the transmitter side. Simulation results show that comparing with the ad hoc networks employing the omni-directional antennas, the ad hoc networks with the directional antennas have not only higher throughput, but also higher energy efficiency. In addition, the authors present the simulation results for the maximum possible savings of power consumption in the same network when an ideal power control scheme is applied. Using the directional antennas could increase the lifetime of nodes and networks significantly. However, in order to utilize the directional antennas, effective algorithms, or MAC protocols are needed to enable nodes to point their antenna directions to the right place at the right time. Spyropoulos and Raghavendra [34] propose a four-step algorithm that coordinates node communications efficiently. The approach is to do energy-efficient routing first, in order to find minimum energy paths. Then, the transmissions of the nodes are scheduled accordingly, in order to minimize the total time it takes for all possible transmitter–receiver pairs to communicate with each other. Simulation results show that the energy savings achieved consist of two major components. The first component, which is the result of using the directional antennas instead of the omni-directional ones, is proportional to the antenna gain. The second component, which is due to the employment of the energy-efficient routing instead of the conventional routing schemes (e.g., minimum hop routing), ranges from 10% to 45%.

The energy consumed per successfully transmitted data bit is an important value in analyzing the performance of the energy-efficient algorithms for the ad hoc networks. Farha and Adve [35] study the performance of the space–time coding for the power reduction of the ad hoc networks. The research is conducted from cross-layer point of view, where the space–time coding in physical layer is combined with a proposed MAC protocol (namely MAC-2 protocol) based on sending the packets at the minimum power needed to achieve a threshold packet error rate (PER) at the receiver. The simulation results prove that the total data delivered per unit of energy consumed increases with the network load for the MAC-2 protocol both with and without Alamouti STC. However, there exists a difference between the cases with and without Alamouti STC for a particular load, which shows that the use of STC at the physical layer can yield additional power savings.

An energy-efficient virtual MIMO communications architecture is studied [36]. The architecture is based on V-BLAST receiver processing. Numerical results show that the rate-optimized 4×4 virtual MIMO system provides about 63% of energy savings for large transmission distance, and the 16×16 virtual MIMO system achieves about 80% of savings over the traditional SISO communications. These results also indicate that while rate optimization over transmission distance may offer improved energy efficiencies in some cases, this is not essential in achieving energy savings as opposed to the Alamouti scheme-based virtual MIMO implementations. In most scenarios, a fixed rate virtual MIMO system with binary phase-shift-keying (BPSK) can achieve performance very close to that of a variable-rate system with optimized rates. However, the results also indicate a trade-off between the achievable energy efficiency and the delay incurred. In other words, as the order of the virtual MIMO architecture grows, the virtual MIMO architecture also leads to larger delay penalties.

In addition, a distributed power control mechanism has been proposed for the ad hoc networks and the WMNs with smart antennas [37]. The basic approach of this protocol is to gather minimum SINR values locally at the nodes during an active link operation and to use this information to estimate the power reduction factors for each activated link. Specifically, the minimum SINR values include the minimum SINR during the RTS, CTS, DATA transmissions, and ACK. Then, by using a dedicated power control algorithm, the appropriate power reduction is decided. Simulation results show that the distributed power control protocol can be adapted to different physical situations and the capacity of the system achieved is very close to the best static local power control schemes.

The power saving performances of multiple antenna techniques for the ad hoc networks and the WMNs are discussed above. Since broadcast/multicast is an important communication mode in the ad hoc networks and the WMNs, we discuss the energy efficiency of the broadcast/multicast for the ad hoc networks and the WMNs, when multiple antennas are employed. The networks need a broadcast/multicast mechanism to update their states and maintain the routes between nodes. Before any data transmission, a broadcast/multicast tree rooted at the source node will be built to cover all destination nodes. Once the tree is known, a node in the tree will deliver each broadcast/multicast packet to all of its one-hop children nodes. Tong and Ramanathan [38] study energy-efficient multicast in the ad hoc networks with MIMO in multipath environments. The goal is to find the optimal antenna patterns for the transmitters and receivers so that the transmit power can be minimized while the broadcasted packet could be correctly received by all the children nodes in the multicast tree. Two computationally efficient heuristics are proposed and the simulation results show that the two can provide considerable power saving as compared to the single-antenna case. In addition, when energy efficiency is considered, the ad hoc networks or the WMNs will require a power-aware metric for their energy-efficiency broadcast/multicast routing algorithms.

Typically, the optimization metrics for routing in the ad hoc networks and the WMNs could be classified into either maximizing the broadcast/multicast lifetime or minimizing the total power assigned to all nodes in the broadcast/multicast tree. Guo and Yang [39] present the performance of various algorithms meeting the former optimization metric of a given multicast connection in the ad hoc networks and the WMNs that use the directional antennas [39]. Simulation results show that with the directional antennas, the static-weight directed prim multicast tree (S-DPMT) algorithm is optimal for unicast, while the dynamic-weight directed prim multicast tree (D-DPMT) algorithm outperforms all the other algorithms for multicast and broadcast. Furthermore, it is proved that the minimal total energy consumption does not guarantee maximum lifetime for the ad hoc networks and the WMNs either for broadcast or multicast.

11.4.4 Other Performance Improvement

In addition to the above-mentioned performance improvement for the ad hoc networks and the WMNs, the multiple antenna techniques bring about the performance improvement in many other aspects, such as connectivity and QoS support, location estimation, and so on.

The improved link quality by using multiple antenna techniques results in better connectivity. Bettsletter et al. [40] study how much the connectivity of the ad hoc networks could be improved by using random or opportunistic beamforming technique. Simulation results show that the percentage of connected node pairs is increased by 60% to 450% depending on the network topologies.

QoS can be defined, in a general way, as the effect of service performance which determines the degree of satisfaction of a service user. Various QoS-related topics are studied for the ad hoc networks and the WMNs with multiple antennas. Both the analytical and the simulation results prove that the networks could provide better QoS support when using different multiple antenna techniques. When considering the QoS improvement for the ad hoc networks or the WMNs with multiple-antenna techniques, cross-layer design or optimization should be taken into account. This is because the multiple-antenna techniques are mainly in the physical layer, while the QoS-related topics are concerned mainly at the MAC and above layers. If the MAC layer cannot manage free resources in an efficient way, it is impossible to offer QoS connections inside the network. A cross-layer structure of ad hoc spatial reuse time division multiple access (STDMA) networks utilizing smart antennas is presented by Martinez and Altuna [41]. In this architecture, the information shared between nodes includes the MAC layer metrics and the network layer metrics, where the network layer metrics contain the network topology, path gains, and transmission power, and the MAC layer metrics contain the MAC layer buffer state. Appropriate scheduling algorithms are demanded to manage different services in the ad hoc networks and the WMNs using cross-layer information, smart antennas, and STDMA as the multiple access scheme. Moreover, Mundarath [42] illustrates a QoS aware protocol, termed as Q-NULLHOC, for the ad hoc networks with MIMO. The basic idea of Q-NULLHOC is that the maximum available degrees of freedom available for nulling is not fixed, but is adaptively selected based on the QoS requirements of the node. The network layer feeds the difference between the service received and the service required to Q-NULLHOC. In turn, Q-NULLHOC selects maximum degrees of freedom to be used for antenna nulling based on the information from the network layer. At the physical layer, antenna nulling technique is used to mitigate the cochannel interferences. As a result, the energy savings increase as the number of degrees of freedom increase.

Furthermore, Saha et al. [43] and Ueda et al. [44] study the priority-based QoS in mobile ad hoc networks, where a zone reservation, adaptive call blocking scheme, and a distributed feedback control mechanism are proposed, respectively. Both studies utilize the directional antennas to improve the QoS performance. The main idea [32] is to take a control-theoretic approach to adaptively control the low-priority flows so as to maintain the high-priority flow rates at their desired level, thus guaranteeing QoS to high-priority flow. Comparably, zone–disjoint routes are used to avoid mutual interference in the ad hoc networks with the employment of the directional antennas [44]. Furthermore, by using game theory, Baccarelli et al. [45] study spatial power allocation multiple antenna game under the best efforts and contracted QoS policies. The various theories originated from computer science will be promising in the study of the ad hoc networks and the WMNs.

Location estimation is an important topic in the ad hoc networks and the WMNs. In the presence of mobility, the MAC protocol should incorporate mechanisms by which a node can efficiently locate and track its neighbors. It should be noted that the solution to location determination with the omnidirectional antennas is not applicable to the directional antennas since the radiation patterns are different and the received power is dependent on angle and distance. Malhotra et al. [46] try to estimate the location of the nodes by measuring the received signal strength from just one or two anchors in a two-dimensional (2D) plane with the directional antennas. The location estimation technique is implemented by using Berkeley MICA2 sensor motes and shows up to three times more accurate than triangulation using the omni-directional antennas. Simulation results show that the error is reduced from 27.5% in an omni-directional-based system to 11.6% with two-directional anchor nodes, and the accuracy of location determination increases with the node density. In addition, an integrated neighbor discovery and polling-based MAC protocol is studied [47] for the ad hoc networks using the directional antennas. The simulation results [47] show that an extremely high per-node channel utilization of up to 80% in static scenarios and up to 50% in mobile scenarios could be achieved.

The ad hoc networks and the WMNs are often implemented for tactical military scenarios. Moreover, direct sequence spread spectrum (DSSS) is an important and popular technique for military communications due to its low intercept probability. Therefore, the performance of the DSSS ad hoc networks and WMNs are intensively studied, mainly for military purpose. In the DSSS ad hoc networks, when each node uses the same 11-chip pseudo-noise (PN) spreading code, a 10.4 dB processing gain could be achieved [48]. Multiple antenna techniques are utilized to further increase the performance of the DSSS ad hoc networks and WMNs. For example, spatial diversity technique is used to mitigate the narrowband and the broadband interferences in the DSSS ad hoc networks and WMNs [49]. Since accurate array designs that embed spatial characteristics of the channel and radiation patterns are necessary to quantify the performance benefits, a spatial diversity model is designed for performance simulation. When comparing the performance with and without the spatial diversity for channels affected by the narrowband and the broadband interferences, equal-gain combining is used for the signals received by a five-element antenna array. Performance of throughput, delay, and maximum throughput performance improvement gain (MPG) are studied. Simulation results show that with the narrowband interference the average network throughput performance is 38.3% better with spatial diversity than without, whereas with the broadband interference it is 47.62%. The highest MPG with the narrowband interference is 2.06 and occurred with 100 nodes whereas for the broadband interference the MPG is 1.95 and occurred with 100 nodes. Furthermore, the average delay improvement for the narrowband was 56.69% with spatial diversity and 38.98% for the broadband.

Swaminathan et al. [50] study the effect of the receiver blocking problem on the performance of the DSSS ad hoc networks and WMNs in which a subset of the nodes employs multiple directional antennas. The studied scenario suits the infrastructure/backbone or hybrid WMNs well, where a number of mesh routers are placed. The mesh routers are relatively easier or more practical to be equipped with multiple antennas. Receiver blocking problem occurs if the receiver is blocked by an overheard CTS from a node that lies within the receiver's range, but not within the transmitter's range [51]. It is shown that the presence of the directional antennas in the ad hoc networks and the WMNs increases the network's vulnerability to the receiver blocking problem with a correspondingly severe impact on network performance. An advanced MAC protocol that employs negative CTS (NCTS) signaling is developed to solve this problem for the DSSS ad hoc networks and WMNs.

11.5 CONCLUSIONS AND DISCUSSIONS

Although the current research on the WMNs mainly focuses on the MAC protocol and the network layer, the physical layer techniques are more fundamental for the performance of the WMNs. However, the physical layer techniques for the conventional cellular systems could not be used directly in the WMNs, due to the architecture differences.

Multiple antenna techniques play important roles in physical layer techniques. This chapter provides a survey of multiple antenna techniques for the WMNs. By using multiple-antenna techniques, the capacity and throughput of the WMNs could be remarkably enlarged, and the energy efficiency and routing performance of the WMNs could be greatly improved. Furthermore, implementing the multiple antenna techniques could provide better connectivity and more accurate location estimation for the WMNs.

The beamforming and the directional antennas have been intensively studied, due to the low complexity of the signal processing and the multihop aspect of the WMNs. In addition, the other conventional multiple-antenna techniques, such as space–time coding and spatial multiplexing, have been exploited for the WMNs combining with the new MAC protocols from cross-layer approaches. However, the emerging multiple-antenna techniques, such as the virtual MIMO and the cooperative diversity schemes, which utilize the distributed character of the nodes, are also promising for the WMNs and need further research.

REFERENCES

1. R. Bruno, M. Conti, and E. Gregori, "Mesh Networks: Commodity Multihop Ad Hoc Networks," *IEEE Commun. Mag.*, vol. 43, pp. 123–131, March 2005.
2. I.F. Akyildiz and X. Wang, "A Survey on Wireless Mesh Networks," *IEEE Commun. Mag.*, vol. 43, pp. S23–S30, September 2005.
3. J.H. Winters, "Smart Antennas for Wireless Systems," *IEEE Pers. Commun.*, vol. 5, No. 1, pp. 23–27, February 1998.
4. G.J. Foschini and M.J. Gans, "On Limits of Wireless Communications in a Fading Environment When Using Multiple Antennas," *Wireless Pers. Commun.*, vol. 6, pp. 311–335, March 1998.
5. G.J. Foschini, "Layered Space–Time Architecture for Wireless Communication in a Fading Environment When Using Multiple Element Antennas," *Bell Labs Tech. J.*, pp. 41–59, Autumn 1996.
6. L. Zheng and D.N.C. Tse, "Diversity and Multiplexing: A Fundamental Tradeoff in Multiple-Antenna Channels," *IEEE Trans. Info. Theory*, vol. 49, pp. 1073–1096, May 2003.
7. Ming-Ju Ho, G.L. Stber, and M.D. Austin, "Performance of Switched-Beam Smart Antennas for Cellular Radio Systems," *IEEE Trans. Veb. Technol.*, vol. 47, pp. 10–19, February 1998.
8. Z. Zhang, M.F. Iskander, Z. Yun, and A. Host-Madsen, "Hybrid Smart Antenna System Using Directional Elements Performance Analysis in Flat Rayleigh Fading," *IEEE Trans. Antennas Propagat.*, vol. 51, No. 10, pp. 2926–2935, October 2003.

9. P. Viswanath, D.N.C. Tse, and R. Laroia, "Opportunistic Beamforming Using Dumb Antennas," *IEEE Trans. Info. Theory*, vol. 48, pp. 1277–1294, June 2002.

10. H.L. Hu, M. Weckerle, J. Luo, and E. Schulz, "Organized Beam-Hopping Technique with Adaptive Transmission Mode Selection for Mobile Communication Systems," *Euro. Trans. Telecommun.*, in press.

11. H.L. Hu and J.K. Zhu, "Performance Analysis of Distributed-Antenna Communication Systems Using Beam-Hopping under Strong Directional Interference," *J. Wireless Pers. Commun.*, vol. 32, pp. 89–105, January 2005.

12. V. Tarokh, N. Seshadri, and A.R. Calderbank, "Space–Time Codes for High Data Rate Wireless Communication: Performance Criterion and Code Construction," *IEEE Trans. Info. Theory*, vol. 44, pp. 744–765, March 1998.

13. S. Sanayei and A. Nosratinia, "Antenna Selection in MIMO Systems," *IEEE Commun. Mag.*, vol. 42, pp. 74–80, October 2004.

14. G. Li, L. Lily Yang, W.S. Conner, and B. Sadeghi, "Opportunities and Challenges for Mesh Networks Using Directional Antennas," www.cs.ucdavis. edu/~prasant/WIMESH/p13.pdf.

15. S. Yi, Y. Pei, and S. Kalyanaraman, "On the Capacity Improvement of Ad Hoc Wireless Networks Using Directional Antennas," *Proceedings of ACM MOBI-HOC 2003*, pp. 108–116, June 2003.

16. A. Spyropoulos and C.S. Raghavendra, "Capacity Bounds for Ad-Hoc Networks Using Directional Antennas", *Proceedings of IEEE ICC 2003*, vol. 26, pp. 348–352, May 2003.

17. K. Dyberg, L. Farman, F. Eklof, J. Gronkvist, U. Sterner, and J. Rantakokko, "On The Performance of Antenna Arrays in Spatial Reuse TDMA Ad Hoc Networks," *Proceedings of IEEE MILCOM 2002*, pp. 270–275, October 2002.

18. S. Ye and R.S. Blum, "On the Rate Regions for Wireless MIMO Ad Hoc Networks," *Proceedings of IEEE VTC 2004 Fall*, vol. 3, pp. 1648–1652, 2004.

19. S. Bellofiore, J. Foutz, R. Govindarajula, et al., "Smart Antenna System Analysis, Integration and Performance for Mobile Ad-Hoc Networks (MANETs)," *IEEE Trans. Antennas Propagat.*, vol. 50, pp. 571–581, May 2002.

20. L. Thomson, "Performance of Systems Using Adaptive Arrays in Mobile Ad Hoc Networks," *Proceedings of IEEE MILCOM 2003*, 2003.

21. R. Radhakrishnan, D. Lal, J. Caffery, Jr., and D.P. Agrawal, "Performance Comparison of Smart Antenna Techniques for Spatial Multiplexing in Wireless Ad Hoc Networks," *Proceedings of ISWPMC 2002*, pp. 614–619, October 2002.

22. T. Ueda, K. Masayama, S. Horisawa, M. Kosuga, and K. Hasuike, "Evaluating the Performance of Wireless Ad Hoc Network Testbed with Smart Antenna," *Proceedings of IMWCN 2002*, pp. 135–139, September 2002.

23. M. Hennhoefer, M. Haardt, and G.D. Galdo, "Increasing the Throughput in Multi-hop Wireless Networks by Using Spatial Multiplexing," *Proceedings of IEEE VTC 2004 Spring*, May 2004.

24. M. Park, S. Nettles, and R.W. Heath, Jr., "Improving Throughput and Fairness for MIMO Ad Hoc Networks Using Antenna Selection Diversity," *Proceedings of IEEE GLOBECOM 2004*, pp. 3363–3367, November 2004.

25. B. Chen and Michael Gans, "Limiting Throughput of MIMO Ad Hoc Networks," *Proceedings of IEEE ICASSP 2005*, March 2005.

26. M. Hu and J. Zhang, "MIMO Ad Hoc Networks: Medium Access Control, Saturation Throughput, and Optimal Hop Distance," Special Issue on Mobile Ad Hoc Networks, *J. Commun. Networks*, pp. 317–330, December 2004.

27. M.M. Islam, R. Pose, and C. Kopp, "Multiple Directional Antennas in Suburban Ad Hoc Networks," *Proceedings of IEEE ITCC 2004*, pp. 385–389, 2004.

28. Sung-Ho Kim and Young-Bae Ko, "A Directional Antenna Based Path Optimization Scheme for Wireless Ad Hoc Networks," *Proceedings of MSN 2005*, LNCS, pp. 317–326, 2005.

29. T. Joshi, H. Gossain, C. Cordeiro, and D.P. Agrawal, "Route Recovery Mechanisms for Ad Hoc Networks Equipped with Switched Single Beam Antennas," *Proceedings of 38th Simulation Symposium*, pp. 41–48, April 2005.

30. A.K. Saha and D. Johnson, "Routing Improvements Using Directional Antennas in Mobile Ad Hoc Networks," *Proceedings of IEEE GLOBECOM 2004*, pp. 2902–2908, 2004.

31. Y. Li, H. Man, J. Yu, and Y. Yao, "Multipath Routing in Ad Hoc Networks Using Directional Antennas," *Proceedings of IEEE SAWWC 2004*, pp. 119–122, April 2004.

32. Y.T. Hou, Y. Shi, H.D. Sherali, and J.E. Wieselthier, "Online Lifetime-Centric Multicast Routing for Ad Hoc Networks with Directional Antennas," *Proceedings of IEEE INFOCOM 2005*, pp. 761–772, March 2005.

33. A. Nasipuri, K. Li, and U.R. Sappidi, "Power Consumption and Throughput in Mobile Ad Hoc Networks Using Directional Antennas," *Proceedings of IEEE ICCCN 2002*, October 2002.

34. A. Spyropoulos and C.S. Raghavendra, "Energy Efficient Communications in Ad Hoc Networks Using Directional Antennas," *Proceedings of IEEE INFOCOM 2003*, April 2003.

35. R.F. Farha and R.S. Adve, "Combining Space–Time Coding with Power-Efficient Medium Access Control for Mobile Ad-Hoc Networks," *Proceedings of IEEE GLOBECOM 2003*, vol. 22, pp. 92–96, December 2003.

36. S.K. Jayaweera, "An Energy-efficient Virtual MIMO Communications Architecture Based on V-BLAST Processing for Distributed Wireless Sensor Networks," *Proceedings of IEEE SECON 2004*, October 2004.

37. N.S. Fahmy, T.D. Todd, and V. Kezys, "Distributed Power Control for Ad Hoc Networks with Smart Antennas," *Proceedings of IEEE VTC 2002*, pp. 2141–2144, 2002.

38. L. Tong and P. Ramanathan, "Energy Efficient Multicasting Using Smart Antennas for Wireless Ad Hoc Networks in Multipath Environments," *Proceedings of IEEE GLOBECOM 2004*, November 2004.

39. S. Guo and O. Yang, "Improving Energy Efficiency for Multicasting in Wireless Ad-Hoc Networks with Adaptive Antennas," *Proceedings of IEEE WiMob 2005*, August 2005.

40. C. Bettstetter, C. Hartmann, and C.S. Moser, "How Does Randomized Beamforming Improve the Connectivity of Ad Hoc Networks," *Proceedings of IEEE ICC 2005*, May 2005.

41. I. Martinez and J. Altuna, "A Cross-Layer Design for Ad Hoc Wireless Networks with Smart Antennas and QoS Support," *Proceedings of IEEE PIMRC 2004*, vol. 1, pp. 589–593, September 2004.

42. J. Mundarath, "QoS and MAC Protocols for Adaptive Antenna Array Based Wireless Ad Hoc Networks in Multipath Environments," Masters Research Report, National Institute of Technology, Calicut, India, August 2003.

43. D. Saha, S. Roy, S. Bandyopadhyay, T. Ueda, and S. Tanaka, "A Distributed Feedback Control Mechanism for Priority-based Flow-Rate Control to Support QoS Provisioning in Ad Hoc Wireless Networks with Directional Antenna," *Proceedings of IEEE ICC 2004*, vol. 27, pp. 4172–4176, June 2004.

44. T. Ueda, S. Tanaka, S. Roy, D. Saha, and S. Bandyopadhyay, "A Priority-based QoS Routing Protocol with Zone Reservation and Adaptive Call Blocking for Mobile Ad Hoc Networks with Directional Antenna," *Proceedings of IEEE GLOBECOM 2004*, pp. 50–55, 2004.

45. E. Baccarelli, M. Biagi, C. Pelizzoni, and Giuseppe Razzano, "Contracted QoS Policies for Distributed Space Division Multiple Access for Ad-Hoc Networks," *Proceedings of IEEE QSHINE 2004*, pp. 36–43, October 2004.

46. N. Malhotra, M. Krasniewski, C. Yang, S. Bagchi, and W. Chappell, "Location Estimation in Ad Hoc Networks with Directional Antennas," *Proceedings of IEEE ICDCS 2005*, pp. 633–642, June 2005.

47. G. Jakllari, W. Luo, and S.V. Krishnamurthy, "An Integrated Neighbor Discovery and MAC Protocol for Ad Hoc Networks Using Directional Antennas," *Proceedings of IEEE WoWMoM 2005*, pp. 11–21, June 2005.

48. IEEE Standard 802.11b: Supplement to ANSI/IEEE Std 802.11 Edition, Wireless LAN Medium Access Control (MAC) and Physical Layer (PHY) Specifications, IEEE, New York, 1999.

49. S. Furman and M. Gerla, "The Design of a Spatial Diversity Model to Mitigate Narrowband and Broadband Interference in DSSS Ad Hoc Networks," *Proceedings of IEEE ICC 2003*, pp. 1201–1205, 2003.

50. A. Swaminathan, D.L. Noneaker, and H.B. Russell, "The Receiver Blocking Problem in a DS Mobile Ad Hoc Network with Directional Antennas," *Proceedings of IEEE MILCOM 2004*, pp. 920–926, 2004.

51. V. Bharghavan, A. Demers, S. Shenker, and L. Zhang, "MACAW: A Medium Access Protocol for Wireless LANs," *Proceedings of ACM SIGCOMM 1994*, pp. 212–225, 1994.

12

STANDARDIZATION OF WIRELESS LAN MESH NETWORKS IN IEEE 802.11s

Shah I. Rahman

CONTENTS

One of the technologies that may change our lives most over the next few years is wireless mesh networking (WMN). Wireless mesh technologies have been around almost as long as wireless local area networks (WLANs) have, but they could never take the wireless industry by storm as they have in recent times. As the popularity of WMNs is growing, end users are demanding higher bandwidth, greater coverage, and improved reliability. To date, there is no official standard for WMNs. However, a fertile field for standards to emerge has been created with market demands and industry interests. The industry has come together at various IEEE 802 workgroups to standardize WMNs with the right ingredients and framework. These standards may fuel widespread adoption of WMNs around the world. This chapter examines the standardization activities of WLAN mesh networking Task Group at IEEE code-named 802. 11s for amendments to the IEEE 802.11 base standard. WLAN mesh networking has been the forerunner in topic research, product development, and network deployment of wireless mesh. At the time of writing this chapter, the Task Group has created a baseline draft based on mandatory WLAN mesh requirements, followed by more advanced and optional features. The roadmap ahead is to develop a full, official extended service set (ESS) mesh standard within the next 2 y [1].

12.1 INTRODUCTION

Several emerging and commercially interesting applications for high-speed networks based on WLAN mesh have been deployed recently. WLAN mesh technology is envisioned as an economically viable networking paradigm to build up broadband, municipal, public safety, tactical, and large-scale wireless commodity networks often called hot zones [2]. Original mesh architectures emerged from mobile ad hoc networks (MANETs) used for military networks. IETF MANET Work Group has been developing various MANET protocols for almost a decade [3–6]. Mesh networks are different from MANETs in that there is more infrastructure communication rather than direct, peer-to-peer communication. This results in a natural hierarchy in the networks. As mesh networks became popular and many vendors started building products for mesh networks, the need for standardization and interoperability became evident around 2003. IEEE 802.15.5 Task Group was formed in November 2003 followed by 802.11s Task Group in September 2004.

IEEE 802.11s started with a charter to extend WLAN for ESS mesh networking. Existing IEEE 802.11 standards specify WLAN access network operations between WLAN clients [stations (STAs)] and access points (APs). In order to extend IEEE 802.11 standards for mesh, backhaul (infrastructure WLAN links) and gateway (infrastructure WLAN to wired-LAN links) operations must be amended to the existing standards (see Figure 12.1). These operations are in the areas of medium access control (MAC), power saving, routing and forwarding, interworking with 802 other networks, security, quality of service (QoS), management and configuration of a WLAN mesh network. The current baseline draft was created from a merged proposal from two WLAN industry groups: SEE-Mesh and WiMesh, consisting of more than 25 companies and universities from a wide variety of backgrounds such as researchers, chip vendors, system developers, and network integrators. The draft forms the foundation for fulfilling the vision of a self-configuring, self-healing, and self-monitoring WLAN mesh standard and addresses many aspects of practical WLAN mesh

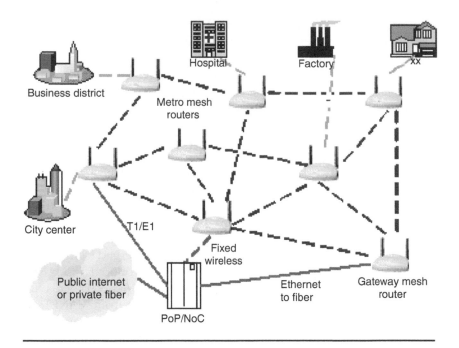

Figure 12.1   A wireless LAN mesh network.

networking. Much work is needed to make the standard truly success-ful and complete.

12.2   WLAN MESH PRIMER

A WLAN mesh network is a fully IEEE 802.11-based wireless network that employs multihop communications to forward traffic en route to and from wired Internet entry points. Recently standardized IEEE 802.11a [7] and 802.11g [8] WLAN standards have substantially increased data rates of WLAN networks by using spectrally efficient modulation schemes (up to 54 Mbps). Upcoming IEEE 802.11n stand-ard based on Multiple-input Multiple-output (MIMO) techniques promises to increase data rates further (up to 200–300 Mbps) [1]. A WLAN mesh network uses 802.11-based physical layer (PHY) device and medium access (MAC) for providing the functionality of an ESS mesh network. The 802.11 AP (known as mesh point [MP] when used in WLAN mesh) establishes wireless links among each other to enable automatic topology learning and dynamic path configuration. The MP-to-MP links form a wireless backbone known as mesh backhaul, which provides users with low-cost, high-bandwidth, and seamless multihop interconnection services with a limited number of Internet entry points and with other users within the network. Each MP may optionally provide wireless access connections to users known as mesh access. These devices are called mesh access points (MAPs). Figure 12.1 shows a typical WLAN mesh network and its various components. WLAN mesh networks are defined [1] as:

> A WLAN mesh is an IEEE 802.11-based wireless distribution system (WDS), which is a part of a DS, consisting of a set of two or more MPs interconnected by IEEE 802.11 links and communicating through the WLAN mesh services. A WLAN mesh may support zero or more entry points (mesh portals [MPPs]), automatic topology learning, and dynamic path selection (including across multiple hops).

12.2.1   WLAN Mesh Topologies

WLAN mesh networks are targeted primarily for home, commercial, neighborhood, community, municipality, rural broadband, emergency and first responder, public safety, small to medium business, large enterprise, and military networks. Each of these markets represents purely one or a combination of the two main mesh topologies: (1) ad hoc and (2) infrastructure. Ad hoc topologies breed from original MAN-ETs; however, they are augmented with a few Internet entry points known as MPPs whenever possible. These networks require mobility as an integral part of the mesh backhaul links as well as dynamic establishment and tearing of these links. Public safety, emergency and first responder, and military networks are examples of ad hoc mesh networks. Infrastructure topologies are multihop WLAN networks extending a single-hop, access network with wireless backhaul mesh to a single or multiple portals. These networks are mostly fixed and static in nature. They do not require mobility of backhaul links except roaming of APs due to radio frequency (RF) or other types of link failures. Home, community, municipality, rural broadband, small and medium businesses (SMB), and enterprises require infrastructure mesh networks.

A third type of topology that combines the flexibility of ad hoc networks and robustness of infrastructure networks is gaining popularity. This is hybrid mesh topology, where ad hoc devices connect to the infrastructure as necessary. For example, a fire truck ad hoc network on a service call may connect to the city WLAN mesh infrastructure as necessary for communicating with the central control room located in another part of the city.

12.2.2   WLAN Mesh Standardization

The IEEE 802.11 working group [1] is an umbrella organization that contains several Task Groups developing technologies for 802.11 WLAN environment. The standardization activities currently underway at the 802.11s Task Group [1] promise to lead to the availability of a highly interoperable WLAN mesh standard. The scope of the group is to amend to the IEEE 802.11 MAC protocol such that an 802.11 WDS that supports both broadcast/multicast and unicast frame delivery at the MAC layer using radio-aware metrics. The 802.11 PHY is not expected to change in any manner. The Task Group started to hear proposals from March 2005 and downselected two full proposals by November 2005. The two surviving full proposals agreed to merge into a single, joint proposal at the interim Task Group meeting at Hawaii in January 2006. The joint proposal was confirmed by the Task Group at Denver in March 2006. The baseline draft is ready for presentation at Jacksonville meeting in May 2006. The release of a complete standard is expected by the end of 2006 or early 2007.

The rest of the chapter examines each major area IEEE 802.11s Task Group is developing for amendment to the IEEE 802.11 base standard, and concludes by showing how WLAN mesh standard would be effective and efficient in building large, scalable WMNs for different applications. The standard is expected to be a major catalyst in popularizing WLAN mesh much like WLAN access was popularized by various IEEE 802.11 standards.

12.3   WLAN MESH BASIC SERVICES

12.3.1   WLAN Frame Formats

WLAN mesh frame formats reuse IEEE 802.11 MAC frame formats defined [9] and extend them appropriately for supporting ESS mesh services. MAC frame header is appended with a mesh forwarding control field, which is a 24-bit field that includes a time to live (TTL) field for use in multihop forwarding to eliminate the possibility of infinite loops and a mesh E2E sequence number for use in controlled broadcast flooding and other services. This field is present in all frames of type extended with subtype mesh data (+ CF-ACK).

There are two new control frames proposed by the 802.11s first draft: request to switch (RTX) and clear to switch (CTX) for backhaul channel change operations.

The exchange of 802.11 management frames shall be supported between neighboring MPs. All 802.11 management frames are extended to include mesh-specific information elements (IEs). A non-exhaustive list of these IEs includes:

1. Mesh ID
2. Mesh capability
3. Neighbor list
4. MPP reachability
5. Peer request
6. Peer response
7. Active profile announcement

Neighbor discovery, HWMP routing, congestion control, beaconing and synchronization, and mesh deterministic access (MDA) use 802.11 management action frames and encode IEs defined by each mechanism [1]. For example, mesh routing-specific route request (RREQ), route response (RREP), route response acknowledgment (RREP-ACK), route error (RERR) IEs are defined each and encoded into a specific action frame to be used for mesh routing and forwarding [1].

12.3.2 Backhaul Channel Selection

This is a mandatory feature of 802.11s draft standard. A WLAN mesh network topology may include MPs with one or more radio interfaces and may utilize one or more channels for communication between MPs. When channel switching is not supported, each radio interface on an MP operates on one channel at a time. But the channel may change during the lifetime of the mesh network according to dynamic frequency selection (DFS) requirements. The specific channel selection scheme used in a WLAN mesh network may vary with different topology and application requirements. A set of MP radio interfaces that are interconnected to each other by a common channel are referred to as a unified channel graph (UCG). The same device may belong to different UCGs. An MP logical radio interface that is in simple unification mode selects a channel in a controlled way such that it enables the formation of a UCG that becomes merged and hence fully connected. The MP logical radio interface thus establishes links with neighbors that match the mesh ID and mesh profile and selects its channel based on the highest channel precedence value.

12.3.2.1 Simple Channel Unification Protocol

At boot time, an MP logical radio interface that is configured in simple channel unification mode performs passive or active scanning to discover neighboring MPs. If an MP is unable to detect any neighboring MPs, it adopts a mesh ID from one of its profiles, and selects a channel for operation as well as an initial channel precedence value. The initial channel precedence value may be initialized to the number of microseconds since the boot time of the MP plus a random value. In the event that an MP logical radio interface that is configured in simple channel unification mode discovers a disjoint mesh, i.e., the list of candidate peer MPs spans more than one channel, it selects the channel that is indicated by the candidate peer MP which has the numerically highest channel precedence indicator to be the unification channel.

12.3.2.2 Channel Cluster Switch Protocol

The MP that determines the need to switch the channel of its cluster first chooses a channel cluster switch wait timer. It sets a local timer with this wait time and then sends a channel cluster switch announcement frame to each peer MP that has an active association in the UCG. It copies the value of the new candidate channel and new candidate channel precedence indicator, and sets a channel switch wait time.

If an MP receives a channel cluster switch frame with a channel precedence value larger than the current channel precedence value of the logical interface on which the frame was received, the MP sets a channel cluster switch timer equal to the channel switch wait timer value of the frame and then sends a channel cluster switch announcement frame to each peer MP that has an active association on the logical radio interface, copying the values from the received channel cluster switch announcement frame.

It is possible that more than one MP in the UCG may independently detect the need to switch channels and send separate channel cluster switch announcement frames. If an MP receives more than one channel cluster switch announcement frame, it acts upon the frame only if the channel precedence value is larger than the channel precedence value of a previously received channel cluster switch announcement frame. In case a newly received channel cluster announcement frame has the same channel precedence value as a previously received frame, the new frame is acted upon only if the source address is smaller than the source address from the previously received frame. If the MP acts upon the newly received channel cluster switch frame, it updates its candidate channel and candidate channel precedence indicator, sets its channel cluster switch wait timer to the channel switch wait value of the frame, and then sends a channel cluster switch announcement frame to each peer MP that has an active association on the logical radio interface, copying the values from the received channel cluster switch announcement frame.

If a channel switch wait timer has been set on an MP, the MP does not originate a new channel cluster switch announcement frame during the duration of the channel switch wait timer. When the channel cluster switch wait timer expires, the MP switches its radio interface to the candidate channel and updates its channel precedence indicator to the candidate channel precedence indicator.

12.3.3 Mesh Link Operations

These are mandatory features of 802.11s draft standard. An MP must be able to establish at least one mesh link with a peer MP, and may be able to establish many such links simultaneously. It is possible that there are more candidate peer MPs than the device is capable of being associated with simultaneously. In this case, the MP must select MPs to establish peer links based on some measure of signal quality, such as gathered during the discovery phase, or other statistics received from candidate neighbor MPs. An MP will continue to look for received beacons on any of the UCGs it is operating on. On receipt of a beacon from an unknown neighboring MP, but containing a matching mesh ID, an MP will attempt to create a peer link to the neighboring MP.

The purpose of the local link-state discovery procedure is to populate the r and e_{pt} fields for each peer MP in the neighboring table. These are subsequently used by the route establishment algorithm to determine the most efficient available routes. All mesh links are asymmetric, in that a pairwise link consists of one node designated as superordinate and the other designated as subordinate. These labels are illustrative and represent no hierarchical relationship. In order to ensure that the measured link state is symmetric, the superordinate node is responsible for making a determination as to the link quality. It may use any method, but must make a determination as to the following two parameters:

- r: current bit rate in use, i.e., the modulation mode
- e_{pt}: packet error rate at the current bit rate for a data frame with a 1000 byte payload

A superordinate node makes this determination for a link in the superordinate, down, state, and at future intervals at its option. On making such a determination, it includes the information in a local linkstate announcement frame and transmits it to the subordinate node. On successful transmission of the frame, it updates the values in its neighboring MP table with the new values, changing the state from superordinate down to superordinate up if this is an initial assessment. A subordinate node shall update the values in its neighboring MP table whenever a local link-state announcement message is received.

12.3.4   Mesh Beaconing

This is an optional feature of 802.11s draft standard. Synchronization and beacon generation services in a WLAN mesh are based upon the procedures defined in clause 11.1 of 802.11 standard [9] for infrastructure and independent basic service set (IBSS) modes of operations.

12.3.4.1   Mesh Synchronization Protocol

An MP supporting synchronization may choose to be either synchronized or unsynchronized based on its own requirements or those of its peers. MPs synchronization behavior is communicated through the synchronization capability field within the WLAN mesh capability element. The synchronization behavior of the two classes is defined as follows:

1. Unsynchronized mode
2. Synchronized mode

An unsynchronized MP maintains an independent time synchronization function (TSF) timer and does not update the value of its TSF timer based on time stamps and offsets received in beacons or probe responses from other MPs. An unsynchronized MP may start its TSF timer independently of other MPs. A synchronized MP updates its timer based on the time stamps and offsets (if any) received in beacons and probe responses from other synchronized MPs. Synchronized MPs should attempt to maintain a common TSF timer called the mesh TSF timer.

12.3.4.2   Beacon Generation

Any MP may choose to beacon as defined either in the IBSS mode or in the infrastructure mode of operation in IEEE 802.11 [9] and 802.11e [10]. Both synchronized and unsynchronized MPs generate beacons according to the beacon generation procedures defined in clause 11.1.2.1 in 802.11 standard [9].

Unsynchronized MPs choose their own beacon interval and TSF independent of other MPs. Unsynchronized MPs may implement beacon collision avoidance mechanisms to reduce the chances that it will transmit beacons at the same time as one of its neighbors. The value of a beacon period attribute used by synchronized MPs equals a submultiple of the mesh delivery traffic indication message (DTIM) interval. Synchronized MPs use and advertise a nonzero self-TBTT offset value using the beacon timing element. Each synchronized MP can select its own beacon interval, but all synchronized MPs need to share a common mesh DTIM interval. The beacon intervals selected by an MP must always be submultiples of the mesh DTIM interval. A synchronized MP that establishes a mesh selects its beacon interval and the MP DTIM period, and determines the common mesh DTIM interval of the mesh. The mesh DTIM interval equals the product of the beacon interval and the mesh DTIM period. A Synchronized MP that joins an existing mesh needs to adopt the mesh DTIM interval of the mesh.

Another mode of beaconing is defined for MPs lacking beacon generation capability to support the ability to designate other MPs to broadcast beacons on their behalf. The designated beacon broadcaster method enables a designated MP to perform beaconing for a defined period of time while all other MPs defer from sending beacons. The beacon broadcaster role must be changed periodically.

12.3.4.3 Mesh Beacon Collision Avoidance Protocol

MPs implementing mesh beaconing may optionally adjust their TSF timers to reduce the chances that they will transmit beacons at the same time as one of their neighbors. Individual MPs may take steps either before or during association in order to select a target beacon transition time (TBTT) that does not conflict with its mesh neighbors. An MP may adjust its TSF timer if it discovers that its TBTT may repeatedly collide with the TBTT of a neighbor. Options of an MP for adjusting its TSF include advancing or suspending the TSF for a period of time. An MP may collect and report information about the TBTT of neighboring synchronized and unsynchronized MPs using a variety of techniques. In addition, 802.11k beacon reports may be used by MPs to exchange beacon timing information of their neighbors [11].

As another option, synchronized MPs may occasionally delay their beacons after their TBTTs for a random time. The random delay is chosen so that the transmission time is interpreted by the MP as not colliding with other beacons. This behavior further helps discovering neighbors through beacons in case they choose colliding offsets. The minimum-blocking channel assignment (MBCA) mechanism may then be used for choosing noncolliding offsets, in case any colliding offsets are observed.

12.4 WLAN MESH MEDIUM ACCESS CONTROL

MAC of the IEEE 802.11 is contention based, using distributed coordination function (DCF) mechanisms, or contention-free, using point coordination function (PCF) mechanisms. Contention-based MAC protocols are robust against environmental interference and noise, making them suitable for use in WLAN mesh networks. This section describes how the MAC is used and adapted in a WLAN mesh in IEEE 802.11s with appropriate amendments. Mandatory changes to 802.11 MAC are not specified [1], but two optional mechanisms for more suitably adapting 802.11 MAC services in a WLAN mesh network [1] are:

1. Multichannel MAC
2. MDA

12.4.1 Multichannel Medium Access Control

Thus far the 802.11 MAC has been based on a single channel, i.e., all the devices on the network share the same channel. Using a multichannel MAC, where transmissions can take place simultaneously on orthogonal channels, the aggregate throughput can be increased considerably. Multichannel MAC protocols are traditionally developed for multiradio devices. However, the common channel framework (CCF) described here enables the operation of a single-radio device in a multichannel environment. Known methods for channel access, e.g., DCF or enhanced distributed coordination function (EDCF), can be used within this framework. MPs can utilize the common channel to select an available channel as shown in Figure 12.2. This, in essence, is a dynamic channel allocation scheme. The destination channel information (channel n) is exchanged using the RTX and CTX frames followed by data frame transmission on the destination channel n. While the data frame transmission is ongoing on channel n, another transmission can be initiated on another destination channel m. A single-radio MP on the common channel cannot communicate with MPs on other channels. At the same time, single-radio MPs on other channels cannot know the network status on the common channel

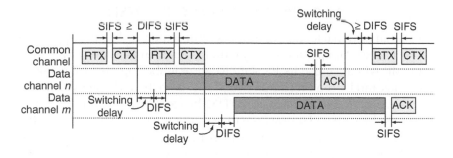

Figure 12.2   Common channel framework.

and vice versa. A multichannel MAC protocol designed for single radio should therefore:

1. Facilitate connectivity among arbitrary MPs that may be on different channels
2. Facilitate protection of the ongoing transactions

In order to address the foregoing issues the concept of a channel coordination window (CCW) is available in CCF. At the start of CCW, CCF-enabled MPs tune to the common channel. This enables arbitrary MPs to establish communication with each other. Secondly, at the start of CCW, the channel occupancy status is reset and MPs can renegotiate channels. CCW is repeated with a period P, and the duration of CCW is usually a fraction of P. A channel coordination mechanism is used with the help of a common control channel (CCC).

12.4.2   Mesh Deterministic Access

MDA is an optional access method that allows supporting MPs to access the channel with lower contention than otherwise in selected times. The method sets up time periods in mesh neighborhoods when a number of MDA-supporting MPs that may potentially interfere with each others' transmissions or receptions are set to not initiate any transmission sequences. For each such time period, supporting MPs that set up the state for the use of these time periods are allowed to access the channel using MDA access parameters (*CW Min, CW Max*, and *AIFSN*). In order to use the MDA method for access, an MP must be a synchronized MP. The MDA method is described in detail [1].

12.5 POWER SAVING IN MESH

This is an optional feature of 802.11s draft standard. An MP supporting power save (PS) operations may either operate in an active or PS state. The MP will advertise its PS state to all neighboring MPs by using its beacons and by sending a null data frame with the PS bit active. MPs in PS mode periodically listen for DTIM beacons. An MP waking up to receive a beacon will stay awake for a minimum period of ad hoc traffic indication map (ATIM) window as indicated in their beacons, before returning to sleep. MPs in PS mode will also wake up according to any negotiated schedule as part of traffic specification (TSPEC) setup with other MPs. The MP will remain awake until the end of the service period.

An MP wishing to communicate with MPs that are in PS mode could buffer the traffic targeted for these MPs in one of the following ways:

1. Send traffic to these MPs only on agreed schedules as negotiated as part of automatic power save delivery (APSD) TSPEC setup (see [10] for details on APSD and TSPEC).
2. Send directed or broadcast ATIM frames to MPs in PS mode during ATIM window in order to signal them to remain awake and wait for further traffic.
3. Send a single null data frame to MPs in PS mode during their ATIM window in order to reactivate a flow that has been suspended or to signal PS state change.

Any MP that wishes to communicate with an unsynchronized MP, and enters PS mode will wake up for the BSS DTIM beacon of the MP. If such an MP wishes to communicate with more than one unsynchronized MP, it will wake up for the BSS DTIM beacons for each such MP in addition to any mesh DTIM TBTT that may be scheduled for its synchronized MP neighbors.

12.6 NETWORK DISCOVERY IN MESH

This is a mandatory feature of 802.11s draft standard. Mesh formation requires that the members of a mesh network have sufficient information about themselves and the available connections between them. This process requires detection of mesh members through beacons or active scanning using mesh probe requests, followed by the exchange of routing information, which may include link-state information. Mesh formation is a continuous process that entails monitoring of neighbor nodes and their connectivity so as to detect and react to changes in mesh memberships and in connectivity between mesh members. Sections 12.6.1 and Section 12.6.2 describe the two subelements of network discovery:

1. Topology discovery
2. Neighbor discovery

12.6.1 Topology Discovery

802.11s draft standard supports topology discovery using profiles. A device must support at least one profile. A profile consists of:

- Mesh ID
- Path selection protocol identifier
- Path selection metric identifier

The path selection protocol and path selection metrics in use may be different for different profiles.

12.6.2 Neighbor Discovery

An MP performs passive or active scanning to discover neighboring MPs. In case of passive scanning, a device is considered a neighbor MP if and only if all of the following conditions are met (similar conditions must be met with probe responses in case of active scanning):

1. A beacon is received from that device.
2. The received beacon contains a mesh ID that matches the mesh ID of at least one of the profiles on the MP.
3. The received beacon contains WLAN mesh capabilities: Version, MP-active indication, and path selection protocol identifier with a matching metric identifier.

A neighbor MP is considered a candidate peer if and only if:

1. The beacon contains a WLAN mesh capability IE with nonzero peer link available value.
2. The MP attempts to discover all neighbors and candidate peer devices, and maintains the neighbor MP information indicating the MAC address of each device, the most recently observed link-state parameters, the received channel number, and the state equal to neighbor or candidate peer as determined by the rules in this section.

When MP devices are discovered, the path selection protocol and metric should be checked for a match with the profile. If there is no match, the newly discovered device should be ignored. If an MP is unable to detect any neighbor MPs, it adopts a mesh ID from one of its profiles, and proceeds to the active state, which, in the case of an MAP is the AP initialization state. This will occur when the MP is the first device to power on (or multiple MPs power on simultaneously). Any peer MP links will be established later as part of the continuous mesh formation procedures.

12.7   MESH ROUTING AND FORWARDING

Wireless routing has been a topic of research interest for over a decade. Routing frames from a wireless host by the backbone network to the appropriate intra-, inter-, or gateway-node through dynamic RF environments has been a major challenge to date. WLAN mesh routing protocol adapts earlier works of IETF MANET Work Group as well as wireless spanning tree protocols (STP) are specified [1]. The terms mesh path selection and mesh forwarding are used to describe selection of single-hop or multihop paths and forwarding of data frames across these paths between MPs at the link layer. Data messages use the 802.11 standard four-address format [9], with 802.11e extensions for 802.1Q tag transfer [12], and some additional mesh-specific information. Simple client STA nodes associate with one of the mesh AP devices as normal, since the mesh AP devices are logically a collection of APs that are part of the same ESS. Mesh path selection services consist of baseline management messages for neighbor discovery, local link-state measurement and maintenance, and identification of an active path selection protocol. Each WLAN mesh uses a single method to determine paths through the mesh, although a single device may be capable of supporting several methods. This section defines a path selection protocol targeted for small to medium, unmanaged mesh networks.

12.7.1   WLAN Mesh Routing Framework

IEEE 802.11s draft specifies an extensible framework to enable flexible implementation of path selection protocols and metrics within the 802.11s standard. It specifies a default mandatory protocol and metric for all implementations, to ensure baseline interoperability between devices from different vendors. However, it also allows any vendor to implement any protocol and/or metric in the 802.11s framework to meet special application needs. An MP may include multiple protocol implementations (e.g., default protocol, optional protocols, and future standard protocols), but only one protocol will be active on a particular link at a time. Different WLAN meshes may have different active path selection protocols, but a particular mesh will have one active protocol at a time.

WLAN mesh routing framework is largely based on layer-2 or MAC layer routing. From a Transport Control Protocol/Internet Protocol (TCP/IP) perspective, this framework works closely with MAC and logical link control (LLC) layers while most wired routing protocols, such as Border Gateway Protocol (BGP) and Open Shorter Path First (OSPF), operate in layer-3 or IP layer. This raises the question of how effective would WLAN mesh routing framework be in relatively larger mesh networks. The IEEE 802.11s Task Group charter limits itself to a maximum of 32 routing nodes, which implies that the framework will probably not scale beyond this upper bound. In other words, a single WLAN mesh network can be up to 32 nodes and must be augmented with 802 LANs and/or IP layer routing in order to build larger WLAN mesh networks. Layer-2 based routing alone may not be practical in many large networks, such as a citywide wireless mesh network.

12.7.2 Path Selection Metrics

IEEE 802.11s draft standard allows, in principle, a WLAN mesh to be implemented with any path selection metrics. This section defines a default radio-aware path selection metric to enable baseline interoperability. In order to compute the unicast forwarding table from the cached link-state information generated by each node, the MP must first calculate the link cost for each pairwise link in the mesh. This section defines a default link metric that may be used by a path selection protocol to identify an efficient radio-aware path. The extensibility framework allows this metric to be overridden by any routing metric as specified in the active profile.

The cost function for establishment of the radio-aware paths is based on airtime cost. Airtime cost reflects the amount of channel resources consumed by transmitting the frame over a particular link. This measure is approximate and designed for ease of implementation and interoperability. The airtime cost for each link is calculated as:

$$C_a = \left(O_{ca} + O_p + \frac{B_t}{r} \right) \times \left(\frac{1}{1 - e_{pt}} \right)$$

where O_{ca}, O_p, and B_t are constants listed below, and the input parameters r and e_{pt} are the bit rate in Mbps and the frame error rate for the test frame size B_t, respectively. The rate r is dependent on local implementation of rate adaptation and represents the rate at which the MP would transmit a frame of standard size (B_t) based on current conditions. Estimation of e_{pt} is a local implementation choice and is intended to estimate the e_{pt} for transmissions of standard size frames (B_t) at the current transmit bit rate that is used to transmit frames of that size (r). Packet drops due to exceeding TTL should not be included in this estimate as they are not correlated with link performance.

- O_{ca}: 75 μs (802.11a) and 335 μs (802.11b)
- O_p: 110 μs (802.11a) and 364 μs (802.11b)
- B_t: 8224 (802.11a and 802.11b)

12.7.3   Hybrid Wireless Mesh Protocol

This is the mandatory routing protocol of the 802.11s draft standard. Hybrid wireless mesh protocol (HWMP) is a combination of ad hoc and spanning tree-based routing protocols that incorporates both proactive and reactive components for targeting all of ad hoc, infrastructure, and hybrid WLAN mesh markets. HWMP brings the best of both worlds (ad hoc and infrastructure) together by blending the flexibility of on-demand route discovery with extensions that enable efficient proactive routing to MPPs. This combination allows MPs to perform the discovery and maintenance of optimal routes themselves or to rely on the formation of a tree structure based on a root node (logically placed in an MPP). In both cases, neighbor node selection is based on a predefined metric. HWMP uses a single set of protocol primitives derived from ad hoc on-demand distance vector (AODV) [4]. If a mesh network has no root node configured (e.g., an ad hoc network), on-demand route discovery is used for all routing in the mesh network. A tree-structured network is enabled by configuring an MP (typically an MPP) as a root node. In that case, other MPs proactively maintain routes to the root node and a proactive, distance vector routing tree is created and maintained. Some of the key benefits of the HWMP hybrid routing approach are:

- Flexibility to adapt to the requirements of a wide range of scenarios, ranging from fixed to mobile mesh networks
- MPs discover and use the best metric path to any destination in the mesh with low complexity

In addition, when a root node is configured in the mesh:

■ Flooding of route discovery packets in the mesh is reduced if the destination is outside the mesh
■ The need to buffer messages at the source while on-demand route discovery is in progress is reduced
■ Nondiscovery broadcast and multicast traffic can be delivered along the tree topology
■ On-demand routes have the topology tree to fall back on should an on-demand route become unavailable or during route rediscovery

HWMP has a unique hybrid routing feature. If a proactive tree exists, it may be used by default while on-demand route discovery is in progress for intramesh destinations. An MP may choose to rely solely on the routing tree for all intramesh as well as gateway traffic routing.

12.7.3.1 Tree-Based Routing in HWMP

If an MP (typically an MPP) in a WLAN mesh is optionally configured as a root node, other MPs proactively maintain routes to the root node using topology discovery primitives. HWMP topology formation begins when the root portal announces itself with the root announcement message, which contains the distance metric and a sequence number. The value of the metric is zero. Any MP hearing the announcement directly updates its route table as directly connected child of the root and the metric associated with the link. It then rebroadcasts the root announcement with an updated distance vector metric. Thus, the topology builds away from the root as each MP updates the distance vector to root and readvertises to its neighbors the cumulative cost to the root portal.

When a node wants to send a frame to another node, and if it has no route to that root (mapping its address to a given MP) it may send the frame to the root. The root looks up the routing and bridging tables to see if the packet is intended for a node within the mesh or outside. It forwards the message appropriately back to the mesh or its uplink. If it finds the entry inside the mesh, it sends the frame to the destination parent MP using an additional tunnel encapsulation. When the packet reaches the destination parent MP and checks the tunnel encapsulation, it knows that the address is within the mesh and may initiate an RREQ back to the source. This hybrid routing mechanism allows the initial frame to be forwarded on the tree topology path followed by establishing an optimal, on-demand route between the source–destination pair for all subsequent frames among them. Any frame sent from the root follows the optimal path to any other MP in the network by the spanning tree properties.

When there are multiple portals in a mesh network, a single portal takes the root role either by provisioning or by a dynamic procedure. All other portals assume nonroot roles. In presence of multiple portals, the root forwards all frames with unknown addresses outside the mesh to its own uplink as well as other portals for forwarding on their uplinks. All nonroot portals forward frames from outside the mesh to the root portal for further forwarding within the mesh network.

12.7.3.2   On-Demand Routing in HWMP

On-demand routing in HWMP uses an RREQ and route reply RREP mechanisms of AODV [4] to establish routes between two MPs. Original AODV was developed for IP-based networks and routing at layer-3. HWMP adapts original AODV to work at layer-2 and changes all IP and IP-addressing references to MAC and MAC addresses. Apart from this adaptation, it uses the following mechanisms of original AODV:

- Route discovery
- Destination-only and reply-and-forwarding
- Route maintenance
- Best candidate route caching
- Sequencing
- Route acknowledgment
- Route errors

In order to be compatible with legacy STA devices, MAPs generate and manage messages on behalf of the legacy STAs that are associated with them. The functionality is similar to the situation when an MAP has multiple addresses. The associated STA addresses may be thought of as alias addresses for the MAP. However, STA handoffs due to roaming may cause a route to become stale. RERR mechanism needs to be modified in order to make sure that backhaul routing continues to function when there are many legacy STAs in a network and/or there are frequent and rapid STA handoffs.

12.7.4   Radio-Aware Optimized Link-State Routing Protocol

This is the optional routing protocol of the 802.11s draft standard. Radio-aware optimized link-state routing (RA-OLSR) is a unified and extensible proactive, link-state routing framework for WMNs based on the original OLSR [13] protocol with extensions from fisheye state routing (FSR) [14] protocol. RA-OLSR enables the discovery and maintenance of optimal routes based on a predefined metric, given that each MP has a mechanism to determine the metric cost of a link to each of its neighbors. In order to propagate the metric information between MPs, a metric field is used in RA-OLSR control messages. In disseminating topology information over the network, RA-OLSR adopts the following approaches in order to reduce the related control overhead:

- It uses only a subset of MPs in the network, called multipoint relays (MPRs), in flooding process.
- It can control (and thereby reduce) the message exchange frequencies based on the fisheye scopes.

The current RA-OLSR protocol specifications also include association discovery protocol to support legacy 802.11 stations. The MAPs select paths among MAPs and MPs by running RA-OLSR protocol and complement routing information among MAPs and MPs with the information of legacy 802.11 stations associated with them. OLSR is an optimization over the classical link-state routing protocol, tailored for MANETs. It inherits the stability of a link-state routing protocol and has the advantage of having routes immediately available when needed due to its proactive nature.

12.7.5 Forwarding in WLAN Mesh

In an 802.11s mesh network, path selection and forwarding operations are implemented as layer-2 mechanisms. When data frames are forwarded in such a multihop mesh network, multipath routing (either due to load balancing or dynamic route changes) can easily result in arrival of out-of-order and duplicate frames, the destination MP. The probability of having out-of-order and duplicate frames increases as the rate of topology changes, load level variations, and/or wireless channel fluctuations increases. The Sequence Control field in 802.11 data frame headers is used on a hop-by-hop basis to detect duplicates or missing frames at each hop and is changed by each intermediate MP. Hence, it cannot be used to detect out-of-order or duplicate frame delivery in an end-to-end fashion. The mesh E2E sequence number in the mesh control field is added to uniquely identify the data frames sent from a given source MP. By the pair of source MP address

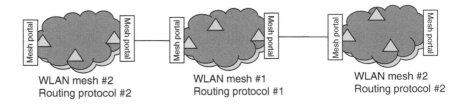

Figure 12.3 **WLAN mesh networked with 802 LANs using bridging.**

1. A node in the mesh
2. A node outside the mesh that is reachable without traversing the mesh
3. A node outside the mesh that is reachable through the mesh (by other MPPs)

A frame being sent by a node in the WLAN mesh has two possible final destinations:

1. A node inside the mesh
2. A node outside the mesh

The MPs collocated with MPPs may participate in transparent layer-2 bridging, allowing users to build networks that include a WLAN mesh in combination with other layer-2 networks. As such, each MP that is collocated with an MPP may participate in the STPs and maintain a node table to determine through which port each node in the logical network can be reached. The MPP may maintain 802.1d bridge tables, learning, and support VLANs.

12.8.2 Interworking with Higher Layers

Figure 12.4 illustrates a network in which the two WLAN mesh LANs are internet worked with other 802.3 LAN segments using layer-3 routing (e.g., IP). In this example, the MPPs perform the functionality of IP gateway routers, resulting in a network with multiple intercon-nected subnet LANs. Rest of the IP interworking mechanisms are beyond the scope of 802.11s Task Group.

12.8.3 Support of Multiple Portals

If there are more than one MPP in a WLAN mesh, each MPP contains layer-2 bridging functionality and may participate in an IEEE 802.1d

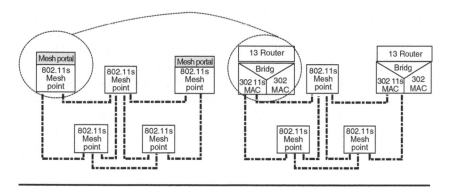

Figure 12.4   WLAN mesh networked with 802 LANs using routing.

STP running in wired LAN. The mesh network appears as a single loop-free segment to STP as bridge protocol data unit (BPDU) frames are transparently bridged across the mesh from portal to portal as data packets. Arbitrarily assuming that MPP-1 has a higher bridge port priority than MPP-2, MPP-1 is elected by the STP as the designated bridge for the mesh LAN segment and MPP-2 is blocked by STP. In this scenario, the mesh connectivity to the wired LAN is reduced to a single portal only.

As STP control packets are transparently bridged across the mesh from portal to portal as data packets, the cost of mesh links are effectively free as far as STP is concerned. If multiple MPPs are connected to different bridged LAN segments, the root bridge may reside on either of them and the STP sees two links connecting them:

1. Wired bridged link
2. Meshed link between two portals

Depending on the total cost of these two paths, either can be selected as the active path and the other will be blocked. In other words, for any LAN segment with nonroot bridge, if the wired bridged path cost to the root is lower than the path cost through the mesh, the bridge connecting the LAN segment to the root will be elected as the designated bridge for this LAN segment and the portal connected to this LAN segment will be blocked. This means that the mesh network can no longer use this portal for data forwarding. If the STP path cost through the mesh LAN segment is lower than the cost of the path through the wired bridged LAN, the portal connected to this LAN will be elected as the designated bridge for this LAN segment and the bridge connecting the wired LAN will be blocked. This means that the wired LAN is fragmented and traffic between this LAN segment and others will be bridged through the mesh network. This can cause interportal traffic starving the actual mesh traffic. 802.11s Task Group is working with 802.1, 802.15, and 802.16 Task Groups in order to resolve these issues and reach solutions that are mutually beneficial for 802.11s and other 802 LANs. These issues are critical for adoption of WLAN mesh networks in larger 802-based networks. 802.21 media-independent interworking Task Group is working toward developing protocols for all different types of 802 LANs so that they can coexist.

12.9   WLAN MESH SECURITY

This is a mandatory feature of 802.11s draft standard, which utilizes 802.11i-based security mechanisms to enable link security in a WLAN mesh network. 802.11i [16] provides link-by-link security in a WLAN mesh network. End-to-end security may be layered on top of WLAN mesh security, e.g., IP security (IPSec/VPN).

12.9.1   Security Framework

The link access protocol is based on 802.11i Robust Security Network Association (RSNA) security and supports both centralized and distributed IEEE 802.1x-based authentication and key management [16,17]. In a WLAN mesh, an MP performs the roles of both the supplicant and the authenticator, and may optionally perform the roles of an authentication server (AS). The AS may be collocated with an MP or be located in a remote entity with which the MP has a secure connection (this is assumed and specified by the 802.11s proposal). Figure 12.5 shows the security framework in a WLAN mesh network. A node establishes RSNA in one of the three ways:

1. Centralized 802.1x authentication model
2. Distributed 802.1x authentication model
3. Pre-shared key authentication model

The first two use 802.1x EAP-based authentication followed by 802.11i 4-way handshake. An authenticator is used in the first model, whereas MP–MP perform mutual authentication in the second model. The pre-shared model does not quite scale to meshes where multihop routing is required. In particular, it is infeasible to secure routing functionality when a pre-shared key is used in a

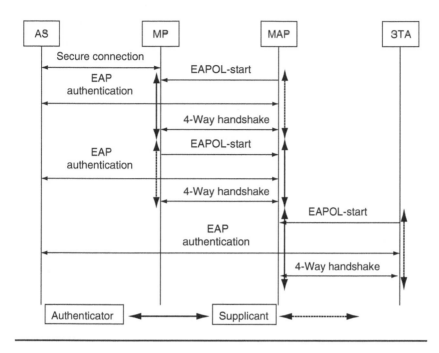

Figure 12.5 WLAN mesh security framework.

mesh with more than two nodes, because it is no longer possible to reliably determine the source of any message. IEEE 802.11s Task Group is discussing more robust security for WLAN mesh and is expected to make some substantial changes to the draft security specification.

12.9.2 Management Frame Security

The objective of management frame security in a WLAN mesh is to assure authenticity, integrity, and privacy (where appropriate) of the management frames sent and received among MPs on a link-by-link basis. The 802.11i-based link level authentication model is used to support authentication, key distribution, and encryption of management frames. There are no separate management frame-specific authentication and encryption architecture. Management frames have the same level of security and use the same mechanisms as data frames. Wherever possible, the security mechanisms defined by the Task Group 802.11w [18] will be utilized. Section 12.9 defines management frame security among MPs. WLAN mesh management frame protection is used for the following purposes in a WLAN mesh network:

1. Forgery protection
2. Confidentiality protection
3. Compatibility with 802.11i key hierarchy
4. Incremental inclusion of new management frames
5. Protection only after key establishment
6. Fragmentation support for management frames

When considering security, the mesh management frames and 802.11 standard [1] management frames can be classified in two broad categories: (1) those sent before authentication and (2) those sent after authentication. The management frames sent before authentication are beacon, probe request/response, authentication request/response, and the 4-way handshake. When 802.1x EAP is used, the data frames used are not protected at the link layer. The management frames sent and received after authentication are beacon, reassociation request/response, ATIM, disassociation, and mesh-specific management frames.

12.10 QUALITY OF SERVICE IN WLAN MESH

QoS is an important element of a network and is mainly required for mission-critical communication, such as voice, real-time data, and video traffic. Over a multihop, contention-based network, providing QoS guarantee is a big challenge. IEEE 802.11s draft standard informatively includes mechanisms so that appropriate QoS can be provided by a WLAN mesh backhaul network, thereby, supporting voice and other QoS-dependent applications. IEEE 802.11e and enhanced distributed channel access (EDCA) as the basis for achieving QoS in a WLAN mesh network are specified [1]. The Task Group is working toward developing a set of recommendations to further optimize EDCA and network allocation vector (NAV) settings for a WLAN mesh. Beyond IEEE 802.11e-based QoS, there are two more areas that focus on ensuring QoS in a WLAN mesh network [1]:

1. Backhaul and access traffic separation
2. Backhaul congestion control

12.10.1 Backhaul and Access Traffic Separation

Since an MP is only a logical entity, it is possible to physically collocate it with an AP or implement it on a device that also acts as an application end point, i.e., in addition to participating in the mesh and forwarding frames on behalf of other MPs, it also generates its own application traffic. In both cases, one single device has to forward a mixture of mesh traffic (with four-address frame formats) and BSS traffic (with three-address frame formats). Handling of these two different kinds of traffic within a single device can have a profound impact on QoS availed to each traffic type. On any given MP, backhaul traffic has traveled farther than any local BSS traffic. Hence, prioritizing backhaul traffic over local BSS traffic makes sense for reducing latency of overall backhaul traffic. It is also possible that an aggressive STA with heavy traffic backlog in the BSS can potentially starve the neighboring MPs in the network.

Traffic prioritization may have different implications from the point of view of fairness and the prioritization policies may depend on the mesh network deployment scenario and the business model in use. There are many implementation choices on how to best support traffic prioritization within a single device, e.g., an MAP. One may choose to employ multiple radios to separate the BSS traffic and mesh forwarding traffic into different radios operating at different channels. The draft standard does not specify any particular mechanism for backhaul/BSS traffic separation. But it recommends that backhaul/BSS traffic separation is considered whenever possible. When it is not possible, interaction between the traffic types should be regulated in order to provide QoS guarantee to all traffic streams in a WLAN mesh network.

12.10.2 Congestion Control in WLAN Mesh

The original 802.11 MAC and all its recent enhancements (e.g., 802.11e [10], 802.11i [11], 802.11k [16]) are designed primarily for single-hop wireless networks. Multihop data forwarding is central to WLAN mesh networks. Neither 802.11 DCF nor 802.11e EDCA provides any QoS over a multihop WLAN network. Each MP contends for the channel independently, without any regard for what is happening in the upstream or downstream nodes. One of the consequences is that a sender with backlogged traffic may rapidly inject many packets into the network, which would result in local congestion of nodes downstream, thereby deteriorating QoS of downstream nodes. Local congestion is defined as the condition when an intermediate MP receives more packets than it can transmit in a predefined time window. The result of local congestion is that the local buffer gets filled up quickly, and eventually the buffer may become full and packets will have to be dropped from the buffer. The situation is exacerbated by the presence of hidden and exposed nodes on the same channel causing extensive back-off and retransmissions.

One of the recommendations is to use transport layer QoS in order to achieve QoS over a multihop path. But most multimedia applications (video and voice) use user datagram protocol (UDP) transport, which does not have any form of congestion control or QoS provisioning. Congestion control for UDP may not be as critical in a wired network as it is in a wireless network, because each individual hop in the wired network is isolated from other hops. Research [19] shows that TCP congestion control does not work well across a multihop wireless network largely due to its susceptibility to high packet loss. Hence, simply relying on TCP is not a viable solution either.

A simple hop-by-hop congestion control mechanism is described to address the problem. This mechanism must be implemented at each MP, and it includes three basic elements:

1. Local congestion monitoring
2. Congestion control signaling
3. Local rate control

The basic idea is that each MP will actively monitor its local channel utilization condition so that it can detect local congestion when it happens. Three new mesh action frames are defined for this purpose:

1. Congestion control request
2. Congestion control response
3. Neighborhood congestion announcement

Upon receiving congestion control request from a downstream MP, the upstream neighbors employ local rate control to help relieve the congestion being experienced downstream, and upon receiving neighborhood congestion announcement from a neighbor MP, the neighbors employ local rate control to help relieve the congestion being experienced in the neighborhood.

12.11 MANAGEMENT AND CONFIGURATION

WMNs were originally envisioned to be self-configuring, self-healing, and self-monitoring networks. The need for management and configuration is reduced to a very minimum set as by the 802.11s draft standard. As far as management information bases (MIBs) and other management entities are concerned, the draft has not specified these yet. They are expected to be worked upon over the next several months.

12.12 IEEE 802.11s AND PRACTICAL MESH NETWORKS

The participants of 802.11s are working toward developing a comprehensive and effective WLAN mesh standard in order to position and popularize WLAN mesh much like they did with 802.11 WLANs. Most consumer, commercial, enterprise, public safety, and tactical networks and applications are able to build mesh networks by taking full advantage of the 802.11s standard and using standardized products, solutions, and technologies. The mechanisms specified [1] allow enabling basic mesh services in a WLAN network by implementing the amendments to IEEE 802.11 protocol and processing rules without changing any part of the PHY. Most of these amendments are simple extensions of existing 802.11 MAC such as power saving and beaconing. Routing, forwarding, interworking, security, and QoS are the five areas where the draft standard deviates significantly from 802.11 base standard.

Routing and forwarding enable forming multihop routes and frames to be forwarded correctly within and in and out of a mesh network. An efficient routing protocol is a fundamental building block of a communication network, and 802.11s attempts to specify a suitable protocol based on earlier MANET work. Interworking is essential when a mesh network must work and coexist with 802 other networks such as Ethernet [20] or WiMAX [21]. Mesh security is essential to ensure that data privacy and confidentiality is maintained while traffic passes over each backhaul link and that the network cannot be tempered with or compromised by adversaries. 802.11i standard is a single-hop security solution, and if a network only deploys 802.11i, traffic will be encrypted between STA and MAP, but not between MPs. Mesh QoS is important to ensure that required QoS by various applications is maintained as traffic flows through the backhaul. 802.11e standard is a single-hop QoS solution and has no provision for ensuring QoS over backhaul links.

Some of the known limitations of 802.11s, which may or may not be addressed by the Task Group include:

1. Mesh routing is layer-2 based, which is traditionally known for not scaling all that well.
2. Hybrid mesh routing is new and innovative, which has not gone through rigorous testing yet.

Wireless mesh networking (WMN) is a promising architecture to converge the future generation wireless networks with the dynamic self-organization, self-configuration, and self-healing characteristics. Due to the inherent flexibility, scalability, and reliability advantages of WMN topology, the IEEE 802.16 has standardized the mesh networking mode. This chapter will comprehensively describe the mesh mode operation procedures in the IEEE 802.16 WiMAX. In particular, we present the standard activities in the IEEE 802.16, the physical layer (PHY) basics, the frame format in point-to-multipoint (PMP) and mesh modes, the entry process (or topology control), energy management mechanism, quality of service (QoS), distributed/centralized scheduling algorithm, and authentication and security management. This chapter will be helpful in understanding the IEEE 802.16 WiMAX mesh networking principle and will serve as guidance for the future research issues in the related context.

13.1   INTRODUCTION

The IEEE 802.16 working group is originally organized to develop standards and recommend practices to support the development and deployment of fixed broadband wireless access (BWA) [3,5–7,9–11]. By adding the mobility capability, the IEEE 802.16 working group targets at designing a high-speed, high-bandwidth, and high-capacity standard for both fixed and mobile BWA [3,5,7,11]. The alternative in Europe is the standard High Performance Radio Metropolitan Area Network (HIPERMAN) created by the European Telecommunications Standards Institute (ETSI). HIPERMAN is able to provide the BWA in the 2–11 GHz bands. The equivalent standard in Korea is the Wireless Broadband (WiBro). WiBro defines the specifications in the licensed radio spectrum, offering the data throughput of 30–50 Mbit/s and covering a service radius of 1–5 km for portable Internet usage over the 2.3 GHz spectrum. In addition, WiBro is designed to provide all IP services and offers QoS schemes to differentiate the loss-sensitive data and real-time stream video multimedia services.

Worldwide interoperability for microwave access (WiMAX) is a certification mark for products that pass conformity and interoperability tests for the IEEE 802.16 standards [4]. In the framework of WiMAX, there are three standard candidates: IEEE 802.16, HIPERMAN, and WiBro. To ensure the interoperability with the aim to reduce deployment cost, the WiMAX Forum is working on approaches that enable 802.16 and HIPERMAN interwork seamlessly; and requires that products developed by the WiMAX Forum members should pass the certification process. Furthermore, WiBro and WiMAX have agreed on the interoperability.

WMN refers to the network architecture where the nodes can communicate with each other via multihop routing or forwarding [45]. It is characterized by dynamic self-organization, self-configuration, and self-correction to enable flexible integration, quick deployment, easy maintenance, low cost, high scalability, and reliable services, as well as to enhance the network capacity, connectivity, and throughput. Due to these inherent advantages, WMN is believed to be a highly promising technology converging the future generation wireless mobile networks.

Wireless Metropolitan Area Network (WMAN) provides broadband wireless access system as an excellent alternative to cabled access networks, such as fiber optic links, coaxial systems, and digital subscriber line links. This system has advantages including low-cost building and maintenances. As stated, the three variant standards can operate seamlessly within the WiMAX framework. Hence, without loss of generality, we focus on the IEEE 802.16 specifications and applications.

13.1.1 Standardization Activities

The first version IEEE 802.16.1 defined the PMP mode, and addressed the line-of-sight (LOS) problem employing the Orthogonal Frequency Division Multiplexing (OFDM) technique with the spectrum range 10–66 GHz and up to 134 Mbps data rate. Due to the characteristics of this radio spectrum, the system is inapplicable in the non-line-of-sight (NLOS) environments. This version was completed in December 2001 and can only be used in scenarios with fixed nodes. The next temporary version 802.16.2 attempted to minimize the interference between coexisting WMAN systems.

Owing to the attractive benefits of WMN, the subsequent version IEEE 802.16a introduced and defined the key operation procedures for the mesh networking mode. This version, approved in January 2003, supported the NLOS capability, operational in both licensed and unlicensed spectrum ranging from 2 to 11 GHz. It can support the data rate up to 75 Mbps and maximum range of 50 km. Mobility capability was still absent in this version. All the aforementioned versions serving the fixed BWA and their characteristics were incorporated into the finalized standard 802.16d-2004.

In December 2005, the latest version IEEE 802.16e was approved by adding the mobility capability, including the components supporting PMP and mesh modes, and seamless handover operation. It may achieve data rate up to 15 Mpbs in 5 MHz channel bandwidth.

13.1.2 Point-to-Multipoint and Mesh Networking Modes

Figure 13.1 compares the PMP and mesh topologies. In PMP mode, a Base Station (BS) performs the centric role to coordinate and relay all communications. The Subscriber Station (SS) under the management of the BS has to communicate with BS before transmitting data with other SSs. This architecture is exactly similar to cellular networks. Under certain situations, e.g., emergency or disaster scenarios, PMP mode may not be suitable for timely and efficient deployment while mesh mode serves as an excellent candidate for temporary network construction. Unlike the PMP mode, there is no separate downlink (DL) and uplink (UL) in the mesh mode. Every SS can directly communicate with its neighbors without the help of BS. In typical installation, one or several nodes play the role of BS to connect the mesh network to the external backhaul link, e.g., Internet or telecommunication networks. Such nodes are called mesh BS while the other nodes are called mesh SS.

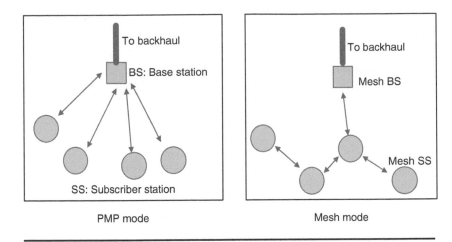

Figure 13.1 PMP mode and mesh mode.

In this chapter, we will comprehensively describe the mesh networking mode in the IEEE 802.16 WiMAX and explain the fundamental operation mechanisms. Section 13.2 introduces the PHY fundamentals. The WMAN-OFDM modulation scheme is deliberately explained since this air interface is used in both PMP and mesh networking modes. Following the PHY overview, the Medium Access Control (MAC) layer overview is also provided. Subsequently, the frame structure, energy management, and security management in the PMP mode is presented. In Section 13.4, following the explanation of the mesh frame structure and the functionalities of each subframe, we elaborate the entry process of a new node before it is entitled to transmit data in the mesh network. In addition, the scheduling algorithms for determining the transmission opportunities of the control messages and the data subframe are presented. In Section 13.5, a new priority scheme is proposed to differentiate diverse QoS. Section 13.6 presents the simulation environment and illustrative examples to demonstrate the achievable throughput in WiMAX mesh networks. Finally, in Section 13.7, open issues are identified for further study due either to unrealistic assumptions or to undefined specifications in the latest standard.

13.2 PHYSICAL LAYER AND MAC LAYER OVERVIEW

IEEE 802.16 defines three different PHY specifications for PMP mode in the 2–11 GHz frequency band. Each of these air interfaces is able to work together with the MAC layer to provide a reliable end-to-end link.

■ WMAN-SCa: a single carrier (SC) modulation technology
■ WMAN-OFDM: a 256-carrier OFDM modulation, in which Time Division Multiplex Access (TDMA) mechanism is utilized for multiple access—this air interface is mandatory for license-exempt bands
■ WMAN-OFDMA: a 2048-carrier OFDM modulation, in which multiple access is achieved by allocating a subset of the available carriers to an individual

Although these three candidates are designed for NLOS environments and applied in the frequency band 2–11 GHz. The two OFDM-based modulation schemes are more attractive for NLOS situations owing to the implementation simplicity of the signal equalization. In addition, compared to the 2048-based WMAN-OFDMA scheme, the 256-carrier-based WMAN-OFDM needs less strict requirement for frequency synchronization and fewer fast fourier transforms (FFTs). Therefore, WMAN-OFDM is preferred and specified by the WiMAX forum. The same PHY specification WMAN-OFDM with time division duplex (TDD) mode is also defined for mesh mode in the 2–11 GHz frequency band. Hence, WMAN-OFDM modulation is discussed in the following sections.

13.2.1 OFDM Symbol

Figure 13.2 shows the OFDM symbol in the time domain and the frequency domain. In Figure 13.2a, T_b denotes the useful symbol time referred as the OFDM waveform time duration. A copy of the last part T_g in the useful symbol time is termed cyclic prefix (CP). This portion is used to combat multipath while maintaining the orthogonality of the tones. The summation of CP and the useful symbol time is referred as the symbol time T_s. Figure 13.2b shows the OFDM symbol frequency description. An OFDM symbol is made up of subcarriers, the number of which determines the FFT size N_{FFT}. There are three subcarrier types:

■ Data carriers: for data transmission
■ Pilot carriers: for various estimation purposes
■ Null carriers: no transmission, for guard band and the DC carrier

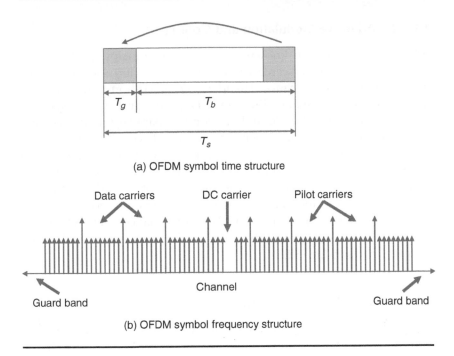

(a) OFDM symbol time structure

(b) OFDM symbol frequency structure

Figure 13.2 (a) OFDM symbol time structure; (b) OFDM symbol frequency structure.

The purpose of the guard bands is to enable the signal to naturally decay and create the FFT "brick wall" shaping. The transmitted signal during OFDM symbol is specified as

$$s(t) = Re\left\{ e^{j2\pi f_c t} \sum_{k=-N_{used}/2;\, k \neq 0}^{N_{used}/2} c_k \cdot e^{j2\pi k \Delta_f (t - T_g)} \right\}$$

where $Re\{\cdot\}$: the real part of a complex number; t: time; f_c: carrier frequency; N_{used}: number of used subcarriers, which is the summation of data and pilot carriers and equal to 200; c_k: a complex number specifying a point in a quadrature amplitude modulation (QAM) constellation. This value represents the data to be transmitted on the carrier whose frequency offset index is k; Δ_f: carrier spacing, which is equal to the ratio between the sampling frequency f_s and the number of FFT size N_{FFT} ($= 256$), i.e., $\Delta_f = f_s / N_{FFT}$; and T_g: CP time.

13.2.2 Adaptive Modulation and Coding

Wireless link has been known as one of the principle performance bottlenecks due to the scarce wireless resources, and also because of the remarkable performance degradation due to short-term multipath fading, long-term shadowing, and Doppler and time-dispersive effects. To improve the spectral efficiency, adaptive modulation and coding (AMC) has been popularly employed to select appropriate transmission mechanisms and parameters with the time-varying channel conditions. The IEEE 802.16 standard also defines a variety of combinations of AMC schemes to trade off the data rate and robustness depending on the channel and interference situations. Figure 13.3 summarizes the available combinations. For efficiently employing the AMC property, one of the key issues is the link adaptation algorithm [32].

13.2.3 MAC Layer Overview

MAC layer is comprised of three sublayers: service specific convergence sublayer (SSCS), MAC common part sublayer (MAC CPS), and privacy sublayer (PS). SSCS is defined as the interface to higher layers and provides mapping function from the transport layer diverse traffic to the flexible MAC. MAC CPS provides the core MAC function including access control, collision resolution, control or data scheduling, bandwidth request, and allocation. PS ensures the secure connection establishment, provides network access authentication, and generates key exchange and encryption for data privacy [21].

Modulation	Uncoded block size [byte]	Coded block size [byte]	Coding rate	Information bits/OFDM symbol
QPSK	24	48	1/2	184
QPSK	36	48	3/4	280
16QAM	48	96	1/2	376
16QAM	72	96	3/4	568
64QAM	96	144	2/3	760
64QAM	108	144	3/4	856

Figure 13.3 Modulation and coding in IEEE 802.16.

13.3 POINT-TO-MULTIPOINT MODE IN IEEE 802.16 WiMAX

13.3.1 Frame Format

The OFDM PHY of the system supports the frame-based transmission with the frame length 0.5 ms, 1 ms, or 2 ms. Figure 13.4 illustrates the frame structure for the OFDM physical layer operating in TDD mode. Each frame consists of a DL subframe and an UL subframe. The Tx / Rx transition gap (TTG) is used to separate the DL and UP subframes and allows the terminal to change operation from reception to transmission. Similarly, the Rx/Tx transition gap (RTG) is used to separate the UL and DL subframes and enables the turning around from transmission to reception.

The DL subframe only consists of one DL PHY packet data unit (PDU) starting with a long preamble (2 OFDM symbols), which is used for PHY synchronization. Frame control header (FCH) follows the preamble and has the length 1 OFDM symbol. The FCH is modulated by the most robust scheme Binary phase shift keying (BPSK) with coding rate 1/2. The FCH is followed by one or multiple DL bursts. The first DL burst #1 contains the broadcast MAC management messages, i.e., DL-MAP, UL-MAP, as well as the DL and UL channel descriptor (DCD and UCD, respectively). DL-MAP defines the access strategy to the DL channel while UP-MAP specifies the access scheme to the UL channel. DCD and UCD define the physical channel

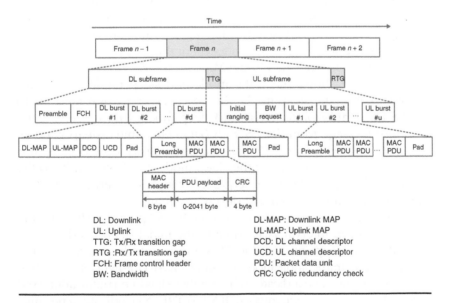

Figure 13.4 PMP frame format with TDD in IEEE 802.16.

characteristics. Each of the other DL bursts starts with an optional preamble to enhance the synchronization and channel estimation. Following the preamble, a number of MAC PDUs are scheduled to transmit in a DL burst. These MAC PDUs may be associated with different service flow/connections or SSs; but all of these PDUs are encoded and modulated using the same PHY mode. In either DL or UL direction, the size of burst is an integer number of the OFDM symbol length to exactly match the OFDM symbol and burst boundaries. To form an integer number of OFDM symbols, unused bytes in the burst payload might be padded by the bytes $0 \times FF$.

The UL subframe consists of contention slots for initial ranging, contention slots for bandwidth request, and one or multiple number of UL PHY transmission bursts. The purpose of initial ranging is for the SSs entry into the system including the functionalities power control, frequency offset adjustment, time offset correction, and basic management request. The bandwidth request interval is used for the SSs to transmit the bandwidth request message. The UL burst structure is similar to the DL burst.

13.3.2 Energy Management in PMP Mode

The amendment 802.16e [7] adds mobility component for WiMAX and defines both physical and MAC layers for combined fixed and mobile operations in licensed bands. Due to the promising mobility capability in IEEE 802.16e, the mechanism in efficiently managing the limited energy is becoming very significant since a Mobile Subscriber Station (MSS) is generally powered by battery. For this, sleep mode operation has been recently specified in the MAC protocol [7,16,34].

Figure 13.5 shows the wake mode and sleep mode of an MSS. Before entering the sleep mode, the MSS sends a request message to the BS for permission to transit into sleep mode. Upon receiving the response message from the BS with parameters initial-sleep window (T_{min}), final-sleep window (T_{max}), and listening window (L), the MSS enters into sleep mode. After the sleep mode, the MSS transits to the wake mode again. As a consequence, the MSS alternatively stays in wake mode and sleep mode during its lifetime.

Now, we focus on the mechanism in the sleep mode. The duration of the first sleep interval T_1 is equal to the initial-sleep window T_{min}. After the first sleep interval, the MSS transits into listening state and listens to the traffic indication message MOB-TRF-IND broadcasting from BS. The message indicates whether there has been traffic addressed to the MSS during its sleep interval. If MOB-TRF-IND indicates a negative

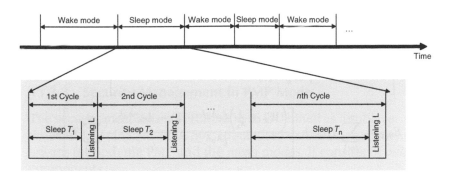

Figure 13.5   Energy management scheme in IEEE 802.16.

indication, the MSS continues its sleep mode after the listening interval L. Otherwise, the MSS returns to wake mode. We term the sleep interval and its subsequent listening interval a cycle.

If the MSS continues its sleep mode, the next sleep window starts from the end of the previous listening window; and it doubles the preceding sleep interval. This process is repeated as long as the sleep interval does not exceed the final-sleep window T_{max}. When the MSS reaches T_{max}, it keeps the sleep interval as fixed T_{max}. That is, the duration of sleep interval in the nth cycle is given by

$$T_n = \begin{cases} T_{\min}, & n = 1 \\ \min\left(2^{n-1}T_{\min}, T_{\max}\right), & n > 1 \end{cases} \qquad (13.1)$$

In case there are frames addressed to a particular sleeping MSS, the MSS exits the sleep mode in the next listening interval. In contrast, if there are external operations, the sleep mode is terminated immediately. As a consequence, the instants in terminating sleep mode are different for different traffics. We assume that the incoming and outgoing frames addressed to the MSS follow the Poisson processes with rate λ_c and λ_g, respectively. Here, the outgoing frames can also be understood as a process constructed by the external operations. Let $\lambda = \lambda_c + \lambda_g$ be the total arrival rate to the MSS. Let E_s and E_L denote the consumed energy units per unit time in the sleep interval and the listening interval, respectively. Suppose $W_n = \sum_{j=1}^{n}(T_j + L)$. The consumed energy during a sleep mode is expressed as [16]

$$\text{Energy} = \sum_{n=1}^{\infty} \sum_{k=1}^{3} E_{n,k}\,\phi_{n,k} \qquad (13.2)$$

message is not sent by the SS. Once the SS is authorized, it will enforce additional security by initiating a separate traffic encryption key (TEK) state machine for each SAID in the authorization reply message. TEK is used for managing the keys to encrypt the actual data traffic. During the TEK exchange, the SS sends a key request message to the BS first. After receiving the request, the BS replies with a key reply message that contains the TEK encrypted with a key encryption key (KEK) derived from AK using 3-DES algorithm. The SS can get TEK using KEK and all data traffic is encrypted with TEK. TEK also has a limited lifetime and the SS should request and receive new TEK before the BS expires the SS's current TEK. The procedure of AK and TEK exchange is shown in Figure 13.7. During the TEK exchange, to verify the hash message authentication code (HMAC)-digest in UL and DL messages, the UL authentication key HMAC-KEY-U and the DL authentication key HMAC-KEY-D are derived from AK. For the UL key request message, the SS calculates the HMAC-digest with the secure hash algorithm SHA-1. The digest is then encrypted using HMAC-KEY-U and sent to the BS with the message. BS authenticates the message using HMAC-KEY-U. For the DL key replay message, the procedure is the same except that the DL authentication key HMAC-KEY-D is used.

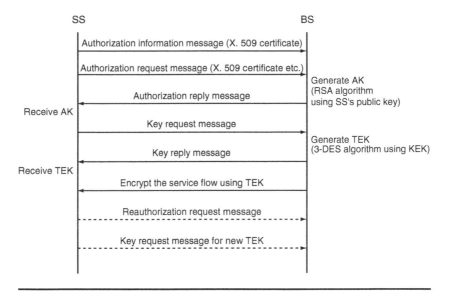

Figure 13.7 AK and TEK exchange procedure.

13.4 MESH MODE IN IEEE 802.16 WiMAX

In the mesh mode, traffic can be routed through other SSs and can occur directly between SSs without being routed through mesh BS. Mesh mode defines the direct communication between SSs in the MAC layer and allows to set up multihop communications. Here, the setup can be achieved using two scheduling schemes: centralized scheduling and distributed scheduling.

13.4.1 Frame Format

Figure 13.8 shows the frame format in the mesh mode. A frame consists of a control subframe and a data subframe. The length of the control subframe is fixed as MSH-CTRL-LEN × 7 OFDM symbols, where the parameter MSH-CTRL-LEN has 4 bits (i.e., value ranges between 0 and 15) and is advertised in the structure Network Descriptor IE (information element). The data subframe is divided into minislots.

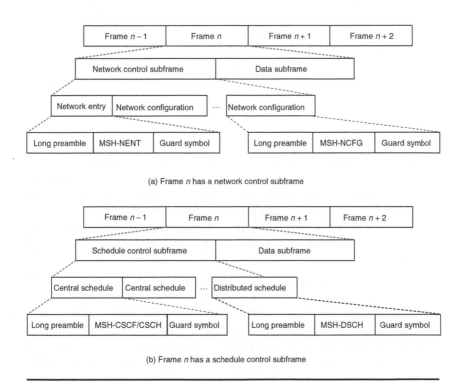

(a) Frame n has a network control subframe

(b) Frame n has a schedule control subframe

Figure 13.8   Frame structure in the mesh mode in IEEE 802.16.

Figure 13.8 illustrates two control subframes: network control subframe in case (a) and schedule control subframe in case (b). The network control subframe occurs periodically, with the period indicated in the Network Descriptor IE. The schedule control subframe occurs in all other frames without network control subframe. In particular, the field scheduling frame in the Network Descriptor IE defines the number of frames having a schedule control subframe between two frames with network control subframe in multiples of 4 frames. For example, if scheduling frame = 4, after a certain number of frames with network control subframe, the following 4×4 frames have schedule control subframe, followed by the next frame with network control subframe.

The network control subframe is defined primarily for new nodes gaining synchronization and joining a mesh network. The first transmission opportunity is the network entry component carrying the information of mesh network entry message MSH-NENT. This part is reserved for new nodes that expect to enter the mesh network system. The remaining (MSH-CTRL-LEN-1) transmission opportunities are the network configuration components carrying the information of mesh network configuration message MSH-NCFG. The portion for network configuration is defined to broadcast network configuration information to all nodes. The length of each transmission opportunity accounts for 7 OFDM symbols. Hence, the length of the transmission opportunities carrying MSH-NENT and MSH-NCFG is equal to 7 OFDM symbols and (MSH-CTRL-LEN-1) $\times$ 7 OFDM symbols, respectively.

The schedule control subframe is defined for centralized or distributed scheduling of the sharing nodes in a common medium, indicating that in Network Descriptor IE there are MSH-DSCH-NUM number of mesh distributed scheduling messages MSH-DSCH. This suggests that the first (MSH-CTRL-LEN-MSH-DSCH-NUM) $\times$ 7 OFDM symbols are allocated for transmitting the mesh centralized scheduling messages MSH-CSCH and mesh centralized configuration message MSH-CSCF.

The data subframe serves the PHY transmission bursts, which start with a long preamble (2 OFDM symbols) serving for synchronization, immediately followed by several MAC PDUs. Each MAC PDU comprises a 6-byte MAC header, a 2-byte mesh subheader with node ID, a variable length MAC payload (0-2039 bytes), and a 4-byte optional cyclic redundancy check (CRC). Consequently, the length of a MAC PDU varies between 12 and 2051 bytes.

13.4.2 Entry Process

The entry process defines the procedure for a new node joining and synchronizing the mesh network before starting normal data transmission. The entry process is the pivotal procedure for mesh network topology formulation and might be viewed as a variant of topology control.

Figure 13.9 shows the scenario when a new node A expects to join the mesh network. Nodes B, C, and D are regular active nodes in the mesh network and also the neighbor nodes from the new node A. Here, a neighbor node is defined as a node that is exactly one hop away from the particular node. The set of all neighbor nodes is called a neighborhood. In addition, the set of all neighbors of a neighborhood is called the two-hop extended neighborhood. Figure 13.10 shows the six-phase entry process for a new node.

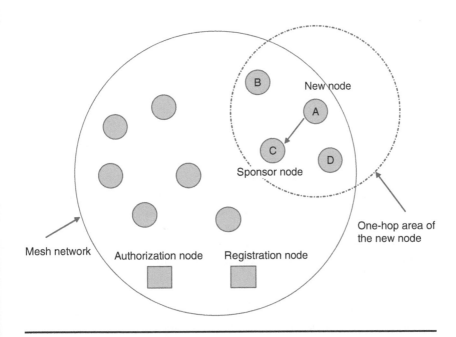

Figure 13.9   An example for a new node's entry process. New node: the node that joins the mesh network; sponsor node: a neighboring node that relays both control and data transmissions to and from the new node/base station; authentication node: the node that performs the authorization in a mesh network; registration node: the node that performs the registration in a mesh network.

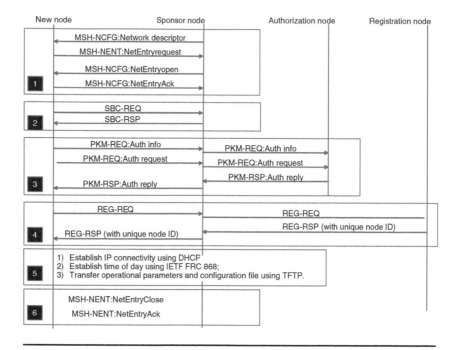

Figure 13.10   Entry process for a new node.

13.4.2.1   Synchronization and Opening Sponsor Channel

MSH-NCFG message is used for advertising the basic network configuration of a mesh network. MSH-NENT message is defined for helping new nodes to synchronize and join the mesh network. Every active node within the mesh network periodically advertises MSH-NCFG message with the Network Descriptor IE to inform the neighborhood of the basic network configuration.

On a new node, initialization or signal loss, it performs scanning and searches for the message MSH-NCFG Network Descriptor to acquire coarse synchronization with the mesh network. Specifically, if the node has storage of the last operational parameters, it will first try to gain coarse synchronization with the network utilizing the saved information. If such information is unavailable, the node will continuously scan every possible channel of the frequency band until it finds a valid network.

After the coarse synchronization, the new node is able to build the physical neighbor list with each record including the information of nearby nodes to facilitate future communication, e.g., 48-bit MAC address, hop count, 16-bit node ID, XmtHoldoffTime, and NextXmt-Time. From the established physical neighbor list, the new node selects a sponsor node having the best signal quality metrics, i.e., received signal strength indicator (RSSI) and carrier-to-interference-and-noise-ratio (CINR). Then, the new node synchronizes its time to the chosen sponsor node with zero propagation delay assumption; and sends out a MSH-NENT:NetEntryRequest message (a MSH-NENT message with type $= 0 \times 02$) to the sponsor node.

Upon receiving this entry request message, the sponsor node will evaluate the request and decide to open or reject a sponsor channel for the new node. If the sponsor node finds the invalid operator authentication value, excess propagation delay, or incapability to support more new nodes, it will reject the entry request and respond with MSH-NCFG:NetEntryReject (a MSH-NCFG message with type $= 0 \times 03$). If the sponsor node accepts the request, it will send out MSH-NCFG:Net-EntryOpen message (a MSH-NCFG message with type $= 0 \times 02$) to the new node with the updated field estimated propagation delay and also open a sponsor channel for the communication between the two nodes. On the basis of the refreshed estimated delay, the new node is able to perform fine time synchronization and additionally corrects its transmission time. Then, the new node acknowledges the acceptance by replying MSH-NENT:NetEntry Ack (a MSH-NENT message with type $= 0 \times 01$).

Now, the new node is able to negotiate the basic capability, authorize its validation, register its presence, and additionally obtain the network setting through the communication with the bridge-like sponsor node.

13.4.2.2 Basic Capabilities Negotiation

This phase is defined to negotiate the basic capabilities between the two nodes through the message pair SS basic capability request (SBC-REQ) and SS basic capability response (SBC-RSP). The new node sends the SBC-REQ message to the sponsor node to request the supporting capability of physical parameters and bandwidth allocation. Upon the reception, the sponsor node replies to the response SBC-RSP, indicating the available or unavailable capability.

13.4.2.3 Node Authorization

This step defines the procedure for the mesh network to verify the new node validness before using the network resources and transmitting data. The new node performs the authorization via the sponsor node. The new node first sends the PKM request (PKM-REQ) message with authentication information PKM-REQ: Auth Info. Subsequently, it sends PKM-REQ message with authorization request PKM-REQ: Auth Request to the sponsor node, with the fields indicating X.509 certificate and supporting cryptographic algorithms. The sponsor node tunnels the received messages over UDP/IP to the authentication node. In response to the authorization request, the authentication node validates the requesting new node's identity, determines the encryption algorithms and authentication key, and sends back the PKM response message with authorization reply PKM-RSP: Auth Reply to the sponsor node. Then, the sponsor node detunnels the authentication reply message to the new node.

13.4.2.4   Node Registration

If the new node is successfully verified by the mesh network, it is entitled to use the network resources and to transmit data packets. However, before such operations, the new node needs to perform registration to obtain a unique node ID. To achieve this, it sends the registration request message REG-REQ to the sponsor node. Following the message tunnel process over UDP/IP, the sponsor node propagates the message to the registration node, which returns the registration response REG-RSP with the generated unique node ID. Upon receipt of the response, the sponsor node forwards the message to the new node.

13.4.2.5   Supplementary Information Acquirement

To perform higher layer operation, the new node needs an IP address, synchronized time, and network configuration parameters. The new node acquires an IP address using dynamic host configuration protocol (DHCP) [1], retrieves the time using the time protocol defined in IETF RFC 868 [1], and then downloads a parameter file using the trivial file transfer protocol (TFTP) defined in IETF RFC 1350 [1]. This configuration file includes IP addresses of authentication node and registration node. This address information is very important for this new node because, after the successful entry process, the new node becomes an active node in the mesh network. Consequently, it may be chosen as a potential sponsor node by other new nodes that want to join the mesh network. In such a case, the new node should be aware of the authorization node and registration node addresses and thus be able to tunnel/detunnel authentication requests and registration requests for other new nodes.

13.4.2.6   Entry Process Termination

When all the above procedures are successful, the new node terminates the entry process by sending a MSH-NENT:NetEntryClose message (a MSH-NENT message with type $= 0 \times 03$) to the sponsor node, which will acknowledge the termination by replying MSH-NENT:NetEntryAck. After this, the new node becomes a potential sponsor node and can accept the MSH-NENT:NetEntryRequest message to help other new nodes enter the mesh network.

13.4.3   Scheduling MSH-NENT in Control Subframe

In the mesh mode, transmission opportunities in the control subframe and minislots in the data subframe are separated. Each node competes the control channel access. The contention consequence in the control subframe does not have effect on the data transmission during the data subframe of the same frame. There are two different messages in the control subframes: MSH-NENT and MSH-NCFG. Accordingly, there are two scheduling algorithms defined for these two messages. In this section, we will elaborate the scheduling algorithm for the MSH-NENT message, and in Section 13.4.4, the MSH-NCFG scheduling.

The scheduling algorithm to transmit MSH-NENT message defines the mechanism for a non-fully functional new node to communicate with the fully functional nodes of the mesh network. In particular, a new node sends the network entry message MSH-NENT in the network entry transmission opportunity following two steps:

■ Step 1: On hearing a MSH-NCFG message from the targeted sponsor node with the sponsored MAC address 0×000000000000, the new node will send the MSH-NENT: NetEntryRequest message in the immediate following MSH-NENT transmission opportunity. This random-access, contention-based situation may cause collision if there are two or more new nodes competing for the same slot.
■ Step 2: On hearing a MSH-NCFG message from the targeted sponsor node with the field netentry MAC address equal to the new node's MAC address, the new node will send the MSH-NENT:NetEntryRequest message in the immediate following MSH-NENT transmission opportunity after the received MSH-NCFG message.

13.4.4   Scheduling MSH-NCFG in Control Subframe

Distributed election scheduling is defined to determine the next transmission time NextXmtTime of a node's MSH-NCFG during its current transmission time XmtTime.

There are two fields NextXmtMx and XmtHoldoffExponent in MSH-NCFG to determine the next eligibility interval. Here, the eligibility interval refers to the duration in which the node can transmit in any slot and is given by

$$2^{\text{XmtHoldoffExponent}} \cdot \text{NextXmtMx} < \text{NextXmtTime}$$
$$< 2^{\text{XmtHoldoffExponent}} \cdot (\text{NextXmtMx} + 1) \tag{13.3}$$

The length of the eligibility interval is equal to the difference between the upper bound and the lower bound, i.e., $2^{\text{XmtHoldoffExponent}}$. After the eligibility interval, the node has to wait for a holdoff time XmtHoldoffTime before a new transmission with

$$\text{XmtHoldoffTime} = 2^{\text{XmtHoldoffExponent}+4} \tag{13.4}$$

For example, if NextXmtMx = 2 and XmtHoldoffExponent = 4, the node is eligible for the next MSH-NCFG transmission between the 33th and 48th transmission opportunity. After the eligibility interval of 16 transmission opportunities, the node waits for 256 transmission opportunities before the next transmission.

The node chooses the temporary transmission opportunity TempXmtTime equal to the first transmission slot after the holdoff time XmtHoldoffTime. Then, the node determines the set of all eligible nodes competing for this slot TempXmtTime. The set of eligible competing nodes includes all nodes in the extended neighborhood satisfying any of the following properties:

■ NextXmtTime includes TempXmtTime
■ NextXmtTime is unknown
■ EarliestSubsequentXmtTime occurs no later than the TempXmtTime, where

$$\text{EarliestSubsequentXmtTime}$$
$$= \text{NextXmtTime} + \text{XmtHoldoffTime}$$
$$= \text{NextXmtTime} + 2^{\text{XmtHoldoffExponent}+4} \tag{13.5}$$

After building the set for the specific node, a pseudorandom mixing function will calculate a pseudorandom MIX value for each node. If the specific node generates the biggest MIX value, it wins the competition and the next transmission time NextXmtTime is set as TempXmtTime. Then, the node broadcasts to the neighbors in the MSH-NCFG message. Otherwise, the specific node fails competing in this slot. The node sets the TempXmtTime as the next transmission slot and repeats the competing procedures until it wins.

The design of the distributed election scheduling algorithm has taken into account distribution, fairness, and robustness. In terms of distributed algorithm, this protocol requires no centralized control on coordinating the transmission opportunity allocation. Fairness refers to the strategy that the algorithm treats all nodes equally in competing for the transmission opportunities. In the sense of robustness, the seed in the pseudorandom algorithm varies in each frame and this mechanism is able to resolve the persisting collision.

13.4.5 Scheduling Data Subframe

The IEEE 802.16 mesh mode has defined three different scheduling schemes to manage the minislots in the data subframe, i.e., centralized scheduling, uncoordinated distributed scheduling, and coordinated distributed scheduling.

13.4.5.1 Centralized Scheduling

In centralized scheduling, there is a mesh node, named the mesh BS, to coordinate the data subframe scheduling, resource allocation, and grant for other nodes in the mesh network. The MSH-CSCH and MSH-CSCF play the most important roles in determining the scheduling. Each SS estimates traffic demands by itself and also its children. Then, the SS sends the MSH-CSCH: Request message to its parent. After receiving the resource request, the parent summarizes the total resource requirement by itself and its children, and then sends the MSH-CSCH: Request message to its own parent. This process is repeated until the MSH-CSCH reaches the mesh BS.

In response, the mesh BS determines the amount of granted resources for each link and broadcasts the message MSH-CSCH: Grant to all its neighbors. All the intermediate nodes forward the MSH-CSCH: Grant message to their own neighbors.

13.4.5.2 Uncoordinated Distributed Scheduling

Uncoordinated distributed scheduling serves the temporary communication between two nodes. The algorithm is based on a three-way-handshake mechanism (Figure 13.11).

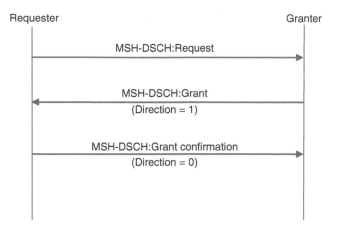

Figure 13.11   Three-way-handshake process.

In uncoordinated distributed scheduling, mesh distributed schedule (MSH-DSCH) message, which transmits in the data subframe, is the most important message in the scheduling process. An MSH-DSCH message carries the following fields: (1) availabilities IE, indicating the starting frame number, the starting minislot within the frame, and the number of available minislots for the granter to assign; (2) Scheduling IE, showing the next MSH-DSCH transmission time NextXmtTime, and XmtHoldoffExponent of the node as well as of its neighbor nodes (note that scheduling IE is useful in the coordinated distributed scheduling discussed in Section 13.4.5.3); (3) Request IE, having the resource demand of the node; (4) grants IE, conveying the granted starting frame number, the granted starting minislot within the frame, and the granted minislots range. Every node sends its available channel resource table to neighbor nodes using MSH-DSCH messages. Hence, each node has the knowledge of idle slots throughout its extended neighborhood. The requester randomly selects an idle slot during data subframe, and sends out the MSH-DSCH: Request message to acquire resource. In case of collision in the random-access, the node performs the random back-off algorithm and then sends out the MSH-DSCH: Request again.

On receiving the MSH-DSCH: Request message, the granter evaluates the request through a *slot allocation algorithm*. If the algorithm returns successfully, the granter replies with the bandwidth grant message MSH-DSCH: Grant to the requester carrying the updated Grants IE. All neighbor nodes of the granter are able to hear the broadcasted MSH-DSCH: Grant and are aware of the reserved minislots.

In confirmation, the requester copies the grant information and acknowledges the resource grant by sending back the MSH-DSCH: Grant Confirmation message to the granter. Then, all neighbor nodes of the requester are able to hear the broadcasted MSH-DSCH: Grant Confirmation and are aware of the reserved minislots. Following this, all neighbor nodes of the requester and the granter are informed about the transmission between the two nodes.

Even the latest version, IEEE 802.16e, has not defined a specific slot allocation algorithm. This provides implementation flexibility with respect to diverse requirements under different scenarios. However, the algorithm's absence also gives rise to significant research challenge to accommodate diverse traffic demands and to achieve QoS.

13.4.5.3 Coordinated Distributed Scheduling

Similar to uncoordinated distributed scheduling, MSH-DSCH message plays a significant role in the whole scheduling process. Unlike the MSH-DSCH message taking place in data subframe in uncoordinated distributed scheduling, the MSH-DSCH in coordinated distributed scheduling occurs in control subframe.

The transmission opportunity of MSH-DSCH in coordinated distributed scheduling follows a similar distributed election algorithm as in Section 13.4.4. That is, the distributed election scheduling algorithm is applied to determine the next transmission time NextXmtTime of the message MSH-DSCH in coordinated distributed scheduling. After determining the transmission time of MSH-DSCH, the node employs the three-way-handshake process (Figure 13.11) to build the communication link with a neighbor node. When the link is constructed successfully, the two nodes can transmit data in the reserved minislots. Similarly, the slot allocation algorithm has not been defined in coordinated distributed scheduling.

13.4.6 Security Management in Mesh Mode

In the mesh mode, the SSs can communicate with each other without the help of BS. Traffic can be routed through other SSs or occurs directly between SSs. Thus the security mechanism for the PMP mode in Section 13.3.3 has to be modified to accommodate this change, mainly in the TEK exchange. In the mesh mode, once a node is authorized, for each neighbor, it starts a separate TEK state machine for each SAID identified in the authorization reply message and maintains these TEK state machines. The TEK state machines send key request messages to the node's neighbors periodically to refresh the TEK for each SAID.

After receiving the key request message, a neighbor sends back a key reply message to the node that contains new TEK encrypted using the node's public key. The TEK is used to encrypt the data traffic between the node and its neighbor. At all times, for each neighbor, the node maintains two active sets of TEKs per SAID. The two generations of TEKs have overlapped lifetimes for seamless key transition. For the mesh mode, a new attribute, operator-shared secret, is added to the authorization reply. It is known to all nodes in the network. Each node maintains two active operator-shared secrets. During the TEK exchange with its neighbors, a node uses the operator-shared secret to calculate the HMAC-digest for the key request and key reply messages. The HMAC-digest is encrypted with a key HMAC-KEY-S derived from the operator-shared secret for authentication purpose.

13.5 QoS SCHEME FOR WiMAX MESH MODE

In the mesh mode, the transmission opportunities in the control sub frame and the minislots in the data subframe are separated. Each node competes the control channel access. The contention consequence in the control subframe does not affect the data transmission during the data subframe of the same frame. Hence, the contention process in the control subframe is elaborated for deriving the performance metrics.

In the distributed scheduling, MSH-DSCH message plays a significant role in the whole scheduling process. An MSH-DSCH message carries the field Scheduling IE showing the next MSH-DSCH transmission time NextXmtTime, and XmtHoldoffExponent of the node and of its neighbor nodes. As stated, the distributed election scheduling in Section 13.4.4 is followed to determine the next transmission time NextXmtTime of a node's MSH-DSCH during its current transmission time XmtTime.

For different types of services, the MSH-DSCH transmission interval between its current transmission time and the next transmission time should be different. For instance, real-time voice-over IP (VOIP) should experience a short transmission interval whereas non-real-time e-mail service can tolerate a long transmission delay. In IEEE 802.16 PMP mode, the standard defines four connection-based QoS classes: unsolicited grant service (UGS), real-time polling service (rtPS), non-real-time polling service (nrtPS), and best effort (BE). Comparatively, for the mesh mode, no similar terms or priority schemes have been defined. In this section, we propose a simple but effective scheme to prioritize various kinds of traffic as well as enable the QoS differentiation.

Firstly, the eligibility interval and its length in Equation 13.3 are generalized. For the sake of presentation, we denote $x =$ XmtHoldoff-Exponent as the transmission holdoff exponent. The original base value 2 is generalized into a real number α in determining the eligibility interval and the length of this interval. That is, the eligible next transmission time NextXmtTime becomes

$$\alpha^x \cdot \text{NextXmtMx} < \text{NextXmtTime} \leq \alpha^x \cdot (\text{NextXmtMx} + 1) \quad (13.6)$$

where the upper and lower bounds should be rounded to the nearest integer. The node can transmit in any slot during the eligibility interval. As a consequence, the length of the eligibility interval V is given by the difference between the lower bound and the upper bound.

$$V = \alpha^x \quad (13.7)$$

Secondly, we introduce another real number holdoff base value β and holdoff exponent y to determine transmission holdoff time XmtHoldoffTime H. Then, H is given as

$$H = \beta^{y+4} \quad (13.8)$$

In Equation 13.3 and Equation 13.4, the base value is a constant integer 2 and the exponent XmtHoldoffExponent is same and fixed. In our scheme, the real number base value α may be different from β; and the exponent x may be unequal to y.

For a different node in a mesh network, the set of parameters $P = (\alpha, x, \beta, y)$ is different to differentiate the services. Suppose there are N nodes in the mesh network. Let N represent the set of all nodes. For a particular node k ($k \in N$), the set of parameters is denoted as $P_k = (\alpha_k, x_k, \beta_k, y_k)$.

Let S_k denote the number of slots in which the node fails during the distributed election scheduling before it wins. Denote τ_k as the interval between two consecutive MSH-DSCH transmission opportunities. Then, τ_k is the summation of the holdoff transmission time H_k and S_k.

$$E(\tau_k) = H_k + E(S_k)$$
$$= (\beta_k)^{y_k+4} + E(S_k) \quad (13.9)$$

where $E(X)$ represents the expected value of a nonnegative random variable. The collocated scenario, i.e., all nodes are one-hop neighbors of each other, is considered to indicate the effectiveness of the

Table 13.1 Simulation System Setting

Parameter	Setting
WiMAX frame duration	10 ms
Number of slots per frame	250
Number of control slots per frame	36
Number of data slots per frame	214
MSH-NCFG holdoffExponent	Variable
MSH-DSCH holdoffExponent	Variable
Data scheduling mechanism	Coordinated distributed scheduling
WiMAX MAC ARQ	Yes
Ad hoc routing protocol type	AODV
Physical mode type	802.11a
Physical data rate	24 Mbps

such that each node is only within the radio range of its adjacent nodes. Thus, SS1 can only hear SS2, whereas SS2 can hear both SS1 and SS3. In such a scenario, for SS1 to communicate with SS6, SS1 will have to utilize SS2, SS3, SS4, and SS5 as intermediate nodes. In the simulation conducted, route setup will be done automatically at the network layer, using the AODV ad hoc routing protocol. For performance evaluation, constant bit rate (CBR) traffic flow is used and the evaluation metrics selected are throughput and delay. Simulations were conducted using the WiMAX system parameter settings as shown in Table 13.1. The parameters of the CBR flow are shown in Table 13.2.

The results obtained from the simulation evaluation for single-flow transmission over various hops, for both throughput and delay, using two different holdoffExponent settings, are as shown in Figure 13.15 and Figure 13.16 respectively. On the basis of these results, by changing the value of the holdoffExponent, the performance of the system

Table 13.2 CBR Parameters

Parameter	Setting
Source node	SS1
Destination node	SS2, SS3, SS4, SS5, SS6
Number of packets	500
Packet size	2000 bytes
Interval	10 ms

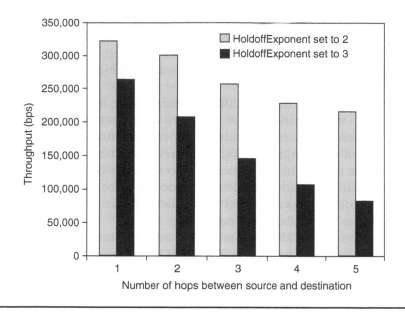

Figure 13.15   Throughput evaluation under multihop scenario.

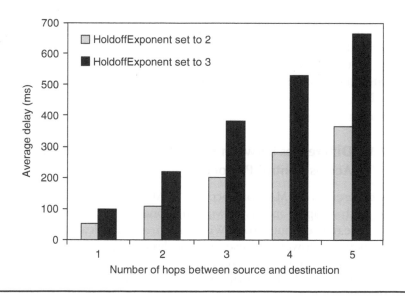

Figure 13.16   Average delay evaluation under multihop scenario.

not been considered in the standard and the consequence is still unclear. One potential solution for the coexistence of heterogeneous networks may be the dynamic spectrum mechanism [44].

13.7.5 Scalability

The proposed mesh mode in IEEE 802.16e cannot support a large number of mobile nodes. In widely imagined single-radio $(1 + 1)$ mesh network architecture, the performance degrades drastically several hops away with an increasing number of nodes [2]. Here, $(1 + 1)$ represents the two-layer network architecture. The mesh routers are themselves organized as a mesh network and the mesh routers also provide the access points under their radio coverage.

To address scalability, the potential approaches include: (1) new network architecture, e.g., multiradio and multichannel (MRMC) system, in which the routing and dynamic channel allocation scheme become the important components to achieve higher capacity and throughput; (2) localization technique with the aid of position, velocity, and direction information, i.e., in route discovery or resource reservation, the control packets do not need to propagate to the whole network; (3) hierarchical routing or cluster-based scheme by schematically deploying several fixed nodes to perform the role of relay candidates.

13.7.6 Topology Control

During the new node entry process described in Section 13.4.2, the new node should first choose a candidate sponsor node to obtain the necessary network configuration and to relay data. The suggested criteria in selecting the sponsor node is the best signal quality, i.e., RSSI and CINR. It is intuitively understood that different criteria will result in various mesh network topologies. Wei et al. [12] proposed an interference-aware sponsor node selection scheme. After introducing the concept of blocking metrics, the study specified the algorithms for choosing a sponsor node to minimize the interference. Evaluation was conducted using matrix laboratory (MATLAB) simulation for throughput performance comparison against the basic 802.16 scheme (which is random selection of sponsoring node). The result indicates that the proposed scheme is able to achieve much better performance compared with the scheme ignoring the interference consideration.

In addition to the interference management, in a realistic scenario, mobility characteristics also have a great impact on the sponsor node selection. Under certain instances, a new node may prefer to choose a sponsor node who has a slow mobility or a highly predictive mobility to avoid frequent physical link breakdown. It is suggested to choose a sponsor node taking into account the mobility, interference-awareness, channel-awareness, and CINR simultaneously.

13.7.7 Slot Allocation Algorithm

As stated in Section 13.4.5 and Section 13.4.6, the slot allocation algorithm is not specified in the standard. In a recent work [14], Liu et al. proposed a simple slot allocation algorithm. Specifically, after receiving a resource request, the granter will first look up the resource table and then determine the reasonable transmission time based on the demand level and demand persistence fields in the message MSH-DSCH: Request. As the authors reported, this simple slot allocation algorithm is insufficient to support QoS. On observing this, they proposed two checkpoint-based algorithms to identify the network's good or congested condition. Cao et al. [13] proposed an analytical model for the distributed scheduler. In the last part about the future work, the authors hint at a data reservation scheme by adaptively adjusting the holdoff exponent according to the number of competing nodes.

It is envisaged that more efficient slot allocation algorithms should be implemented to trade off fairness, QoS, and throughput maximization, taking into account the scalability, mobility, and realistic fading channel environment.

13.8   CONCLUSION

In this chapter, we presented the principle of the mesh mode in the emerging WMAN standard IEEE 802.16. We discussed the critical processes in detail, including the entry process to gain synchronization, the scheduling algorithms for control messages competing for a transmission opportunity; and the centralized/distributed scheduling for data subframe. More importantly, we specified unrealistic assumptions or undefined specifications in the standard that motivate the proposed open research challenges and potential solutions. We hope that this chapter will be helpful in understanding the key procedures in joining, managing, and coordinating the IEEE802.16 mesh networking, and be additionally instructive for the future research topics in the context of multihop broadband wireless access.

REFERENCES

1. http://rfc.net
2. http://www.meshdynamics.com
3. Wireless MAN Working Group, http://wirelessman.org/
4. http://www.wimaxforum.org/
5. IEEE 802.16–2001, "IEEE Standard for Local and Metropolitan Area Networks-Part 16: Air Interface for Fixed Broadband Wireless Access Systems," April 2002.
6. IEEE 802.16a-2003, "IEEE Standard for Local and Metropolitan Area Networks Part 16: Air Interface for Fixed Broadband Wireless Access Systems Amendment 2: Medium Access Control Modifications and Additional Physical Layer Specifications for 2–11 GHz," April 2003.
7. IEEE 802.16e/D5-2004, Part 16: Air Interface for Fixed and Mobile Broadband Wireless Access Systems—Amendment for Physical and Medium Access Control Layers for Combined Fixed and Mobile Operation in Licensed Bands, November 2004.
8. J. Dickman, K. Rath, and L. Kotecha, Proposal for 802.16 Connection Oriented Mesh, IEEE 802.16 Standard Proposal, March 2003.
9. D. Beyer, N. Waes, and C. Eklund, Tutorial: 802.16 MAC Layer Mesh Extensions Overview, IEEE 802.16 Standard Proposal, March 2002.
10. C. Eklund, R. Marks, K. Stanwood, and S. Wang "IEEE standard 802.16: A technical overview of the WirelessMAN air interface for broadband wireless access," *IEEE Communications Magazine*, vol. 40, no. 6, pp. 98–107, June 2002.
11. A. Ghosh, D.R. Wolter, J.G. Andrews, and R. Chen, "Broadband wireless access with WiMax/802.16: current performance benchmarks and future potential," *IEEE Communications Magazine*, vol. 43, no. 2, pp. 129–136, February 2005.
12. H. Wei, S. Ganguly, R. Izmailov, and Z. Haas, "Inteference-aware IEEE 802.16 WiMAX mesh networks," *Proceedings of the 61st IEEE Vehicular Technology Conference (VTC'2005-Spring)*, May 2005.
13. M. Cao, W. Ma, Q. Zhang, X. Wang, and W. Zhu, "Performance analysis of the distributed scheduler in IEEE 802.16 mesh mode," *Proceedings of the 6th ACM International Symposium on Mobile Ad Hoc Networking and Computing (MobiHoc'2005)*, pp. 78–89, May 2005.
14. F. Liu, Z. Zeng, J. Tao, Q. Li, and Z. Lin, "Achieving QoS for IEEE 802.16 in mesh mode," *Proceedings of the 8th International Conference on Computer Science and Informatics*, July 2005.
15. S. Xu and T. Saadawi, "Does the IEEE 802.11 MAC protocol work well in multihop wireless ad hoc networks?" *IEEE Communications Magazine*, vol. 39, no. 6, pp. 130–137, 2001.
16. Y. Zhang and M. Fujise, "Energy Management in the IEEE 802.16e MAC," *IEEE Communications Letters*, vol. 10, no. 4, pp. 311–313, April 2006.
17. S. Sengupta, M. Chatterjee, S. Ganguly, and R. Izmailov, "Exploiting MAC flexibility in WiMAX for media streaming," *Proceedings of the IEEE International Symposium on a World of Wireless, Mobile and Multimedia Networks (WOWMOM 2005)*.

18. S.-M. Oh and J.-H. Kim, "The analysis of the optimal contention period for broadband wireless access network," *Proceedings of the 1st International Workshop on Pervasive Wireless Networking (PWN05)*.

19. D.-H. Cho, J.-H. Song, M.-S. Kim, and K.-J. Han, "Performance analysis of the IEEE 802.16 wireless metropolitan area network," *Proceedings of the 1st International Conference on Distributed Frameworks for Multimedia Applications (DFMA'2005)*.

20. C. Hoymann, "Analysis and performance evaluation of the OFDM-based metropolitan area network IEEE 802.16," *Computer Networks*, vol. 49, no. 3, pp. 341–363, 2005.

21. G. Nair, J. Chou, T. Madejski, K. Perycz, D. Putzolu, and J. Sydir, IEEE 802.16 medium access control and service provisioning. *Intel Technology Journal*, vol. 8, no. 4, pp. 213–228, August 2004.

22. H. Wang, W. Li, and D. Agrawal, "Dynamic admission control and QoS for 802.16 wireless MAN," *Proceedings of the Wireless Telecommunications Symposium*, 2005.

23. H. Lee, T. Kwon, and D.-H. Cho, "An efficient uplink scheduling algorithm for VoIP services in IEEE 802.16 BWA systems," *Proceedings of the 60th IEEE Vehicular Technology Conference (VTC'2004-Fall)*, September 2004.

24. S. Redana and M. Lott, "Performance analysis of IEEE 802.16a in mesh operation mode," in *Proceedings of IST SUMMIT 2004*, June 2004.

25. S. Redana, M. Lott, and A. Capone, "Performance evaluation of point-to-multi-point (PMP) and mesh air interface in IEEE standard 802.16a," *Proceedings of the 60th IEEE Vehicular Technology Conference (VTC'2004-Fall)*, September 2004.

26. K. Wongthavarawat and A. Ganz, "IEEE 802.16 based last mile broadband wireless military networks with quality of service support," *Proceedings of the IEEE Military Communications Conference (Milcom'2003)*, vol. 2, pp. 779–784, October 2003.

27. G. Chu, D. Wang, and S. Mei, "A QoS architecture for the MAC protocol of IEEE 802.16 BWA System". *IEEE International Conference on Communications Circuits and System and West Sino Expositions*, vol. 1, China, pp. 435–439, 2002.

28. J. Chen, W. Jiao, and H. Wang, "A fair scheduling for IEEE 802.16 broadband wireless access systems," *Proceedings of the IEEE International Conference on Communications (ICC'2005)*.

29. J. Chen, W. Jiao, and Q. Guo, "Providing integrated Qos control for IEEE 802.16 broadband wireless access," *Proceedings of the 62nd IEEE Vehicular Technology Conference (VTC'2005-Fall)*, September 2005.

30. J. Chen, W. Jiao, and Q. Guo, "An integrated QoS control architecture for IEEE 802.16 broadband wireless access systems," *Proceedings of the 48th IEEE Global Telecommunications Conference (GLOBECOM'2005)*, December 2005.

31. L.F.M. de Moraes and P.D. Maciel, "Analysis and evaluation of a new MAC protocol for, broadband wireless access," *Proceedings of the International Conference on Wireless Networks, Communications and Mobile Computing*, 2005.

32. S. Ramachandran, C.W. Bostian, and S.F. Midkiff, "Link adaptation algorithm for IEEE 802.16," *Proceedings of the IEEE Wireless Communications and Networking Conference*, vol. 3, pp. 1466–1471, March 2005.

33. D. Johnston and J. Walker, "Overview of IEEE 802.16 security," *IEEE. Security and Privacy*, vol. 2, no. 3, pp. 40–48, May 2004.

34. J.-B. Seo, S.-Q. Lee, N.-H. Park, H.-W. Lee, and C.-H Cho, "Performance analysis of sleep-mode operation in IEEE 802.16e," *Proceedings of the 60th IEEE Vehicular Technology Conference (VTC'2004-Spring)*, May 2004.

35. H. Shetiya and V. Sharma, "Algorithms for routing and centralized scheduling to provide QoS in IEEE 802.16 mesh networks," *Proceedings of the 1st ACM workshop on Wireless Multimedia Networking and Performance Modelling (WMuNeP 2005)*, October 2005.

36. S. Choi, G.-H. Hwang, T. Kwon, A.-R. Lim, and D.-H. Cho, "Fast handover scheme for real-time downlink services in IEEE 802.16e BWA system," *Proceedings of the 61st IEEE Vehicular Technology Conference (VTC'2005-Spring)*, vol. 3, pp. 2028–2032, May 2005.

37. H.S. Alavi, M. Mojdeh, and N. Yazdani, "A quality of service architecture for IEEE 802.16 standards communications," *Proceedings of the Asia-Pacific Conference*, pp. 249–253, October 2005.

38. F. Yang, H. Zhou, L. Zhang, and J. Feng, "An improved security scheme in WMAN based on IEEE standard 802.16," *Proceedings of the International Conference on Wireless Communications, Networking and Mobile Computing*, vol. 2, pp. 1145–1148, September 2005.

39. H. Lee and D.-H. Cho, "Reliable multicast services using CDMA codes in IEEE 802.16 OFDMA system," *Proceedings of the 61st IEEE Vehicular Technology Conference (VTC'2005-Spring)*, vol. 3, pp. 2349–2353, May 2005.

40. P.C. Ng, S.C. Liew, and C. Lin, "Voice over Wireless LAN via IEEE 802.16 Wireless MAN and IEEE 802.11 Wireless Distribution System," *Proceedings of the International Conference on Wireless Networks, Communications and Mobile Computing*, vol. 1, pp. 504–509, June 2005.

41. S. Maheshwari, S. Iyer, and K. Paul, "An efficient QoS scheduling architecture for IEEE 802.16 Wireless MANs," Master Thesis, India Institute of Technology, 2005.

42. B. Louazel and S. Murphy, "Implementation of IEEE 802.16a in Glomosim/QualNet," Master Thesis, University of Dublin, August. 2004.

43. D. Kim and A. Ganz, "Architecture for 3G and 802.16 wireless networks integration with QoS support," *Proceedings of the 2nd International Conference on Quality of Service in Heterogeneous Wired/Wireless Networks*, pp. 28–28, August 2005.

44. X. Jing and D. Raychaudhuri, "Spectrum co-existence of IEEE 802.11b and 802.16a networks using the CSCC etiquette protocol," *Proceedings of the 1st IEEE International Symposium on New Frontiers in Dynamic Spectrum Access Networks (DySPAN 2005)*, pp. 243–250, November 2005.

45. I.F. Akyildiz, X. Wang, and W. Wang, "Wireless mesh networks: a survey," *Elsevier Computer Networks*, vol. 47, no. 4, pp. 445–487, March 2005.

46. D.R. Smart, *Fixed Point Theorems*, Cambridge University Press, Cambridge, 1974.

14

COGNITIVE RADIO AND DYNAMIC SPECTRUM MANAGEMENT

Clemens Kloeck, Volker Blaschke, Holger Jaekel,
Friedrich K. Jondral, David Grandblaise,
Jean-Christophe Dunat, and Sophie Gault

CONTENTS

14.1 INTRODUCTION

In order to overcome the fixed and inflexible spectrum allocation new approaches in the spectrum regulation and in the development of enhanced wireless devices become mandatory. At present, regulation authorities assign fixed frequency bands to different subscribers or service providers who have to operate in the licensed band only. This procedure was already formulated in 1920 and may not be able to meet tomorrow's needs [1]. Due to the fixed allocation of the frequency bands to specific users or operators the utilization of the spectrum is not always very efficient. Recent studies have shown that only about 10% of the allocated spectrum in the United States and Europe are utilized [2–4]. In contrast to a fixed assignment of services and frequency bands the demand on several services varies over time and place that results in a highly temporal and geographical variation of the spectral utilization. Therefore, the interest regarding new ways of using, allowing access to, or allocating spectrum by increasing the spectrum efficiency has been grown in many regulatory bodies and standardization groups.

According to the current allocation structure, present wireless terminals have been designed to provide fixed services. Only upon the introduction of software-defined radios (SDRs) a flexible hardware platform was presented, which offers the possibility to integrate different transmission technologies and services into a single wireless device using the same hardware structure. This identifies an important requirement on the way to a dynamic resource allocation.

In addition to the necessarily greater flexibility in fixed transmission way networks, in order to improve the spectrum efficiency, wireless mesh network (WMN) provides an additional degree of freedom by flexibly choosing the wireless transmission way to improve the overall quality of service (QoS). Thus, there always exists a WMN which outperforms a fixed transmission way network. The controlling mechanisms in a WMN are preferably decentralized, so that each node (terminal) carries out controlling and observing tasks. The node must be aware of its transmission and equipment capabilities to assist in determining possible transmission ways over different nodes and also decide the possible operation bandwidth based on the current spectrum usage. This degree of freedom leads to an enhanced radio resource variation in space and time. Without loss of generality, it can be envisaged that the radio resources belong to a party, e.g., for unlicensed radio resources this can be the network as itself and for licensed resources this can be an operator. To increase the overall utility, a dynamic and decentralized radio resource allocation mechanism is needed, which will be described in this chapter.

Furthermore, the development of appropriate mechanisms for evaluating the present options and requirements of a mobile terminal is needed. This includes the consideration of the conditions and potentials of the radio interface and the implemented hardware structure, as well as aspects and strategies to control the allocation of the unused radio resources.

A greater flexibility of the wireless terminals and the available services leads to an increased number of options and possible settings. Therefore, future wireless devices have to provide the capability to present the choices at the highest level possible without involving the user in technical details. Cognitive radios (CRs), with their capability of decision making and adaptation, offer one opportunity of making these control decisions invisible to the user.

Therefore, CRs build the basic platform for any future wireless system that should provide dynamic adaptation. Driven by the increased flexibility of devices, the handling of the spectrum allocation and the adaptation of radio resources have become an important topic of current research and development. The shift from fixed assignment of services to dynamically allocated frequency bands offers a number of advantages. But several challenges have to be met on the way to flexible, adaptable, and more efficient wireless services.

This chapter presents the topic of dynamic and distributed spectrum allocation. Starting with the device-specific aspects (Section 14.2), the concept of CR is presented. First, the definition and the general attributes of such a mobile terminal are described. Furthermore, the conceptual architecture of a cognitive terminal is considered within cognitive networks.

Section 14.3 presents how opportunistic distributed spectrum allocation for CRs can be achieved with methods like swarm intelligence (SI) and game theory. Section 14.4 describes the algorithmic approaches for a dynamic resource allocation. The discussion includes technical aspects and market perspectives. Finally, a conclusion and an outlook are given.

14.2 COGNITIVE RADIO CONCEPT

CRs which are presented by Mitola [5] adapt themselves without user intervention in order to reduce the interference to other users. They also make use of spatial awareness, and understand and follow the user's behavior. Due to these features the CR concept builds the basic platform for a distributed and dynamic resource allocation.

In order to give an overview of the enhancements provided by a cognitive terminal, the present state of research and development within this topic is briefly summarized.

14.2.1 Definition and Attributes

The introduction of the CR concept has caused a strong interest for future communication applications. Due to the wide range of applications that benefit from a cognitive mobile device, several aspects of this new concept become important. The more general interpretation allows a CR to autonomously exploit locally vacant or unused spectrum in order to achieve new options of spectrum access. This requires the definition of basic rules for spectrum access. At least the radio has to stay in compliance with the local regulations. That means, a CR terminal can only roam across the borders of the local radio operations and adjust itself to the local emission regulations. In this case, the allocation mechanisms are already defined but the possible changes in response to the environment are restricted. In both cases, CR negotiate on behalf of the subscriber in order to connect the terminal to the service provider or the frequency band, respectively, offering the lowest costs. This market-oriented perspective provides the basis for the approaches discussed in Section 14.4.

Due to the different meanings of the term CR several definitions have been coined by the regulation bodies and research organizations. The wide range of expectations and demands made to this concept can be seen in different definitions presented in the literature. A more popular definition can be found [6]. The definition given by Haykin describes the technical background of a CR [7]:

> Cognitive Radio is an intelligent wireless communication system that is aware of its surrounding environment (i.e., outside world), and uses the methodology of understanding by building to learn from the environment and adapt its internal states to statistical variations in the incoming RF stimuli by making corresponding changes in certain operation parameters (e.g., transmit power, carrier frequency, and modulation strategy) in real time, with two primary objectives in mind:
>
> ■ Highly reliable communications whenever and wherever needed
> ■ Efficient utilization of the radio spectrum

In the software-defined radio forum's (SDRF) Cognitive Radio Working Group a definition is currently under consideration which should fix only a minimum of requirements in order to allow a further evolution:

> A Cognitive Radio is a software-defined radio that processes the attributes of being RF and spatially aware with the ability to autonomously adjust to its environment accordingly frequency, power, and modulation.

As it can be easily seen, the different definitions which are still under discussion include a wide range of wireless devices which could be named as cognitive. The terms *cognitive terminal, cognitive engine,* and *cognitive network* are defined as follows:

■ *Cognitive Radio* describes a mobile device that supports cognitive capabilities only on the physical and medium access control (MAC) layer.
■ *Cognitive engine* describes the functional unit on the MAC layer of a mobile node that provides cognitive capabilities.
■ *Cognitive network* is a complete network that supports the ability to adapt its nodes in the sense of an end-to-end reconfiguration and adaptation.

■ *Cognitive terminal* is able to adapt also higher layers above the MAC layer in order to affect the whole protocol stack of an end-to-end connection.

Generally, the basic characteristics of a CR described in the following can also be applied to cognitive terminals. All the above-mentioned definitions include, more or less, the following main attributes for such a device:

Awareness: A CR possesses awareness. The terminal is able to understand the RF environment and the spatial surrounding. It is also aware of its geolocation and the associated spectrum usage policies.

Adjustability: A CR responds to its environment of which it is aware. This includes the change of its emissions like radio frequency, power, modulation, or channel bandwidth in real time without user intervention. The adjustment will be done in order to reduce the interference to other users. This requires knowledge about the interference situation at other users caused by the active CR.

Autonomy: The CR can act and react in order to be adjustable without any user interaction. In order to be self-adjustable the terminal has to perform spectrum exploration. Furthermore, the terminal needs to know the local policy restrictions as well as the own specifications for an exploitation of locally vacant and allocable radio channels.

Adaptivity: The CR is able to understand and follow the user's behavior. This requires the ability to learn from the actions and choices taken by the user in order to become more responsive and to anticipate the user's needs.

Some of these attributes are already natural to traditional wireless terminals on a simpler level. But the integration of these attributes in one terminal causes new challenges. So, the spatial observation can be realized using a global positioning system (GPS) receiver. But the spectrum sensing would be much more difficult due to the different classes and access characteristics of licensed user (e.g., TV broadcast vs. wireless local area network (WLAN)). Furthermore, a comprehensive awareness of the spectrum needs a permanent spectral analysis that would not attempt for saving battery power. Also the hidden terminal problem, already known in WLANs, makes the spectrum sensing complicated [8]. In addition to the awareness of the surrounding, the terminal has to be aware of itself [9]. This capability of knowing its internal structure enables a selective response to the observed situation regarding the possibilities supported by the hardware. The resulting adjustment to the environment should provide an increase of the overall network resources that are wasted due to interferences caused by suboptimal adaptation. In current systems the radio environment is observed as transmitter-centric. This means that the transmitted power is designed to approach a prescribed noise floor at a certain distance from the transmitter. In order to shift toward a real-time interaction between the transmitter and the receiver in an adaptive manner, an interference temperature is specified [7]. This metric, described in detail in Section 14.2.5, is intended to quantify and manage the sources of interference in a radio environment. An interference temperature limit could provide a "worst-case" characterization of the RF environment in a particular frequency band and at a particular geographical location, where the receiver could be expected to operate satisfactorily [7]. Due to large number of sensors necessary for accounting the spatial variation of the RF stimuli, this approach leads to cognitive networks that will be described in Section 14.2.3.

In order to provide an autonomous and adaptive terminal, it has to interact with its environment. Considering mechanisms of machine learning, in addition to the orientation and evaluation of the environment, the *cognition cycle* presented [5] describes the principle of this interaction (Figure 14.1).

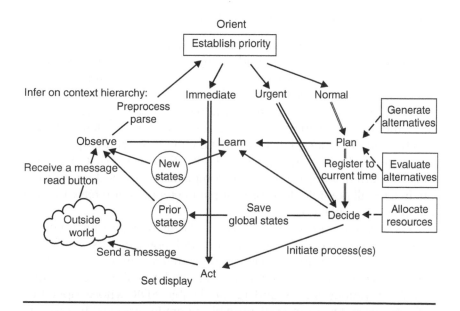

Figure 14.1 Cognition cycle presented by Mitola. (From Mitola, J., and Maguire, G.Q., Cognitive Radio: Making software radios more personal, *IEEE Personal Communications*, 6, 13, August 1999. With permission.)

The observation of the outside world includes the spectral and spacial awareness, as well as the recognition of user's inputs by the user interface. For a self-orientation of the terminal within its surrounding, the relevant parameters are parsed from the observed data. Priorities assigned to the stimuli from the outside world support a rating of the required reaction. For example, interference of other subscribers will invoke an immediate action. The nonrecoverable loss of a signal during transmission might require urgent reallocation of resources. In the normal case of an incoming message or a user request a plan is generated. Several alternatives are generated and evaluated in order to find the optimal reaction to the observation. The decision phase selects the most appropriate strategy that is initiated in the act phase. Information about the present observation, plan, decision, and reaction of the outside world can be used to learn the effectiveness of the chosen communication mode. The achieved knowledge influences the future generation of alternatives in order to become more responsive to the user's behavior. Besides the classical knowledge, processing methods like neuronal or fuzzy logical networks and biologically inspired models of cognition in a mobile terminal have been considered [10].

14.2.2 Cognitive Radio Architecture

Since CRs provide operation in different frequency bands using different modulation forms, the underlying structure is built upon SDRs. SDRs also have the ability to operate on several bands simultaneously which increases the utility for a dynamic resource allocation over different transmission standards. In addition to SDRs, which have to meet the local regulations regarding the new downloaded software, CRs have to be designed in a way that their choices of the operational parameters will meet the regulatory limits. This means that in an SDR only the current software has to be compliant to the local regulations, but in a CR all possible decisions for parameter selection have to meet the limits of regulation.

The definition of an appropriate radio architecture including all aspects of possible operation modes and regulation is still under discussion. In a very abstract view of a CR architecture two subsystems can be pointed out: (1) a cognitive unit which makes the decisions based on various inputs and (2) a flexible SDR whose operating software provides a range of possible operating modes. In order to hide the complexity of the cognitive elements to the applications the SDRF Cognitive Radio Working Group proposed a conceptual model using a layered architecture (Figure 14.2) [11].

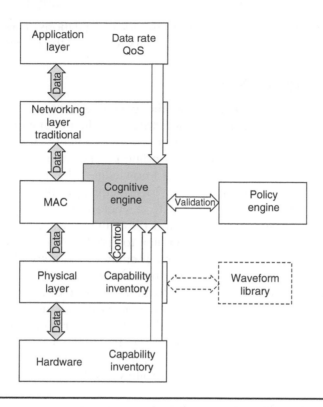

Figure 14.2   Model of cognitive radio architecture. (From Software Defined Radio Forum, Cognitive Radio Working Group (CRWG), www.sdrforun.org. With permission.)

Generally, the SDR is at the base of the model including the hardware and the software-defined physical layer. The cognition layer is directly above the physical layer. In this model the CR engine gets inputs from, and controls, the SDR. Furthermore, status information from the hardware and from the application layer is given to the cognitive engine. The logical position on the MAC layer level points out the basic potential function of the cognitive engine: the mapping of application layer requirements, like data rate, to the radio link layer regarding the available device and spectral resources. Therefore, the cognitive layer knows the available hardware resources, e.g., the capabilities and characteristics of the available modulations. In addition, a Policy engine would watch and control the decisions of radio in order to prevent the violation of any restrictions defined by the operator or regulation bodies. A waveform library provides all transmission modes that adhere to the local regulations. Thus, a minimal user interaction would be necessary to control the physical layer.

14.2.3 Cognitive Networks

In addition to the CR concept, the cognitive network framework describes a complete network structure providing cognition abilities. Within a cognitive network all network elements, the cognitive terminals, observe the network conditions. The information is used to plan, decide, and act by taking knowledge gained from previous actions into account. This behavior is equal to the CR concept described above. But in contrast to a single radio, which can only adapt itself to the environment and the network state, the cognitive networks act simultaneously. Similar to the WMN concept, which requires distributed intelligence, each entity of the network includes learning and decision modules. In addition to a WMN, cognitive networks support even more dynamic resource allocation and adaptation. The entities of cognitive networks change parameters belonging to multiple layers, which is suitable for employing cross-layer optimization. Mitola briefly described the interaction of CRs within the system-level of cognitive networks. Thomas et al. [12] have defined cognition network as follows: [12]

> Cognitive Network is a cognitive process that can perceive current network conditions, plan, decide, and act on those conditions. The network can learn from these adaptations and use them to make future decisions, all while taking into account end-to-end goals.

The network and the overall end-to-end aspects are the main differences compared to the CR definition. According to definition the enhancement cognitive networks contain software adaptable network elements [12]. This means that the network is able to modify one or several layers of the network stack within the network's terminals. The mechanisms necessary for the cognition of the states and the knowledge processing are similar to the CR methods.

14.2.4 Heteromorphic Waveforms for Spectrum Sharing

Heteromorphic waveforms (also named as malleable waveforms in the literature) offer some interesting properties for secondary spectrum usage between different systems. In particular, heteromorphic waveforms based on multicarrier (MC) access technologies have been recently proposed by the research community in support of spectrum sharing. Given the fact that "spectrum holes" can be very narrow and spread fragmentally over the band, heteromorphic waveforms enable to exploit any spectrum hole size. The first approach for MC-based solution is to use orthogonal frequency-division multiplexing (OFDM) modulation [18] with a multiple access scheme. The second approach (quite similar to the previous one) is to implement MC-based multiplexing schemes combined with carrier interferometry (CI) [19]. Solutions based on CI/time division multiple access (TDMA) (multicarrier pulse shape), CI/direct sequence code division multiple access (DS-CDMA) ([19,23]), CI/multicarrier code division multiple access (MC-CDMA) (multicarrier chip shape [20,22]) or frequency division multiple access (FDMA)/DS-CDMA [22]) are considered. MC-CDMA has better spectrum efficiency than FDMA/DS-CDMA, and has also a lower blocking probability for a noninterruptible spectrum sharing [22]. From these observations, MC-CDMA is a more efficient candidate for spectrum sharing than FDMA/DS-CDMA.

The motivations for using MC-based solutions for spectrum sharing are manifold. First, the waveforms can be adapted dynamically (modulation, power control, and diversity gain). Second, it is possible to fill spectrum holes of any radio access technology (RAT), i.e., the bandwidth of one band with an integer multiple of the carrier spacing of the MC-based systems. Third, it provides flexible spectrum-sharing capability in both continuous and noncontinuous bands to adapt dynamically the space time traffic demand.

These MC-based heteromorphic waveforms are very good candidates for: (1) operators to share spectrum in licensed bands by borrowing/renting, (2) for unlicensed systems to share spectrum in unlicensed bands, and (3) for unlicensed systems to use licensed bands in an opportunistic fashion. In case (1), Figure 14.3 illustrates spectrum sharing between two operators using the same MC-based heteromorphic waveform. Figure 14.3a through Figure 14.3c shows the case where spectrum is not shared, shared in continuous fashion, and shared in a fragmented fashion.

Case (3) can be approached with spectrum pooling strategy. This strategy [2] can be implemented by considering the secondary user being enabled with an OFDM-based MC technology [18]. A practical case [21] of spectrum pooling is the coexistence (hot spot) of a cellular system (e.g., GPRS) with a WLAN system (e.g., HiperLAN2) in which HiperLAN2 can reuse the licensed GPRS band in an opportunistic fashion. One of the main challenges of this technique is the capability of the secondary system to detect the activity of the primary system in

(UT) level. The open spectrum supporters [26,27] argue that the spectrum scarcity has been artificially created as a result of the regulators parceling the spectrum band into "can use" and "cannot use" spectrum bands, but this can be changed by switching into a more flexible regulatory framework. A framework, presented here, use distributed collaborative methods between many users to negotiate their spectrum access to several spectrum resources organized in a common spectrum pool. Interesting approaches for that purpose are presented in Section 14.3.1 through Section 14.3.3.

14.3.1   Interest for a Distributed and Collaborative Approach

Several measurements performed by the Shared Spectrum Company have shown that the level of activity was very low in many allocated spectrum bands [2]. This revealed the existence of spectrum holes in spectrum usage, within the fully allocated bands. Unfortunately, there is no model to predict when and for how long a given spectrum band will be used. Accordingly, opportunistic and smart algorithms should be developed to identify these holes and use them as needed, while authorized by the regulators. The objective is to have a much greater flexibility in varying the available spectrum resources for a system, according to its needs.

Unlicensed spectrum bands are a good playground candidate for implementing such innovative algorithms, where the whole spectrum is seen as a common spectrum pool [31] that can be used by several entities according to their needs and assuming it does not interfere with others. Due to the free access of terminals within the unlicensed frequency bands the implemented algorithms are able to handle the highly dynamic allocation of resources which is not centralized or controlled, in order to optimize the QoS.

Throughout nature, an enormous amount of processing takes place at the level of the individual organisms (e.g., ant, wasp, and human beings). Such individuals are autonomous nodes acting according to some rules (from simple to very complex), with imperfect knowledge of their environment (local knowledge), having memory, and are able to interact with other similar entities. Thus, the society composed of all these individuals can be seen as a complex superorganism. This is the way of organizing wireless networks in the future.

In a wireless system, user devices and radio access infrastructure equipments do not have global knowledge of the environment but only of a part thereof. A collaborative approach for spectrum allocation would rely on exchanging information with neighbors to propagate the information, instead of sending it to the central entity, as it is the case in a centralized architecture. The advantages of such an approach are flexibility, scalability, robustness against changes, and self-organization (through interactions).

A common feature of these distributed systems [32] is that organized behavior emerges from the interactions of many simple parts. Dissimilar systems (business, ant colonies, and brains) share fundamental commonalities: individual cells interact to form differentiated body parts, ants interact to form colonies, neurons interact to form intelligent systems, and people interact to form social networks. This is made possible using the CR.

The frequency-planning problem (trying to best allocate a finite spectrum resource to where and when it is best needed in a network and across networks) is a complex optimization problem, especially when considering changes in traffic and channel conditions, temporally and spatially. This approach is described in the Section 14.3.2 and Section 14.3.3, which explore the application of SI and game theory for a smart distributed radio resource allocation, respectively.

14.3.2 Swarm Intelligence

SI is an artificial intelligence technique based on the study of collective behavior in decentralized and self-organized systems, which was introduced by Beni and Wang in 1989, in the context of cellular robotic systems. SI is defined as: *the emergent collective intelligence of groups of simple agents* [33]. Computational intelligence structures at the system level emerge from unstructured starting conditions using powerful interaction mechanisms, while the action of each individual appears random (ants) when overlooked at the system level.

One great result from SI is that even with a population of identical and simple agents (limited capabilities leading to cheap devices), it is still possible to create a very complex and dynamic behavior at the system level. Such a superorganism composed of social artificial agents would have a higher adaptation capability compared to a centralized approach where each of these agents is taught what to do. This structure is very promising for applications to future "intelligent" wireless communication systems. This "intelligence" opens the way to more flexible regulatory rules, new ways of considering frequency reuse and avoiding interference.

SI offers an excellent way of controlling complex distributed systems by defining and ensuring simple local rules at the lowest component level. An interest of the SI is that several parameters can be varied, thus providing a direct mean for controlling the algorithm, with an aim, e.g., to trade between speed of convergence and quality of the final solution. The reason for using the properties found in social insects into the wireless world is that many challenges faced by current and future wired and wireless networks have already been overcome in large-scale biological systems. Accordingly, future wireless networks will definitely benefit from adopting key biological principles and bionics.

Many phenomena around us use a threshold mechanism to trigger actions: below the threshold one action is performed, and above, another. This is simple but very efficient in controlling many interacting entities without a central controller. Social insects have been observed as following a threshold mechanism using a sigmoid function [33]; they react to external stimuli depending on their threshold value vs. the stimulus intensity. Wasps use a variable threshold mechanism (still with a sigmoid function) where a threshold is adapted to reflect the specialization for a given task (time spent in performing it). Reinforcement learning is used to let each individual learn from experience of its actions. As a result wasps are able to perform a highly efficient and fully decentralized adaptive task allocation [33,34] where the number of wasps performing a task is a function of its urgency for the colony and of each individual's specialization in this kind of task. Many other properties of the SI still need to be discovered and modeled, with application to many domains, including the wireless one.

An application of the SI metaheuristic can design an uplink (UL) subcarriers (SCs) allocation between several users using orthogonal frequency division multiple access (OFDMA), without the need for a central controller to dictate the SC choice [35]. Instead of a sigmoid function, a modified Fermi–Dirac distribution was used. The competition to achieve not more than a single user per SC for data transmission was resolved using the SI; each user, on each SC, learnt about its ability to better use this SC than the others or not. As a result, this algorithm maximized the system UL sum capacity, with each user trying to pursuing its own interest. The scalability of the algorithm was demonstrated with respect to the time for negotiating SCs between users and not dependent on the number of users. In addition, the more competitors participate, the better the total sum capacity because of the multiuser diversity. The set of rules applied was very simple and the users could learn from the past actions for the betterment of the entire system.

14.3.3 Game Theory

Game theory is a tool for modeling and analyzing situations with strategic interactions. Since the pioneering work of von Neumann and Morgenstern [36], game theory concepts and methods have been used in various domains, from economics to politics, biology, or telecommunications. Formally, a game is constituted with a set of players, each having the possibility to choose the best strategy in a given space so as to obtain the greatest payoff [37]. The expected outcome of a game is to reach an equilibrium being a kind of stable state. The well-known Nash equilibrium is defined as a profile of strategies such that each player's strategy is an optimal response to the other players' strategies. When a Nash equilibrium is reached, no player can increase one's own utility individually (i.e., by changing one's own strategy) any further. A Nash equilibrium is therefore a stable and targeted operating point, since no player would find it beneficial to deviate from, if the others maintain their strategies. Game theory has several similarities with SI: among others, players do not use any explicit representation of the global environment; moreover, each decision is made in a distributed way. Therefore, it benefits from the common advantages of scalability, light signaling cost, and no computational complexity problem. But it is also based on a distinctive assumption that agents are playing rationally, similar to human beings who are rational with their economic choices, defending their own interest. For all these reasons, game theory is an interesting tool for solving the interactive optimization problem of resource allocation in wireless networks [38] where users compete for power, bandwidth, or SCs in OFDMA-based systems. Although the word "game" suggests a kind of behavior, most situations rather correspond to strong conflicts of interest and competition without any mutually beneficial actions. To make a compromise between user objective and network objective, a branch of game theory aims at designing penalty mechanisms in such a way that efficient operating solutions can be obtained. Pricing is one of those approaches and its principle is to charge locally a player who would tend to play too selfish, by imposing a fine on him. It aims at improving the overall performance by limiting the aggressive competition of the purely noncooperative game. Pricing is also used to control fairness between users. Many references in the literature exist on power control games for systems where users are transmitting in the same band (using, e.g., CDMA [39] or multiuser waterfilling), and a few ones deal with joint OFDMA SC and power allocation [40].

are suitable for wireless infrastructure communication systems capable of managing multihoming. By applying the CR abilities, not only
to the allocation but also to this combined architecture, it is mandatory
to dynamically allocate RRGs by an auction sequence (AS) in order to
exploit the CR abilities [35]. The repetition of auctions will happen
very fast up to milliseconds. A class of auctions being suitable for an
allocation mechanism auction in a determinable time and with an
acceptable signaling effort in comparison with sequential auctions
[36], is the multiunit sealed-bid auction, which is the core of the
enhanced radio auction multiple access.

14.4.3 Enhanced Radio Auction Multiple Access Protocol

MAC mechanisms can be divided into deterministic and statistical
mechanisms. One well-known representation of the last mechanism
is ALOHA and its modified kind-slotted ALOHA [39]. On the other
hand, several deterministic access mechanisms are proposed in order
to avoid access interference. Besides, carrier sense multiple access with
collision detection (CSMA/CD) and carrier sense multiple access
with collision avoidance (CSMA/CA) [41] mechanisms are proposed
based on periodically repeated auctions. The resource auction multiple access (RAMA) [37] is based on sequential single-unit sealed-bid
auction that can be improved by using a multiunit sealed-bid auction
as proposed [54]. In contrast to RAMA for which the bids have no
inherent meaning because of being randomly chosen, dynamic priorities resource auctions multiple access (D-RAMA) [38] allows to weigh
the bids based on the buffer size of the proper service class. The aim is
to reach a better QoS.

These protocols do not allow the users to incorporate their purchase power, their experience of the past, and additional information
of the future. The proposed access and assignment mechanism,
economical RAMA (E-RAMA), allows users to express their needs,
urgency to send, preferences, and purchase power, combined with a
QoS-aware buffer management which categories the data according to
their urgency to be sent in order to fulfill the QoS criteria. E-RAMA
facilitates sophisticated functionalities of the cognitive algorithm
classes as shown in Section 14.4.4.

14.4.4 Auction Sequence and Cognitive Awareness

The demand of RRGs varies with space and time. Consider a big event
like a soccer game, where participants arrive from several directions

and with different means of transportation like cars and trains: the participants will concentrate on a few traffic nodes and edges like train stations and highways and move on to the event location. The demand will thus vary locally and dynamically. Therefore, the RRGs of a BS should be offered to the users attending the cell and the price should be dependent on the demand of the proper cell.

In general, the operator wants to maximize his monetary gain, by charging the participants by their willingness to pay. In order to find this upper limit, negotiations take place between the operator and the users. Negotiations also allow the users to express their urgency to get the RRGs and to incorporate their purchase power in contrast to FPMs. The demand changes very fast, thinking on bursty traffic or moving hot spots. Therefore, the negotiations have to be executed fast and, in order to track the dynamic market, these have to be repeated very shortly. Users cannot permanently address their attention to the auction or, assuming repetition of auction up to a frame duration, users cannot physically react fast. Consequently, agents represent the parties. These algorithms need a proper protocol and rules to negotiate. A negotiation which follows proper instructions is the auction.

In an FPM, the UT need not be intelligent in terms of allocation competition including pricing, allocation, and billing, because the price is fixed and the BS designs the RRGs regardless of the other users. In contrast, within an underlying AS the UTs can gain experience from the history and accordingly modify the bidding strategy to improve the utility by expressing the users' needs with the bids. Collecting data, gaining cognition, and acting according to the learned experience are the major functionalities of the CR. Therefore, the AS allows the CR abilities to increase the operator's gain and provides opportunity to the users to express their needs within the techno-economical environment.

14.4.5 Cognitive Terminal in the Techno-Economical Environment

A terminal is cognitive, if its functionalities, which can be logically located within all open system interconnection (OSI) International Standards Organization (ISO) layers, allow to learn about the environment and act accordingly. Thus, the term cognitive not only includes the learning facility but also the consequent action. The whole processing chain of one cognitive function consists of information extraction, learning process, and decision execution (see Figure 14.5).

The cognitive terminal senses its embedded environment and collects data. The environment is recognized as a set of parameters

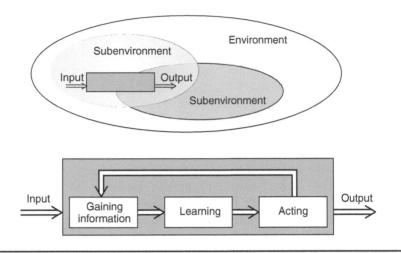

Figure 14.5 Cognitive radio in subenvironments (From Kloeck, C., Jaekel, H., and Jondral, F.K., Dynamic and local Combined pricing, allocation and billing system with Cognitive radios, *IEEE DySPAN*, 2005.)

by the cognitive terminal because of the algorithmic nature. A cognitive terminal has finite resources of computational power, recognition duration, hardware, and energy in general. Consequently, such an entity can at most collect the whole amount of parameters representing the environment. Therefore, a cognitive terminal can only obtain data from a subenvironment, which is defined as a set of parameters by which a cognitive functionality describes its environment. Thus, a cognitive terminal can collect data from a different environment despite the environment the entity will act on (see Figure 14.5), e.g., the CR is included in the cognitive terminal definition.

The AS environment, being one of the aforementioned subenvironments, comprises not only economical aspects that could be assumed at first glance, but also the action of the operators and the users, i.e., their behavior and the technical and physical subenvironments, which include:

■ The economical aspects that include the varying demand and supply, the purchase power of the competitors, the auction mechanism, and the reserve price.
■ The operator mainly strives to optimize his monetary gain by choosing and using a strategy which outputs the reserve price.
■ The user's behavior can approximately be described by preferences, utility characteristics, and budget constraints.

- Besides the finite hardware resources including computational power, the characteristics of the underlying RATs like data rate, control overhead, service provision, and quality represent main parameters.
- In addition to these four virtual subenvironments, the characteristics of the physical circumstances clearly have to be taken into account, because one major influence is the transmission channel characterized by SNR, signal-to-noise and interference ratio (SNIR) and basic encoding rules (BER).

The AS environment is distinguished between the operator's and the user's point of view. Usually, the operator's agent called economic manager (EM) who is logically located close to the BS executes the repeating multiunit sealed-bid auctions and thus gets knowledge of all bids of every user. This information gain can be used to better estimate the current market, users' demands resulting in cell capacity planning to guarantee QoS and future market prediction. The radio auction agent (RAA) which represents the user's interests is responsible for MAC and hence acting within the second layer of the ISO/OSI-layer model. The EM decides the economical information an RAA gets. The only way by which an RAA can get economic information, whereas the bids include the most information content, is indirectly from the EM after an auction. The conveyed information format has to be precisely determined in order to adjust the bidding strategy algorithms.

Besides the economical differences, the physical parameters are clearly different for all users and generally the users do not share the channel information affecting the radio resources needed to get a proper QoS implicating the values of the bids. Again, the BS in an infrastructure network elicits at least the channel state of the mobile terminal that assists in the reserve price adjustment to optimize the operator's gain.

Furthermore, assuming the AS environment to be identical for different users [44], different information extractions do not necessarily extract the same information. Moreover, some data and parameters of the AS environment are only known to one RAT and thus are private information. This leads to different kinds of information provided to different information extractions, whereas there is also common information like the reserve price r and the maximum number of RRGs N_{max} offered per RAT. One major task of the learning process is to estimate the behavior of the other users in order to conclude the most opportune action following game theory [44]. The decision execution transforms the conclusion into an action and, especially in the case of the aforementioned AS environment, this action will affect the same subenvironment, thus creating a recursive behavior and preferring to a dynamic and decentralized market.

14.4.6 Decentralized Multihoming System Structure

The auction repetition can be both periodical and spontaneous. Based on the underlying periodical repetition of logical control protocols in communication systems, e.g., TDMA systems, the implementation of fast auction repetitions up to frame-based auctioning is more suitable than undetermined ASs. In the following periodical auction repetition only is considered.

Generally speaking, a seller can only offer a good if he can surely sell it. Applying this to the auction mechanism and first assuming the user cannot perform multihoming, this mechanism can only offer goods that are not depending on the user's environment. For example, if the EM offers data or data rate and the user comes into an RF shadow, the offered good cannot be provided by the operator. Thus, the RRGs can only be frequency time bits in an FDMA/TDMA system, code time bits in a CDMA system like universal mobile telecommunication system (UMTS), or time bits in a TDMA system like WLAN IEEE 802.11a.

As a second possibility, the user may be able to send the data over more than one RAT simultaneously, i.e., its UT can get data by multihoming. For the sake of simplicity, assuming that multihoming occurs only by RATs of one operator, it seems a UT can bid for data or data rate no matter over which RATs the operator will transmit the data. But, following the same argumentation as before, the operator can only offer RRGs. The RRGs of different RATs need not mandatorily be the same and keeping in mind that each RAT is only specialized for a few services and not all UTs are able to transmit over all RATs, the demand per specific service and resources available can only result in a competition if the UTs bid for the RRGs of each RAT solely. Nevertheless, the traffic splitting over the RRGs of the RATs a user is allowed to send, data can be optimized by a joint radio resource management (JRRM) [45]. It can also be envisaged that if a user won RRGs of a RAT and the connection becomes worse, the JRRM exchanges part or the whole RRGs of another RAT if available and the UT possesses an interface in order to increase the QoS. Therefore, the UT bids for RRGs of each RAT available and gets a certain number of general resource elementary credits (GRECs) per RAT, i.e., kind of vouchers that are exchangeable, increasing the QoS of the UT and not decreasing the revenue of the operator. In Section 14.4.7, a system is proposed in order to realize E-RAMA considering a controlling instance at both the network and the user's side.

14.4.7 System Description

The AS is used for MAC, concluding the function to be logically located in the MAC layer of the ISO/OSI model (see Figure 14.6). The reserve price calculator (RPC) aims to approximate the operator's needs and behavior. Its main goal is to maximize the operator's monetary gain by varying the reserve price of the auction. This leads to coerce the users into bidding their true values. The RPC announces the reserve price to all users attending the radio cell and conveys this value to the auction mechanism. The auction mechanism collects the bid vectors of all users by polling each user. The allocation depends on the underlying RAT, i.e., the bids may have to be distinguished by the QoS classes described by the technology-based constraints like data rate and delay. Moreover, the transmission technique and mode may restrict the RRGs allocation granularity, e.g., in OFDMA IEEE 802.16, UL and DL are separated within a frame, but the border within the frame is dynamically adjustable with the restriction that each link category consists of an integer number of OFDMA symbols. Thus, it may happen that a user with a higher bid and less RRGs will be rejected against a lower bid and more RRGs, in order to fill the

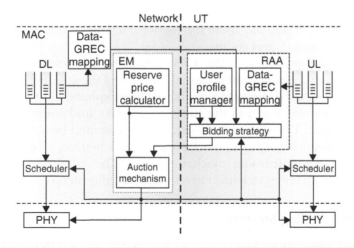

Figure 14.6 System description (From Kloeck, C., Jaekel, H., and Jondral, F.K., Auction sequence as a new resource allocation mechanism, *IEEE VTC Fall 05*, 2005.)

whole frame, increasing the overall capacity and maximizing the operator's gain. Realistically, the auction mechanism is proposed by the operator and thus represents his goals. Nevertheless, the auction mechanism has to be common knowledge, because this is the etiquette of the "game theory." Otherwise, in terms of game theory, the other players have no chance to choose a rational bidding strategy. In a manner of speaking the users are unarmed and have no confidence in the system, resulting in a change of the operator.

In the UT, the RAA is responsible for the bidding process in order to maximize the user's utility subject to the AS environment. Therefore, the RAA needs information from the user about the cost constraints for each service, the relative preference of the services, and the utility depending on the preferences and the fulfillment of the service. The user profile manager (UPM) serves as a user interface and provides information to the RAA entities. The arriving data from the third layer will be splitted into different QoS buffers by a convergence layer. The buffer management categorizes the data for each QoS buffer into critical and uncritical data. Critical data should be urgently sent before the next auction to meet the QoS criteria. By contrast, the uncritical data are waiting in the buffer, but the QoS does not allow to send them within this auction period. The goal of this evaluation is to convey information to the bidding strategy to fulfill the QoS. Consequently, the utility of the critical data has to be higher than the uncritical data. The data GREC mapping (DGM) maps the number of the categorized data to the corresponding number of RRGs needed. Therefore, the DGM needs information about the physical environment, e.g., the channel properties, and about the RAT characteristics like modulation, control overhead, and coding.

The information from the UPM about the user, from the DGM about the data categorization, and the results of the last auction serves as input for the bidding strategy that will be explained in Chapter 15. The bidding strategy and the RPC are adaptive and learn about the environment. These algorithms can be implemented by several artificial intelligence methods like Bayesians and neuronal networks or other analytical learning mechanism [36]. The whole system is a multiagent system specialized to sophistically allocated resources [42].

14.4.8 Bidding Strategy

The bidding strategy [44] possesses an integral and a differential part. The integral part incorporates past auctions and thus the past users' behaviors in order to predict number of GRECs won $N_I^{w,x,y}$ for QoS buffer l of link $y \in \{UL, DL\}$ and data category $x \in \{cri, uncri\}$ depending on the proposed bidding vector **bid**. On the other hand, the differential part changes the bid value b afterwards if the auction conditions like reserve price r and number of demand N_{dem} have been kept stable, but the goals are not fulfilled. The design of the bidding strategy aims at satisfying the user's wishes, i.e., this algorithm tries to act as the user would do. The action goals can be sorted according to the following decreased-ordered priority list:

1. Keeping the budget constraint
2. Fulfillment of the QoS
3. Maximization of the utility
4. Minimization of the costs

As in normal life the bidding strategy will get a budget where the costs must not exceed. The user adjusts this budget constraint with the UPM. This item is the most important restriction to the bidding strategy to keep confidence to the user and is allowed to act without the user's permanent supervision. The budget constraint can be differential or cumulative. A differential budget constraint means that the maximum costs per QoS class, data category, and link are given for one auction. On the other hand, a cumulative budget constraint summarizes all other opportunities like cumulative costs per auction, whereas the sum of the bid per auction must not exceed this limit. In addition, the accumulation over time can be envisaged, i.e., cumulative costs over several auctions have to be below a proper threshold.

Within this economical framework the requested service should be executed as good as possible. A service is defined by a set of parameters that are determined, fixed, or tolerated within an interval to execute the service satisfactorily. In this case, the second layer offers several different services to the third layer. The convergence layer in between has to assign the incoming data to proper service classes. The data in a service class can be categorized according to the urgency to send them with respect to the parameter specification fulfillment. If some or even all parameter specifications are injured, the QoS suffers, i.e., the bidding strategy is responsible for awarding as much RRGs as necessary to transmit the data according to their urgency to fulfill the QoS. The bidding strategy will at first bid for the critical data and afterwards for the uncritical.

After this preselection of the bids that optimally fulfill the QoS, the utility of transmitting the proper data will be determined and maximized. In other words, these data within each category, but over the service classes, will be selected for transmission which maximize the QoS. For this data the bidding strategy will submit bids **bid**.

Rationally, keeping in mind the leeway of the budget constraint for the bids in order to reach the goals of (1) and (2), the bidding strategy must be anxious to minimize the costs in line with the common user's wishes.

14.4.8.1 Input

The bidding strategy gets its input from the QoS buffer management and the UPM. The QoS buffer management determines the critical data D^{cri} that should be urgently sent and correspondingly the uncritical data D^{uncri} within this auction period T, and maps these data to the number of GRECs $N_l^{w,x,y}$ that are the goods to be auctioned per link and QoS and data category.

The UPM provides the bidding strategy with the user relevant data concerning purchase power and preferences. The user can adjust for each QoS class in uplink/downlink a maximum value $\mathbf{c}_i^{x,y}$ that the cost per GREC must not exceed. Thus, he is able to influence the price of the data of different QoS classes mainly representing different applications like real-time audio data for a phone call or best effort data for an e-mail download.

Second, the user can express his preferences by a preference vector α. Its components $\alpha_i^{x,y}$ indicate the relative personal evaluation of the QoS classes, e.g., if a user is more sensitive to a noisy phone call than to a disturbed FTP download, the QoS class mainly containing phone data will get a higher $\alpha_i^{x,y}$ than the one representing FTP.

The input to the BIS is represented as a set of quintuples I:

$$I = \{(\alpha_l^{x,y}, N_l^{b,x,y}, l, x, y) | l = 1 \ldots L, \; x \in \{\mathrm{cri}, \mathrm{uncri}\},$$
$$y \in \{\mathrm{UL}, \mathrm{DL}\}\} \tag{14.1}$$

in which L is the number of QoS classes of only one link.

14.4.8.2 Utility Function

The bidding strategy aims at satisfying the user's utility function as good as possible and contemporarily saving money. The utility function $\eta_i^{x,y}(\alpha_i^{x,y}, N_l^{b,x,y}, R)$ combines the use of resources expressed by the number of GRECs $N_l^{b,x,y}$ needed to satisfy the QoS requirements of each class in uplink/downlink and the user preferences represented by a preference vector α. The variable R serves as a limiter of the utility, i.e., the utility of $N_l^{b,x,y} > R$ remains the same regardless of any additional goods:

$$\eta(\alpha_l^{x,y}, N_l^{b,x,y}, R) = \begin{cases} \eta(\alpha_l^{x,y}, R, R) & N_l^{b,x,y} > R \\ \eta(\alpha_l^{x,y}, N_l^{b,x,y}, R) & N_l^{b,x,y} < R \end{cases} \qquad (14.2)$$

The limitation property can be switched off by converging R to infinity:

$$\eta(\alpha_l^{x,y}, N_l^{b,x,y}) = \lim_{R \to \infty} \eta(\alpha_l^{x,y}, N_l^{b,x,y}, R) \qquad (14.3)$$

The user can also choose his basic utility behavior for each QoS class categorized by the well-known economic expressions: risk-averse, risk-neutral, and risk-encouraged.

If the utility function $\eta_l^{x,y}(\cdot, N_l^{b,x,y}, R)$ is risk-averse in $N_l^{b,x,y}$, the differential utility $d\eta_l^{x,y}$ and in this case the additional utility of more goods will decrease if $N_l^{b,x,y}$ increases. Considering a risk-neutral function, the differential utility $d\eta_l^{x,y}$ is equal to each additional differential good $dN_l^{b,x,y}$. If there exits a QoS class for which the QoS has only fulfilled sending complete datagrams, the user will choose a risk-encouraged function whose differential utility increases in $N_l^{b,x,y}$. All the utility functions have the following two properties in common:

1. $\eta_l^{x,y}(\alpha_l^{x,y}, 0, R) = 0$

2. $\eta(\alpha_1, n_0, R) > \eta(\alpha_2, n_0, R) \leftrightarrow \alpha_1 > \alpha_2$

The first expression means that there is no utility if no goods are available, and the second states the increased utility by an underlying higher preference assuming utility functions of the same category. Especially, for the risk-neutral class another property holds with respect to the derivative η':

3. $\eta'(\alpha_1, \cdot, R) > \eta'(\alpha_2, \cdot, R) \Leftrightarrow \alpha_1 > \alpha_2$

This property mainly reduces the effort of computer in finding the optimized bid vector.

14.4.8.3 Risk-Averse Utility
The differential utility of the critical data is constant in most cases, because a lost critical datum is indistinguishable from its buffer location and temporal position. Consequently, the respective utility function is linear. The critical data of a QoS buffer have to produce a higher utility than the uncritical. The importance and the additional characteristic of the critical data, e.g., to transmit, are apparent in comparison with the uncritical data. All the other characteristics are the same. The data in

the QoS buffer can always be sorted to get a concave utility graph, if the order of relation is the urgency to keep the QoS parameter limits and the differential utility is monotonically decreasing in the urgency. The data will be sorted in a descending order. Thus, the first data are the critical ones, followed by the uncritical. Last but not least to save computation power this concave function can be approximated by two linear functions, a linear utility for each of the critical and uncritical data. This is an approximation of the well-known risk-averse behavior observed in economic studies. The critical and the uncritical utility function, $\eta(\alpha_l^{cri,y}, N_l^{cri,y}, L_l^{cri,y})$ and $\eta(\alpha_l^{ucri,y}, N_l^{ucri,y}, L_l^{ucri,y})$, can be described by the preference-limiter tuple $(\alpha_l^{x,y}, L_l^{x,y})$.

14.4.8.4 Utility Criterion

The main goal of the bidding strategy is to approach the utility function of the GRECs needed and the GRECs won. Therefore, the criterion of the difference function $\Delta_\eta(x, y)$ is defined as:

$$\Delta_\eta(\mathbf{x}, \mathbf{L}) = \sum_i \left[\eta_l^{x,y}(\alpha_{l,i}^{x,y}, x_i, L_i) - \eta_l^{x,y}(\alpha_{l,i}^{x,y}, L_i, L_i) \right] \qquad (14.4)$$

The difference utility function is always nonpositive and the value indicates the utility missed. The quadratic error utility is not chosen, because based on its minimization result, one cannot conclude the utility maximization result. For example, consider two different x_1 and x_2, where both of them affect the same quadratic utility error, but the gained utility is different. Thus, the x has to be chosen with the highest utility to reach utility maximization and not the one with the minimum quadratic utility error. By minimizing the quadratic error utility there is no incentive to favor utility functions to get very close to their utility wanted, but gain less additional utility, over utility functions for which the difference to the utility wanted is higher, but gain more additional utility. Thus, the absolute utility maximization suffers from the quadratic error functions.

14.4.8.5 Bid Representation

The bid vector $\mathbf{bid}_k$ of user k is a vector consisting of quadruple elements:

$$\mathbf{bid}_k = x_i(N_{l,i}^{b,x,y}, b_i, y, l) \qquad (14.5)$$

The information of one quadruple includes the number of RRGs $N_{l,i}^{b,x,y}$ needed for link y and SC l. For each RRG of $N_{l,i}^{b,x,y}$ the RAA bids b_i.

Depending on the RAT, DGM, and auction protocol, the bid vector size can be limited and information can be rejected.

14.4.8.6 Ideal Strategy

Ideally, the bidding strategy possesses complete information of the other users inter alia their bids. The goods a bidder wins are a deterministic function of his own bids **bid** and the bids of the other users $\mathbf{bid}_{-k}$:

$$N^w = AM(\mathbf{bid}_k, \mathbf{bid}_{-k}) \tag{14.6}$$

Besides this, the strategy philosophy has to be expressed in a formal statement and measure in order to design algorithms. The aim of the bidding strategy is to maximize the utility difference of the utility won and the actual utility wanted under condition of a certain bid constellation for the critical data:

$$S_3 = \{\mathbf{bid} | \arg \max_{\mathbf{bid}} \Delta_\eta(N^{w,\mathrm{cri}}, N^{b,\mathrm{cri}})\} \tag{14.7}$$

The set S_3 includes all bids **bid** which maximize the utility difference function for the critical data. The same procedure will be applied for the uncritical data with respect to S_3:

$$S_2 = \{\mathbf{bid} | \arg \max_{\mathbf{bid} \,\in\, S_3} \Delta_\eta(N^{w,\mathrm{ucri}}, N^{b,\mathrm{ucri}})\} \tag{14.8}$$

Till now, this set of bids maximize the utility difference of the uncritical data subject to the maximization of the utility difference of the critical data. After calculating the set S_2, the bid vector $\mathbf{bid} \in S_2$ will be chosen which minimizes the costs:

$$\mathbf{bid}_k \in S_1 = \{\mathbf{bid} | \arg \min_{\mathbf{bid} \,\in\, S_2} c\} \tag{14.9}$$

Especially, in a discriminatory auction the cost c_{dis} depending on the $\mathbf{bid}_i$ and the number of GRECs $N_{l,i}^{b,y}$ can can be expressed by:

$$c_{\mathrm{dis}} = \sum N_{l,i}^{w,x,y} b_i \tag{14.10}$$

14.4.8.7 Strategy for Incomplete Information

The bidding vectors that are intended to be submitted have to be covered by the bidding strategy by not assuming a cooperation among the agents. Thus, the bidding vectors $\mathbf{bid}_{-k}$ are unknown information resulting in using stochastic methods, and the method only approaches the ideal bidding strategy. The number of goods won are no longer deterministic. The $\mathbf{bid}_{-k}$ are characterized by a probability density $b(b_{-k})$. The conditional probability $P\{N^w \mid \mathbf{bid}_k\}$ of the goods won N^w is subject to the own bid vector $\mathbf{bid}_k$.

14.4.8.8 Integral Part

The bidding strategy approaches the ideal strategy, by maximizing the expectation of the utility difference subject to the **bid**. Equation 14.7 and Equation 14.8 are modified to:

$$S_3 = \left\{ \mathbf{bid} \middle| \arg \max_{\mathbf{bid}} E\{\Delta_\eta(N^{w,\text{ucri}}, N^{b,\text{cri}})|\mathbf{bid}\} \right\} \tag{14.11}$$

$$S_2 = \left\{ \mathbf{bid} \middle| \arg \max_{\mathbf{bid} \in S_3} E\{\Delta_\eta(N^{w,\text{ucri}}, N^{b,\text{ucri}})|\mathbf{bid}\} \right\} \tag{14.12}$$

The minimization of the cost is transformed into a worst-case estimation resulting in the minimization of the maximum expected costs given by Equation 14.9:

$$\mathbf{bid}_k \in S_1 = \{\mathbf{bid} | \arg \min_{\mathbf{bid} \in S_2} \max c\} \tag{14.13}$$

specifically for a discriminatory multiunit sealed-bid auction:

$$\max c_{\text{dis}} = \sum N_{l,i}^{b,x,y} b_i \tag{14.14}$$

The selection of the expectation instead of a probability optimization is based on the right balance of a priori calculated losses and the probability that this utility will occur. For instance, choosing a maximum a posteriori or a maximum likelihood estimator, both maximizes probabilities of either the value of the utility difference subject to the bids or vice versa. Consequently, in latter cases the measure of both the utility and the corresponding occurring probability is missed.

14.4.8.9 Allocation Estimation

Assuming that the bids $\mathbf{bid}_{-k}$ of the other users are independent and identically distributed according to the relative frequencies of the histogram $\tilde{b}_x(x|r, N_{\text{dem}})$ based on the $\mathbf{bid}_{-k}$ won, the probability $P\{x < b_i\} = H_k^{k_0}(b_i)$ that given k_0 and other bids at most κ bids are higher than $\mathbf{bid}_i$ can be calculated:

$$H_k^{k_0}(b_i) = \sum_{l=0}^{k-1} \binom{k_0}{l} [H_X(b_i)]^{k_0-l} \cdot [1 - H_X(b_i)]^l \qquad (14.15)$$

where $k = \min\{N_{max}, N_{dem}\}$. If $\kappa > k_0$, Equation 14.15 converges to a binomial progression and thus $H_\kappa^{k_0}(b_i) = 1$. That is, if the demand N_{dem} is smaller than the supply, the user will get $\kappa - k_0 - 1$ GRECs with probability 1. The corresponding probability density $b_\kappa^{k_0}(b_i)$ to Equation 14.15 that will be used in the proposed algorithm is

$$b_\kappa^{k_0}(b_i) = k_0 h_X(b_i) \binom{k_0 - 1}{\kappa - 1} [H_X(b_i)]^{k_0-\kappa}[1 - H_X(b_i)]^{\kappa-1} \qquad (14.16)$$

The expected number of goods N^e won given a proper bid constellation **bid** can be determined by applying Equation 14.15:

$$N^e = E\{N^w|\mathbf{bid}\} = \sum_{i=1}^{2L} N_{l,i}^{e,x,y} \qquad (14.17)$$

$$N_{l,i}^{e,x,y} = \sum_{j=sN_{l-1,i}^y}^{sN_{l,i}^{b,x,y}} H_{N_{max}-j}^{k_0}(b_i) \qquad (14.18)$$

$$sN_l^{b,x,y} = \sum_{u<l} N_u^{b,y} \qquad (14.19)$$

The GRECs are estimated and the corresponding bids are comprised in a tuple $(N_{l,i}^{e,x,y}, b_i)$.

14.4.8.10 Derivation of the Algorithm

In the following it is assumed that the DGM works optimally, i.e., for a given number of goods won, N^w the division to the link, QoS classes, and categories maximizes the utility. Moreover, the bids of the others are assumed to be independent and identically distributed. The own bids underly a differential budget constraint:

$$0 \le b_i \le c_i \qquad (14.20)$$

The goal of the algorithm is to compute the solution of the strategy with incomplete information by Equation 14.11 through Equation 14.13. The expectation of the utility difference subject to bid vector **bid**$_k$ can be written using order statistics:

$$E\{\Delta_\eta|\mathbf{bid}_k\} = \sum_{n=0}^{N_{\max}} E\{\Delta_\eta|N^w = n\} \cdot \left(H_{N_{\max}-n}^{N_{\max}}(b_{i_n}) - H_{N_{\max}-n+1}^{N_{\max}}\left(b_{i_{(n-1)}}\right)\right)$$

$$= \sum_{n=1}^{N_{\max}} \sum_z \eta(\alpha_z, N_z^w, N_z^b) \cdot \left(H_{N_{\max}-n}^{N_{\max}}(b_{i_n}) - H_{N_{\max}-n+1}^{N_{\max}}\left(b_{i_{(n-1)}}\right)\right)$$

$$- \sum_z \eta(\alpha_z, N_z^b, N_z^b) \tag{14.21}$$

$$\sum_z N_z^w = N^w \tag{14.22}$$

The maximization of Equation 14.21 is equivalent to maximization of the first sum in Equation 14.22 with respect to the bid vector. Thanks to the conditional expectation notation, only the order statistics terms are dependent on the bid vector. The cumulative probability function $H_y^{N_{\max}}$ $(x<w)$ is both monotonically increasing in y and w, therefore all summands are positive by keeping in mind the non-negative utility difference η. Now, taking the assumption of the optimal allocation and that a bidding strategy bids at most for N^b RRGs in the whole, the difference between two adjacent utility functions is always nonnegative:

$$\sum_z \eta(\alpha_z, N_z^w, N_z^b)|_{N^b} - \sum_z \eta(\alpha_z, N_z^w, N_z^b)_{N^b-1} \geq 0 \tag{14.23}$$

$$\sum_z N_z^b = N^b \tag{14.24}$$

Reordering Equation 14.22 with respect to the cumulative distribution function gives a sum with nonnegative summands, each of it increases in the bid value b_{i_n}:

$$E\{\Delta_\eta|\mathbf{bid}_k\} = \sum_{n=1}^{N_{\max}} \sum_z [\eta(\alpha_z, N_z^w, N_z^b)|_{N^w=n} - \eta(\alpha_z, N_z^w, N_z^b)|_{N^w=n-1} \cdot H_{N_{\max}-n}^{N_{\max}}(b_{i_n})$$

$$- \sum_z \eta(\alpha_z, N_z^b, N_z^b) \tag{14.25}$$

Thus, the maxima of Equation 14.23 can be found by only considering the probability density function of Equation 14.16. This equation is zero if the single density function $h_x(b_i)$ is zero or the corresponding cumulative distribution function is 0 or 1. The latter two cases are included in the first one. The global maximum of Equation 14.25 for $b_{i_n} \in [0, c_{i_n}]$ is c_{i_n}, if $h_n^{N_{\max}}(c_{i_n})$ is unequal to zero. Otherwise, there exists one single interval, in which the cumulative distribution function is globally maximum. To simplify matters, the kernel equivalence relation π_f of a function $f(x)$, $x \in [a, b]$ is introduced:

$$(x,y) \in \pi_f \iff f(x) = f(y) \tag{14.26}$$

For this equivalence relation the equivalence classes $[x]_{\pi_f}$ are one-to-one.

14.4.8.11 Algorithm

The maximum costs c_{i_n} are in descending order. That is, the definition interval of bid $b_{i_{(n+1)}}$ is a subset of the interval of b_{i_n}. The maximum bid reached by b_{i_n} is at least as high as the one gained by $b_{i_{(n+1)}}$. Thus, bids are also in descending order and this order relation is equivalent to the c_{i_n} order relation. Therefore, the expectation of the utility difference $E\{\Delta_\eta\}$ is maximized by choosing the bids b_{i_n} of $[c_{i_n}]_{\pi H_X}$ and from this select the minimum bid to minimize the maximum expected costs. Consequently, the algorithm starts for the least significant bid $b_{i_{N\max}}$ and steps down from $c_{i_{N\max}}$ to the reserve price until b_X is unequal to zero. That is, given $b_X(x)$ with $x \in [r, c_{i_{N\max}}]$, the first step can be expressed formally as:

$$b_{i_{N\max}} = \min[c_{i_{N\max}}]_{\pi H_X} \tag{14.27}$$

For the second least significant bid the procedure is the same, but the searching space is at least reduced to $c_{i_{N\max}}, c_{i_{N\max}-1}$. If the downstepping search has reached $c_{i_{N\max}}$ without finding b_X is unequal to zero, $b_{i_{N\max}-1}$ is automatically equal to $b_{i_{N\max}}$:

$$b_{i_{N\max}-1} = \frac{\min}{x}\left\{[c_{i_{N\max}-1}]_{\pi H_X}\right\} - \{x | x < b_{i-1(N\max+1)}\} \tag{14.28}$$

The following steps are similar to the second one in finding the corresponding bids. The searching space for the kth bid b_{i_k} is at least to the last maximum cost limit $c_{i(k+1)}$; hence, Equation 14.28 can be generalized to:

$$b_{i_k} = \frac{\min}{x}\left\{[c_{i_k}]_{\pi H_X}\right\} - \{x | x < b_{i(k+1)}\} \tag{14.29}$$

The computation time of the algorithm is independent of the number of users and goods, but at most linear to the number of quantization steps of the bids assuming bids with the same maximum costs are

the operators can better adapt to the market and the users' wishes, and on the other hand the users are now able to express directly their needs by bids. Consequently, a dynamic market model is at least as good as the current FPM for the monetary gain of the operators and the satisfaction of the overall utility.

ACKNOWLEDGMENT

This work is partially funded by the Commission of the European Communities, under the 6th Framework Program for Research and Technological Development, within the project End-to-End Reconfigurability (E^2R). The authors would like to acknowledge the contributions of their colleagues from E^2R consortium.

REFERENCES

1. V. Charkravarthy, A. Shaw, M. Temple, and J. Stephens, "Cognitive Radio—An Adaptive Waveform with Spectral Sharing Capability," IEEE Wireless Communications and Networking Conference (WCNC'05), New Orleans, USA, March 2005.
2. Shared Spectrum Company, "Comprehensive Spectrum Occupancy Measurements over Six Different Locations," www.sharedspectrum.com, August 2005.
3. FCC, "Spectrum Policy Task Force Report, ET Docket No. 02-155," November 2002.
4. J. Juntunen, V. Ranki, and K. Kalliola, "Spectrum Measurements and Spectrum Occupance," 15th Meeting of Wireless World Research Forum, Paris, December 2005.
5. J. Mitola, "Cognitive Radio—An Integrated Agent Architecture for Software Defined Radio," Ph.D. thesis, Royal Institute of Technology (KTH), Kista, Sweden, 2000.
6. Steven Ashley, "Cognitive Radio," Scientific American, www.sicam.com, March 2006.
7. S. Haykin, "Cognitive Radio: Brain-Empowered Wireless Communications," *IEEE Journal on Selected Areas in Communications*, vol. 23, pp. 201–220, February 2005.
8. D. Cabric, S.M. Mishra, and R.W. Brodersen, "Implementation Issues in Spectrum Sensing for Cognitive Radios," 38th Asilomar Conference on Signals, Systems, Computers, vol. 1, pp. 772–776, November 2004.
9. J. Mitola and G.Q. Maguire, "Cognitive Radio: Making Software Radios More Personal," *IEEE Personal Communications*, vol. 6, no. 4, pp. 13–18, August 1999.
10. C.J. Rieser, "Biologically Inspired Cognitive Radio Engine Model Utilizing Distributed Genetic Algorithms for Secure and Robust Wireless Communications and Networking," Ph.D. thesis, Virginia Polytechnic Institute and State University, Blacksburg, VA, August 2004.
11. Software Defined Radio Forum, Cognitive Radio Working Group (CRWG), www.sdrforum.org.

12. R.W. Thomas, L.A. DaSilva, and A.B. MacKenzie, "Cognitive Networks," First IEEE International Symposium on New Frontiers in Dynamic Spectrum Access Networks (DySPAN 2005), Baltimore, MD, pp. 352–360, November 2005.

13. F. Jondral, "Parametrization—A Technique for SDR Implementation," ed., W. Tuttlebee, *Software Defined Radio-Enabling Technology*, John Wiley & Sons, London, pp. 232–256, 2002.

14. F. Jondral, "Parameter Controlled Software defined Radio," *Proceedings of the SDR Forum Technical Conference*, San Diego, CA, November 2002.

15. V. Blaschke and F. Jondral, "A Concept for a Standard-Independent Cognitive Radio Terminal," 12th European Wireless Conference, Athens, April 2006.

16. V. Blaschke and F. Jondral, "An Approach for Providing QoS in Cognitive Radio Terminals," *Proceedings of 4th Karlsruhe Workshop on Software Radios*, Karlsruhe, pp. 139–143, March 2006.

17. E. Perera and A. Seneviratne, "Failover for Mobile Routers: A Vision of Resilient Ambience," IEEE International Conference on Networking (ICN'05), Reunion Island, Work in progress. April 2005.

18. T.A. Weiss and F.K. Jondral, "Spectrum Pooling: An Innovative Strategy for The Enhancement of Spectrum Efficiency," *IEEE Communications Magazine*, vol. 42, no. 3, pp. 8–14, March 2004.

19. S. Hijazi, B. Natarajan, M. Michelini, W. Zhiqiang, and C.R. Nassar, "Flexible Spectrum Use and Better Coexistence at the Physical Layer of Future Wireless Systems via A Multicarrier Platform," *IEEE Wireless Communications Magazine*, vol. 11, no. 2, pp. 64–71, April 2004.

20. S.A. Zekavat and C.R. Nassar, "Spectral Sharing in Multi-System Environments via Multicarrier CDMA," IEEE ICC'03, vol. 3, 11–15 pp. 2223–2228, May 2003.

21. F. Capar, T. Martoyo, T. Weiss, and F. Jondral, "Analysis of Co-Existence Strategies for Cellular and Wireless Local Area Networks," IEEE VTC'03, Orlando, USA, November 2003.

22. A. Pezeshk and S.A. Zekavat, "DS-CDMA vs. MC-CDMA, a Performance Survey in Intervendor Spectrum Sharing Environment," 37th Asilomar Conference on Signals, Systems and Computers, vol. 1, 9–12 pp. 459–464, November 2003.

23. M. Michelini, S. Hijazi, C.R. Nassar, and W. Zhiqiang, "Spectral Sharing Across 2G–3G Systems," 37th Asilomar Conference on Signals, Systems and Computers, 2003, vol. 1, 9–12 pp. 13–17, November 2003.

24. J. Mitola, "Cognitive Radio for Flexible Mobile Multimedia Communications," Mobile Multimedia Communications, 1999. (MoMuC'99), 15–17 pp. 3–10, November 1999.

25. ET Docket No. 02-135, "Spectrum Policy Task Force Group Report," US Federal Communications Commission, 15 November 2002.

26. ET Docket No. 03–108, "Notice of Proposed Rule Making and Order," US Federal Communications Commission, December 2003.

27. D. Grandblaise, "Gestion flexible de spectre," Les Systemes Radiomobiles Reconfigurables, Hermes Science, ISBN: 2-7462-1127-0, 2005.

28. D. Reed, "Why Spectrum is Not Property—The Case for an Entirely New Regime of Wireless Communications Policy," http://www.reed.com/Papers/openspec.html, Work in progress. 2001.

29. K. Werbach, "Open Spectrum: The New Wireless Paradigm, New America Foundation—Spectrum Policy Program," Work in progress. October 2002.

30. F. Capar, et al., "Comparison of Bandwidth Utilization for Controlled and Uncontrolled Channel Assignment in a Spectrum Pooling System," IEEE VTC'02, Work in progress. March 2002.

31. L. Kleinrock, "Distributed Systems," Communications of the ACM, Special Issue, November 1985.

32. C. Ting, S.S. Wildman, and J.M. Bauer, "Government Policy and the Comparative Merits of Alternative Governance Regimes for Wireless Services," in *Proceedings of IEEE DySPAN*, pp. 401–419, November 2005.

33. E. Bonabeau, G. Theraulaz, and J.L. Deneubourg, "Adaptive Task Allocation Inspired by a Model of Division of Labor in Social Insects," in eds., D. Lundh and B. Olsson, *Bio Computation and Emergent Computing*, World Scientific, vol. 28, pp. 36–45, 1998.

34. V. Cicirello and S. Smith, "Wasp-like Agents for Distributed Factory Coordination," Technical Report CMU-RI-TR-01-39, Robotics Institute, Carnegie Mellon University, Pittsburgh, 2001.

35. J.C. Dunat, D. Grandblaise, and C. Bonnet, "Efficient OFDMA Distributed Optimization Algorithm Exploiting Multi-User Diversity," in *Proceedings of IEEE DySPAN*, pp. 233–242, November 2005.

36. J. von Neumann and O. Morgenstern, *The Theory of Games and Economic Behaviour*, John Wiley & Sons, 1944.

37. D. Fudenberg and J. Tirole, *Game Theory*, Cambridge, MA, MIT Press, 1991.

38. A.B. MacKenzie and S.B. Wicker, "Game Theory and the Design of Self-Configuring, Adaptive Wireless Networks," *IEEE Communications Magazine*, vol. 39, pp. 126–131, November 2001.

39. C.U. Saraydar, N.B. Mandayam, and D.J. Goodman, "Efficient Power Control via Pricing in Wireless Data Networks," *IEEE Transactions on Communications*, vol. 50, pp. 291–303, February 2002.

40. Z. Han, Z. Ji, and K.J.R. Liu, "Fair Multiuser Channel Allocation for OFDMA Networks using Nash Bargaining solutions and coalitions," *IEEE Transactions on Communications*, vol. 53, pp. 1366–1376, August 2005.

41. C. Kloeck, H. Jaekel, and F.K. Jondral, "Dynamic and Local Combined Pricing, Allocation and Billing System with Cognitive Radios," IEEE DySPAN 2005 Symposium, 8–11 November 2005.

42. D. Grandblaise, K. Moessner, G. Vivier, and R. Tafazolli, "Credit Tokens Based Scheduling for Inter BS Spectrum Sharing," 4th Karlsruhe Workshop on Software Radios (WSR'06), Karlsruhe, Germany, 22–23 March 2006.

43. D. Grandblaise, et al., "Techno-Economic of Collaborative based Secondary Spectrum Usage—E2R Research Project Outcomes Overview," First IEEE International Symposium on New Frontiers in Dynamic Spectrum Access Networks (DySPAN 2005), Baltimore, MD, 8–11 November 2005. Work in progress.

44. C. Kloeck, H. Jaekel, and F. Jondral, "Auction Sequence as a New Resource Allocation Mechanism," IEEE VTC Fall 05, 2005.

45. N. Amitay, "Resource Auction Multiple Access (RAMA): Efficient Method for Fast Resource Assignment in Decentralised Wireless PCS," *Electronics Letters*, vol. 28, no. 8, pp. 799–801, April 1992.

46. J. Santivanez and J. Roberto Boisson de Marca, "D-RAMA: A New Media Access Protocol for Wireless Multimedia Communications," *Proceedings of*

1997 IEEE Personal, Indoor, and Mobile Radio Communications, Helsinki, Finland, pp. 1043–1048. Work in progress.

47. A.S. Tanenbaum, *Computer Networks,* Englewood Cliffs, NJ, Prentice-Hall, Work in progress. 1981.

48. G.J. Nutt and D.L. Bayer, "Performance of CSMA/CD Networks Under Combined Voice and Data Loads," *IEEE Transactions on Communications,* vol. 30, no. 1, part 1, Work in progress. pp. 6–11, January 1982.

49. A. Colvin, "CSMA with Collision Avoidance," *Computer Communication,* vol. 6, 1983, Work in progress. pp. 227–235.

50. Y. Chevaleyre, P.E. Dunne, U. Endriss, J. Lang, M. Lematre, N. Maudet, J. Padget, S. Phelps, J.A. Rodrguez-Aguilar, and P. Sousa, "Issues in Multiagent Resource Allocation," *Informatica,* vol. 30, 2006, Work in progress. pp. 3–31.

51. P. Cordier, et al., *Cognitive Pilot Channel,* WWRF15 Paris, Work in progress. 2005.

52. T. Mitchell, *Machine Learning,* New York, MacGraw Hill, 1997, ISBN: 0071154671. Work in progress.

53. J. Perez-Giupponi, L. Giupponi, R. Augusti, and O. Sallent, "A Fuzzy Logic Neuronal based Approach for Joint Radio Resource Management in a beyond 3G framework," Quality of Service in Heterogenous Wired/Wireless Networks, Work in progress. 2004.

54. C. Kloeck, H. Jaekel, and F. Jondral, "Auction Sequence as a New Spectrum Allocation Mechanism," IST'05 Dresden, 19–23, June 2005.

15

WIRELESS MESH NETWORKS CASE STUDY: FIRE EMERGENCY MANAGEMENT AND MARKET ANALYSIS

*Carles Gomez, Pau Plans, Marisa Catalan,
Josep Lluis Ferrer, Josep Paradells, Anna Calveras,
Javier Rubio, and Daniel Almodovar*

CONTENTS

Today, public safety and first-responder networks demand solutions based on open standards that perform better than existing emergency networks. Wireless mesh networks (WMNs) constitute a promising technology for these environments. In this chapter, we propose a proof-of-concept public safety WMN for firefighter interventions based on current commercial equipment. We carry out a comprehensive market analysis in order to select a suitable platform to build our prototype. The overall system performance is presented and validated according to specific performance criteria and system requirements for this case. Results show the feasibility of the solution, although further research is needed to fulfill all the requirements. Finally, open issues are discussed.

15.1 INTRODUCTION

Firefighters work in very rough conditions to save people's lives while risking their own. In the United States alone, fire causes more than 4000 civilian deaths, 100 firefighter deaths, 25,000 injuries, and $11 billion in property losses every year [1,2]. Tragic events like the attacks on September 11, 2001 have brought the crucial role played by fire-fighters in emergency situations to the attention of the media.

Firefighters have to move around unknown places to put out fires. They can get lost or even isolated. In a best case scenario, they may have a map of the area (e.g., a burning building). However, smoke, fire, and noise make orientation difficult. In current systems, most radio messages describe a firefighter's location. Ideally, location should be user-transparent, i.e., it should not require human intervention. In general, communication between firefighters and the incident commander, who coordinates the emergency operation at the control center (CC), is vitally important.

Today, communications solutions in emergency scenarios are based on private mobile radio (PMR) systems. These systems use low-frequency signals, which provide good coverage and penetrate walls easily. At low frequencies, channels are narrowband. As a result, they can only offer analog or digital voice capabilities and very limited data services suitable for status messages or short messages. Never-theless, nearly all communications are currently voice-based, despite the surrounding noise (due to the fire, people shouting, sirens blaring, chain saws, etc.). Today's PMR systems are based on the digital tech-nologies of Terrestrial trunked radio (TETRA) [3] and the Association of public safety communications officials (APCO P25) [4]. The TETRA system offers direct mode operation, which allows direct communica-tion between terminals without network infrastructure. TETRA can only be used to transmit voice messages and short messages, due to the very limited data rates available (around 2.4 Kbps, with the highest protection level). With these data rates, the CC cannot be kept informed of the status of each firefighter. In order to increase the available bit rate, higher bandwidth is required, which involves using higher frequencies. If more bandwidth were available, visual information such as maps, in addition to text orders, could be provided to firefighters and viewed on a helmet-mounted display. (Firefighters are too busy to watch a screen for long periods of time, and it is difficult for them to use a terminal while they are wearing gloves.) Likewise, firefighters could send image content to the CC. Video and pictures taken at the scene of a fire are very useful for tactical firefighting response [5]. Other new broadband services that may further improve the efficiency of emergency tasks include robotics, remote control, and data collection from a large number of monitoring sensors.

The US Federal Communications Commission (FCC) allocated 50 MHz in the frequency band from 4940 to 4990 MHz to enable the provision of broadband services for emergencies [6]. This should enable IEEE 802.11a/j and digital short radio communications (DSRC) products to be adapted [7]. The availability of broadband services does not exclude the introduction of voice-over Internet Protocol (VoIP) services in the long term, although the latter may be used as a backup option rather than as a substitute for current voice communications systems [7]. However, integrating services into a single system could be considered for reasons of efficiency and cost. If this were done, IP multicast support could emulate the group voice communications currently used by firefighters. Other US government actions aimed at improving the efficiency of public safety response are led by the SAFECOM program [5]. This program has produced a state of requirements (SoR) document, which contains a set of requirements for an interoperable public safety communications (PSCs) system. The SoR defines an operational organization with different models of PSC networks. The following PSC networks are considered:

- Personal area network (PAN): a network for connecting devices carried by individuals
- PSC user group network (PUGN): an ad hoc network made up of network nodes carried by different individuals
- Incident area network (IAN): a temporary network created to support communications during an incident
- Jurisdictional area network (JAN): a permanent infrastructure covering a wide area (e.g., a city) for supporting communications in emergency situations
- Extended area network (EAN): a network that provides communication between regional and national public safety networks

Communication between the entities deployed in an emergency is provided by an IAN. IANs should be easy to deploy and should support a certain degree of user mobility. Communication with remote locations such as other IANs or dispatch centers is supported by a JAN. JANs should offer backbone capacity. Both IANs and JANs should be robust to any link or node suppression by offering sufficient path redundancy in the event of catastrophe. In view of the features of PUGNs, IANs, and JANs, together with the demand for broadband applications from the public safety community (e.g., firefighters), WMNs appear to be a promising solution for these kinds of PSC networks due to their inherent robustness in terms of available paths, self-configurability, and self-healing properties.

A WMN can use IEEE 802.11a/j-like radio interfaces and thus offer broadband capacity in emergency scenarios. These technologies have range limitations due to restrictions on the power transmission allowed. In the new 4.9 GHz band, power can be boosted to allow better building penetration. Unfortunately, the channel delay spread may still introduce coverage restrictions, especially in the case of outdoor use [7]. The multihop approach of a WMN makes it possible to overcome the aforementioned limitations.

One key factor that may determine the success of WMN technology is interoperability, which is another major problem identified in emergency response operations [5]. Standard technologies must therefore be used. Gateways to interconnect devices using different protocols can only be developed when such protocols have public specifications. Standards also offer other advantages over proprietary solutions. The public availability of standard technologies makes them easy to implement on different platforms. As they are adopted on the mass market, they are also cheap. Wireless local area network (WLAN) technology makes it possible to build WMNs based on commercial off-the-shelf (COTS) technology.

This chapter describes our experience in designing, building, and evaluating a proof-of-concept WMN for fire emergencies. The remainder of the chapter is organized as follows. First, we define the proposed WMN system for fire emergencies. Next, we identify several requirements of the scenario defined. We then discuss the WMN products that are currently available on the market in order to select the platform that best matches our specific requirements and to implement a prototype. We then validate the proposal by implementing and testing a system with open standards that provides voice, video, and data communications between groups at data rates that are higher than those of current emergency systems. Results show the feasibility

include the development of gateways and the availability of low-cost products on the market.

■ *Affordable equipment*: The use of COTS technology is preferred in order to build a WMN at a low price.

15.3.2 Specific Requirements

On the basis of the general requirements, we derived the list of specific requirements shown below.

15.3.2.1 Hardware Requirements

■ *Form factor and weight*: Firefighters should carry several seeds in addition to their own FT. These devices must be small and light enough for them to carry.

■ *Indoor and outdoor operation*: FTs and seeds are expected to be used mainly indoors, while ONs are defined as outdoor devices. Network nodes should therefore be able to operate in both indoor and outdoor scenarios.

■ *Batteries*: The battery life of the nodes should ensure full operation throughout a fire incident.

15.3.2.2 Radio and Network Topology Requirements

■ *Multiple radio interfaces*: The use of multiple radio interfaces and different channels can maximize WMN capacity [9]. Since the system we propose is a hybrid WMN [8] with multiple available radio interfaces, one of them can be devoted to backhaul purposes while alternative interfaces and frequencies can be used for communication between infrastructure nodes (i.e., ONs and seeds) and clients (i.e., FTs). In addition, several radio channels can be used, which makes it possible to avoid using an interfered frequency. Thus, this feature makes the WMN more robust.

■ *Radio interface technology*: The use of a 4.9 GHz spectrum interface conforms to the US public safety band reserved for broadband applications. Additional support for a 2.4 GHz band is useful for connecting laptops or handheld devices (e.g., PDAs) to the WMN, since wireless connectivity of such devices is currently based on IEEE 802.11b technology.

■ *Client relay capability*: In order to provide additional path redundancy to the proposed WMN system, clients should be able to relay data. The system therefore uses the aforementioned hybrid WMN approach, since it includes infrastructure nodes and client relays.

15.3.2.3 Performance Requirements

- *End-to-end latency*: End-to-end delay should be kept to a small or assumable value since it affects the performance of data communications, especially for real-time services such as the remote control of robots and voice communications. Typical one-way delay requirements for VoIP services are below 150 ms.
- *End-to-end bandwidth*: We refer to the available bandwidth in a path without broken links. Reactivity to topology changes is considered in a separate performance parameter. Enough end-to-end bandwidth should be available to allow the use of services such as video streaming or the remote control of robots. For push-to-talk (PTT) voice communications a recommended minimum of 80 Kbps is suggested. Bandwidth requirements for video communications depend on the image quality and frame rate. For video surveillance a bit rate close to 200 Kbps is sufficient.
- *Route change latency*: Connectivity gaps should be minimized, thereby maximizing service availability. We have defined a parameter called route change latency (RCL) as the time difference between the moment in which a link fails and the moment in which an alternative route starts being used [10]. Connection gaps should be lower than 2 s to avoid voice message losses.

15.3.2.4 Routing Protocol Requirements

- *Optimized metrics*: Several works in the literature show that routing metrics that take link quality and/or network congestion into account perform better than the default minimum hop-count metric [11,12].
- *Load balancing*: In order to minimize network congestion, support for load-balancing mechanisms would be a valuable feature [13].
- *Support for multiple interfaces*: To exploit the benefits of using multiple radio interfaces, the routing protocol should be able to deal with more than one interface within the same network node.
- *Multicast support*: If group communications are supported by the WMN system, a routing protocol with multicast support must be used.

15.3.2.5 Other Requirements

- *Location*: In the event of danger (e.g., flashovers, backdrafts, hidden fires, structural collapses, personal hazards), location is a critical element for identifying where any affected firefighters are. Firefighter location requirements are tight and require a location error of less than 1 m.

■ *Security*: Security services such as privacy, authentication, integrity, and preventing denial of service (DoS) attacks are generally desired. However, not all public safety agencies require the same degree of security. For instance, law enforcement teams demand a higher degree of security than fire emergency teams.

15.4   MARKET ANALYSIS

This section examines a comprehensive set of WMN products currently available on the market. Our goal is to choose the most appropriate network node platform for our proposed WMN for fire emergencies. Our analysis takes into account the requirements set out in Section 15.3. Since we expect that further tuning will be necessary to adjust the platform to our needs, we also consider whether the platforms are open. For instance, the possibility of tuning routing protocol parameters is an important issue since network performance is sensitive to routing protocol settings [10,14].

In the last two years, many proprietary and open solutions [15] have emerged to provide wireless mesh connectivity. Some companies focus on extending network coverage and providing broadband wireless access. Other products on the market offer solutions for rapid deployment of temporary networks. Such WMNs are intended for military, disaster recovery, and public safety purposes, among others. All these products may be used as IANs and JANs. However, since our WMN is intended for use in residential fire emergencies, our requirements are tighter, especially in terms of device portability. In order to compare all of the considered WMN solutions, Table 15.1 shows the features of the solutions available on the market, including their degree of compliance with the stated requirements. The information presented was either provided by manufacturers or obtained from public documents [16] in 2005. We have endeavored to keep the information updated. We next identify a number of features that facilitate comparison between products, indicating several representative examples in each case.

15.4.1   Number of Radio Interfaces

The number of radio interfaces available in WMN products ranges from one to a few. This number depends on the architecture of each

Table 15.1 Wireless Mesh Network Solutions: Main Features

Supplier	O.S.	Mesh Routing Protocol [Layer]	Routing Protocol/ Parameters Modification	Air Interface	Client Access Backhaul Separation	Latency	Clients Relay	Location (Accuracy)	Self-Healing	Self-Install	QoS	Multicast	Security	Batteries (Duration)	Dimensions (L(cm) × w(cm) × H(cm))[a]	Weight (kg)[a]	Scalability (Number of Hops)[a]	PUGN Suitability
Tropos	Linux-based	Proprietary PWRP [L3]	No/limited	IEEE 802.11b/g	No	3–5 ms hop	Yes	Associated AP and signal strength. Optional GPS[d]	Yes (few s)	Yes	User prioritization	NA	WEP, WPA, 802.1x, VPN, AES	Yes (8–12 h)	26 × 15.2 × 2.5	1.4	No limit	No
Firetide	FreeBSD Linux	TBRPF [L3][b]	No/no	IEEE 802.11a/b/g 4.9 GHz band	No	2 ms hop	Yes	No	Yes (2–3 s)	Yes	Traffic prioritization.	No. L3 IP multicast	AES, VPN	Yes (2 d)	22.8 × 14.8 × 2.7	0.95	At least 8–10 hops	No
MeshDynamics	Linux	Proprietary [L2]	No/yes	IEEE 802.11a/b/g	Yes	<1 ms hop	NA	Yes. GPS. Triangulation (>9.14 m)	Yes (2 s)	Yes (60 s)	Up to 4 IEEE 802.11e compliant categories	No	WPA/AES, RADIUS, SNMPv3, IEEE802.1x	Yes	20.3 × 15.2 × 5	1.13	Over 8 hops	NA
4G Systems	Linux-based	OLSR [L3]	Yes/yes	IEEE 802.11a/b/g	Yes	1 ms RTT hop	Possible (OLSR clients)	No. Associated AP	Yes (10 s)	Yes (1 min)	OLSR data packets prioritization	No	WPA, Ipsec, Open VPN, secure OLSR	Yes (5 h)	7.4 × 5.3 × 7.6	0.3	5 hops	Yes
Bel Air/ Siemens	Linux	RSTP [L2] OSPF [L3]	No/yes	IEEE 802.11a/b/g	Yes	1.5 s hop	Possible (with PacketHop software)	No	Yes (1 s)	No	VLAN. Supports 802.11e/p categories	NA	WEP. AES. 802.1x	Yes (20–60 min)	20.3 × 13.9 × 34.2	4.54	Over 10 hops	No
Nova Engineering	NA	AODV [L3]	No/yes	915 MHz ISM	No	NA	Yes	GPS[d]	Yes (N.A.)	NA	Differentiated	Yes	IP and MAC filtering, AES	NA	16 × 13.2 × 3.3	0.4	NA	No
Strix Systems	NA	Proprietary [L2]	No/limited	IEEE 802.11a/b/g	Yes	NA	NA	No. Associated AP	Yes (0.5–7 s)	Simple installation	Traffic prioritization	Yes	WEP, WPA, AES, IEEE 802.1x, RADIUS	Yes (15 min)	12.7 × 9.2 ×H[f]	NA	No limit	No
SkyPilot	NA	Proprietary [L2]	No/no	IEEE 802.11a/b/g	Yes	8–10 ms RTT hop	No	GPS	Yes (2 s)	Yes (45 s)	Subscriber rate control, packet prioritization	Yes[e]	VLAN, Packet filtering, node authentication. Encryption	NA	27.9 × 14.2 × 5.1	0.7	Recommends 5 hops	No
Nortel	NA	OSPF variant [L3]	No/no	IEEE 802.11a/b/g	Yes	50 ms hop[c]	No	No	Yes	Yes	Best effort QoS	No	RADIUS, WPA	Street light photoelectric power	26.5 (tall) × 20 (diameter)	2.4	4 hops	No

continued

Interestingly, some solutions like these aforementioned follow a layer-3 routing scheme, while some others perform routing tasks at layer 2. Some examples include products from MeshDynamics [32], SkyPilot, and Strix Systems, which use proprietary schemes as well as Microsoft's [33] link quality source routing (LQSR) [9], a modified version of dynamic source routing (DSR) [34] that operates at layer 2.5. BelAir/Siemens solutions [35] are the only ones that may operate at both layer 2 and layer 3, using standard rapid spanning tree protocol (RSTP) [36] and open shortest path first (OSPF) [37], respectively.

15.4.4 Security

Almost all the considered WMN solutions support security mechanisms, although the capabilities may differ significantly depending on the product. Wired equivalent privacy (WEP) and wi-fi protected access (WPA) mechanisms are provided by a majority of vendors. IEEE 802.1x standard is also supported by a large number of products. Additional mechanisms like RADIUS authentication are provided by MeshDynamics, Strix Systems, Nortel, and Motorola solutions. Virtual private network (VPN) functionality is offered by NexGen City, Tropos, 4G Systems, and Motorola. The 4G Systems solution is remarkable as it allows using a secure version of the OLSR routing protocol.

15.4.5 Quality of Service

Manufacturers propose a variety of ways for providing QoS. Some of the solutions apply traffic or user prioritization. Regarding standard QoS mechanisms, MeshDynamics and BelAir/Siemens products claim to support IEEE 802.11e [38]. The Motorola solution supports Diffserv.

15.4.6 Multicast

A few providers assure multicast support. Nova Engineering and Strix Systems solutions inherently offer it. LocustWorld product requires a special configuration in order to allow multicast communications. SkyPilot presents a solution in which packets are treated as broadcast packets. Firetide product provides IP multicast functionality.

15.4.7 Form Factor and Weight

From the analyzed solutions, the one offered by 4G Systems presents the best combination of dimensions and weight in terms of portability, making it a suitable platform for being carried by firefighters. The rest of the products are designed to offer rather static wireless coverage, assuming no strict portability requirements.

15.4.8 Tunability

We consider a solution to be tunable if it allows software changes as well as protocol configuration parameters customization. Although the operating system of most of the products is Linux-based, a few of them are tunable. For instance, Tropos solution allows only software upgrades, while MeshDynamics product does not allow software changes (both are Linux-based). Regarding protocol customization, solutions from Nova Engineering, BelAir/Siemens, MeshDynamics, 4G Systems, Microsoft, Tropos, and Strix Systems allow routing protocol parameter tuning. The most flexible platform is 4G Systems solution, which allows modifying any software piece of the product due to its open-source nature.

15.4.9 Management and Monitoring

Some WMN products offer management and monitoring capabilities. For example, Firetide, BelAir/Siemens, Nova Engineering, and PacketHop solutions include a mesh management software for basic configuration and monitoring.

15.4.10 Platform Choice

We next draw some conclusions about the suitability of the considered WMN products for our proposed system for fire emergencies. None of the products that are available fulfills all of the requirements identified. We selected the access cube from 4G Systems because it satisfies most of them and can be modified to fulfill the rest of the requirements. The main features of this platform are:

1. Due to its unique form factor and weight, it can be incorporated into firefighter equipment.
2. It runs an embedded Linux distribution, i.e., an open-source operating system that allows full platform tuning.
3. It is built using COTS hardware.
4. It has two mini PCI wireless cards with Atheros chipset model AR5213A [39], which allows IEEE 802.11a/b/g standards to be used. A further advantage is that it can operate in the 4.9 GHz

band because the RF front end of the mini PCI card allows operation between 4900 and 5850 MHz.

5. It uses OLSR, a standard routing protocol. Although the OLSR implementation in the access cube uses a nonstandard extension by default, standard OLSR operation is possible, which facilitates interoperability.

In view of the overall set of related features, we consider the access cube to be suitable for use as an ON, a seed, or an FT.

15.5   PLATFORM DESIGN AND IMPLEMENTATION

In order to validate the proposed WMN for fire emergencies, we implemented a proof-of-concept prototype system. We used the 4G access cube as the core device in our WMN. This section details the hardware and software features of the elements in our system (i.e., FTs, seeds, ONs, and the CC). We included both off-the-shelf and custom-developed functionalities.

15.5.1   Hardware Platform

We only added hardware to the access cubes to build the CC and FTs, as the seeds and ONs could be implemented on the default platform. We developed an FT prototype composed of an access cube connected to a PDA, which acts as a graphical user interface. Figure 15.2 illustrates this setup, in which the Ethernet interface was used to connect the personal digital assistant (PDA) to the access cube. Another option would be to use an IEEE 802.11b interface. Advanced firefighter equipment would include heterogeneous equipment such as sensors to monitor biometric parameters (e.g., heart rate, respiratory rate, body temperature) and environmental parameters (smoke, temperature, CO_2, etc.), as well as cameras (infrared, night vision, etc.) and global positioning system (GPS) devices.

The access cube provides a 400 MHz processor that can process information from different sources. Hence, the access cube may not be devoted solely to routing tasks. It can act as an enhanced ubiquitous computing element by allowing different alarm applications to be implemented, which could be useful in the context of fire emergencies. For instance, the platform can process data from firefighters' sensors or other elements. In a WMN environment, applications can take advantage of the intrinsic distributed architecture of the network,

Figure 15.2 A 4G Access Cube on the left (mesh device), connected via Ethernet to a PDA (processor unit)

which is different from classical approaches that send rough, unprocessed data to a centralized system. This makes it possible for these applications to react swiftly and properly when dangerous environmental conditions are detected (e.g., high CO_2 levels). For example, triggering alarms and transmitting information or alert messages to neighboring nodes would allow a faster response than a centralized scheme. We built a simple functional CC that consists of an access cube connected via Ethernet to a PC, with access to databases with useful information such as maps and group management information.

15.5.2 Software

We next present the software features of the elements in our proposed WMN system. We developed applications and additional software programs that run on the FTs and the CC. Some processes and tools are present in all of the proposed WMN elements. These software programs are responsible for routing, security, and network monitoring. Figure 15.3 describes the functions of each element.

15.5.2.1 FT Applications

The main application developed for the FT is a VoIP application that supports both unicast and multicast communications. It is implemented in the C^{++} language and is based on the session initiation protocol (SIP)

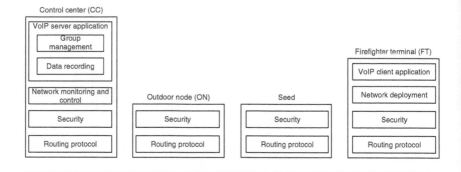

Figure 15.3 Functional description of the elements of the proposed WMN system.

[40]. The PDA runs a VoIP application that implements a PTT service. We also developed a network deployment application in the FT. Its purpose is to assist firefighters during the deployment of seeds in the incident zone. A process in the FT access cube measures the power level from neighboring nodes and triggers either a visual or an audio alarm in the FT when the highest power level received is below a certain threshold. This parameter may depend on the sensitivity of the receiver and could be tuned to achieve the desired degree of path redundancy for the WMN. Thus, firefighters can receive a hint as to when a new seed should be deployed in order to maintain connectivity with the rest of the WMN system.

15.5.2.2 CC Applications

The CC monitors communications by recording all voice calls. Subsequent analysis of this recorded digital data could be useful for future fire brigade interventions. A user interface for creating, modifying, deleting, and monitoring communications groups is also provided, which allows rescue team coordinators to fully manage the groups. Moreover, additional functionalities can be developed and integrated into the CC as modules, depending on the requirements of the brigade. For example, the CC graphical user interface could display a limited number of firefighter video cameras simultaneously, depending on screen size and the number of groups participating in the emergency tasks.

15.5.2.3 Security

A number of security tasks are performed in the access cubes. However, security issues are only partially covered at this stage. The access cubes support the advanced encryption standard (AES) with temporary key integrity protocol (TKIP) included in IEEE 802.11i [41]. These mechanisms allow client authentication and data ciphering in the access interface. However, operation in secure ad hoc mode with AES requires a full 802.11i implementation in the driver, which is not currently supported by the system implementation. As a solution, we adopted a wireless protected access preshared key (WPA-PSK) with preconfigured keys in all of the WMN elements. Thus, we can assume data privacy, integrity, and authentication in the access interfaces. Nevertheless, data privacy and integrity between network nodes require security mechanisms implemented at the application layer. The WPA-PSK can also prevent evil-twin attacks, but avoidance of DoS attacks has not yet been implemented. A simple approach could be MAC filtering at every node, but MAC spoofing attacks cannot be prevented without a more complex intrusion detection system (IDS) that continuously monitors traffic at every node and sends alert messages to the CC. There are therefore a number of unresolved security issues.

15.5.2.4 Routing Protocols

The default routing protocol in the access cubes is an OLSR implementation called olsrd [42], which uses the LQE, a nonstandard feature. This mechanism uses the expected transmission count (ETX) [12] metric to choose the most appropriate route by combining link quality and the number of hops per path. The ETX for each link is obtained as $\frac{1}{NLQ*LQ}$, where LQ is the measured quality of the link between a node and a neighbor in one direction and NLQ is the link quality measured in the opposite direction. Link qualities are obtained as the percentage of successful hello message transmissions. The total ETX of a path is the sum of the ETX values of all of the links in the path. The olsrd implementation allows three different LQE settings to be used, which may affect the system's overall performance:

1. LQE = 0. The extension is disabled.
2. LQE = 1. OLSR only uses this extension to select multipoint relays (MPRs).
3. LQE = 2. OLSR uses this extension to select MPRs and to choose paths for data flows. This is the default setting.

This OLSR implementation provides an easy means of network monitoring through a visualization tool and a dot-draw plug-in, which provides network information in dot file format [43]. This information can be parsed and appropriately displayed (i.e., graphically) in the CC. Images can be created using the graphviz library [44]. This interface can help understand the network configuration and performance at all times (Figure 15.4).

To take advantage of the fact that access cubes provide an open platform, we included another routing protocol in our WMN evaluation. We wanted to study to what extent different routing approaches could provide the required performance. We chose to test AODV as an alternative routing protocol. Thus, we considered two different Internet engineering task force (IETF) standard protocols based on different philosophies: OLSR is proactive, while AODV is reactive. Furthermore, both protocols are popular, according to the number of publicly available implementations and the support for different operating systems [45]. The AODV implementation tested was the AODV-UU version 0.8.1. from Uppsala University [46], which has been used in a number of published research papers in the mobile ad hoc networks (MANET) field [47]. One relevant feature of this implementation is the use of a hello message mechanism to maintain

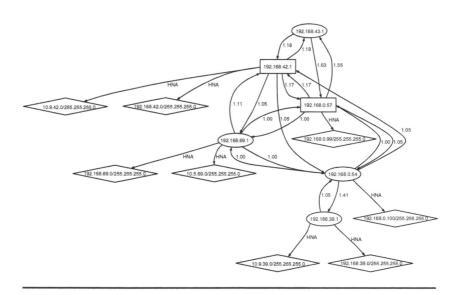

Figure 15.4 An example of a WMN topology view with olsrd. Ellipses and rectangles represent access cubes. The quality of each link is shown. The direction is indicated by an arrow.

local connectivity. In order to allow the possibility of integrating VoIP group communications into our WMN system, we used a multicast routing protocol for MANETs. We tested the multicast AODV (MAODV) patch [48] with the AODV-UU version 0.6. We had to adapt the code to get MAODV to work on the chosen platform. We chose MAODV because it was the only multicast routing protocol available at the time the system was designed and implemented. It is also an extension of one of the tested unicast protocols. However, other currently available multicast routing protocol implementations like multicast OLSR (MOLSR) [49] can also be considered.

15.6 VALIDATION

This section validates the proposed WMN's capability of providing different kinds of services that are useful in fire emergency environments. Examples include interactive voice communications, file transfer, and video streaming. We conducted several experiments by measuring the performance of our WMN in terms of available end-to-end throughput, end-to-end delay, and reactivity to topology changes. However, our main goal was to focus on user perception when fire emergency communications applications were run. The influence of the routing protocol was also evaluated. OLSR and AODV implementations with the default settings were used in all tests, unless explicitly stated otherwise. We also evaluated the proposed system's ability to support multicast in order to emulate a firefighter voice group communications service. We only presented the most significant results based on the performance requirements of our WMN. These experiments were carried out using the IEEE 802.11b radio interface. Unfortunately, 4.9 GHz radio interfaces were not available for use. Although the hardware of the access cube's wireless cards allowed 4.9 GHz spectrum operation, the wireless card driver did not support it. This driver is called MadWifi [50] and the version available at the time of the validation was 13/08/2005. Nevertheless, the functional WMN concept can also be proved using a 2.4 GHz radio interface. It was not possible to use different channels in each access cube radio interface because the driver did not fully support this functionality.

15.6.1 End-to-End Available Throughput

We measured the influence of the end-to-end hop count on the available throughput and the influence of routing protocol signaling on this performance parameter. To do this, we defined a static string

Figure 15.5   String topology scenario.

topology scenario (Figure 15.5). Hence, if a node wants to communicate with a nonneighboring node, it must do so via multihop communication. We conducted experiments using static routes and OLSR and AODV protocols with their default parameter configuration. To obtain a controllable scenario, we emulated network topologies using the *iptables* tool [51]. In each trial, a user datagram protocol (UDP) stream of packets was transmitted between the endpoints of the communication (i.e., nodes S and D in Figure 15.5) at a higher bit rate than that available in one-hop communication. Each trial lasted 2 min. Statistical results were obtained from the average of 15 trials in each case. We considered a number of hops ranging from 1 to 4. The request-to-send/clear-to-send (RTS/CTS) mechanism was not enabled in these tests.

Figure 15.6 illustrates the results obtained. Throughput decreases with the number of hops following a $\frac{1}{N}$ tendency, where N is the hop count of the end-to-end path. This is the expected behavior, since each hop uses the same channel and all nodes are in the same range of transmission. Hence, the wireless medium must be shared among

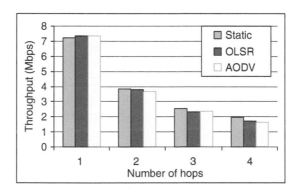

Figure 15.6   Throughput measurements in a string topology scenario.

nodes and the end-to-end available throughput is approximately the throughput achieved in the one-hop string topology divided by the number of hops. Even so, in four-hop communication the throughput obtained is roughtly 1.5 Mbps. This bit rate is sufficient for broadband services such as video or file transmissions. Higher end-to-end bit rates are expected if more than one frequency in the path or IEEE 802.11a/g interfaces are used. Empirical results in an ad hoc network testbed show a $\frac{1}{n^{1.68}}$ tendency, where n is the number of nodes in the network [52]. By comparing the results obtained with OLSR and AODV protocols with those obtained with static routes (i.e., no routing protocol), we conclude that the effect of the control message overhead of both protocols is almost negligible.

15.6.2 End-to-End Delay

To evaluate the effect of routing protocols on end-to-end delay, we performed a set of trials with the topology and conditions mentioned before. The end-to-end delay was obtained from RTT measurements using the *ping* tool. We expected reactive protocols to present an additional route discovery delay if no route was available when data needed to be transmitted. However, no additional latency would be introduced when proactive protocols were used.

The results show that the average RTT is close to 1.5 ms in the one-hop topology with either protocol once a route is known. Each additional hop in the end-to-end path contributes roughly 1 ms to the RTT. These values were obtained with no additional traffic in the network. If background traffic is present in the network, the average values grow. For instance, if background traffic is set at around 80% of the available bandwidth capacity, the end-to-end delay is almost twice the delay without background traffic. We also confirmed that OLSR does not influence end-to-end delay, while AODV incurs additional delay due to the route discovery procedure. However, this delay increase is negligible for the fire emergency applications envisaged, for two reasons: (i) the average route discovery delay we measured was below 40 ms and (ii) it only takes place at the first transmission of a data flow. Performance will not be degraded due to the AODV route discovery increase. Furthermore, the expected number of hops in a WMN should be small (i.e., up to 3 or 4), since performance decreases significantly beyond such figures [53]. Thus we conclude that end-to-end latency requirements are met by our prototype system.

15.6.3 Reactivity of the Routing Protocol to Topology Changes

This section focuses on the reactivity of the system to topology changes. We evaluated two situations by emulating conditions that could easily occur in fire emergencies: (i) an active node of the communication is suddenly lost (e.g., due to battery drain or node damage) and (ii) the mobility of nodes or client relays leads to smooth topology changes. We assume that firefighter teams move at pedestrian speeds in the emergency zone. We measured the RCL in such situations to evaluate the system's self-healing capability. We also tested user perception in a scenario intended to represent realistic fire emergency conditions inside a building.

15.6.3.1 Sudden Topology Changes

To evaluate the reactivity of the protocols to sudden changes, we used a square topology (Figure 15.7). This topology presents two possible routes between the source and destination nodes (i.e., through node A and through node B). A UDP flow was sent from the source to the destination. Once traffic was routed through one route, we disconnected the intermediate node. The routing protocol was needed to detect the link failure and find the alternative available path. We measured connectivity gap durations of close to 20 s with OLSR and 3 s with AODV. The RCL can be reduced by increasing the control message sending rate, at the expense of additional available bandwidth and power consumption. Using different routing protocol parameter settings, we obtained RCL values of roughly 10 and 2 s, respectively. For a

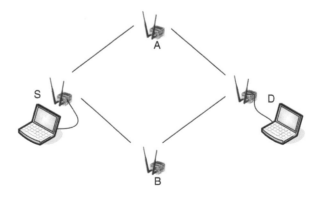

Figure 15.7 Square topology used to evaluate routing protocol reactivity to topology changes.

detailed description of the impact of OLSR and AODV parameters on the performance of ad hoc networks, see [10,14]. These figures do not fulfill RCL requirements. Usage of cross-layer mechanisms such as link layer notification may improve reactivity to topology changes [61].

15.6.3.2   Smooth Topology Changes

We analyzed reactivity to smooth topology changes in an emulated emergency scenario inside a building. We positioned our WMN testbed in a realistic way, with natural multihop communications (i.e., without using *iptables* to emulate coverage). Figure 15.8 illustrates a scenario in which a firefighter walks into a burning building following the path from A to D. The FT of the moving firefighter communicates with node D while the firefighter follows the aforementioned path. At the beginning of the experiment, the data flow follows a four-hop path, while the hop count of the path decreases as the firefighter gets closer to the destination. When the firefighter reaches the destination, the path has only one hop.

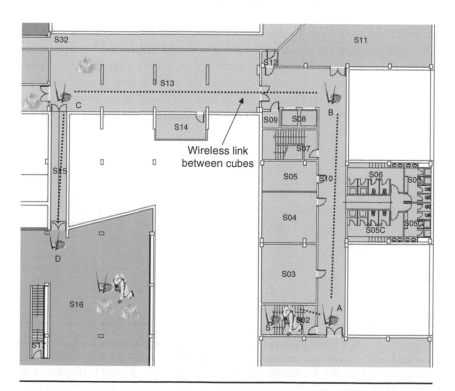

Figure 15.8   Emulated emergency scenario inside a building.

In this scenario, we performed experiments with three types of applications: FTP file transfer, interactive voice communication using the *linphone* [54] application, and video transmission. We also used a Wifibot [55], a robot that includes an access cube and acts as an additional, remotely controlled, WMN node. The Wifibot moved at pedestrian speed in all the tests. Our evaluation focused mainly on user perception, in order to validate the capability of the proposed WMN system to offer the aforementioned services. We also performed detailed measurements for comparison with previous tests.

We first analyzed the behavior of a file transfer in two situations: (i) using AODV and (ii) using OLSR. Figure 15.9 depicts the throughput obtained at the receiver in both situations. Figure 15.9a corresponds to a trial with AODV, while Figure 15.9b was obtained with OLSR. Both figures clearly illustrate the effect of the number of hops on available path throughput. Contrary to the results obtained in the sudden topology change trials, OLSR performed better than AODV, since OLSR gave a higher transmission control protocol (TCP) throughput. However, this comparison is unfair, since this OLSR implementation includes a mechanism for adaptively estimating the link qualities of a network. As the firefighter moves away from the neighboring node, the link quality between the firefighter and the neighboring node decreases. As alternative paths with better quality appear, OLSR performs an almost ideal route change, thereby avoiding connectivity gaps. However, AODV cannot perform the route change until a neighbor has been completely lost. Therefore, we only considered OLSR for the rest of the trials.

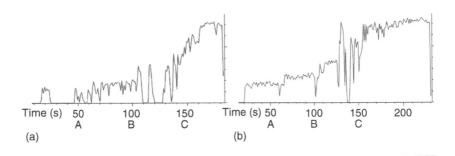

Figure 15.9 TCP throughput received in the emulated emergency scenario. The vertical axis depicts the normalized instantaneous received throughput. The instants in which the firefighter passes next to nodes A, B, and C (Figure 15.8) are plotted on the horizontal axis. Two cases are considered: (a) with AODV; (b) with OLSR.

We performed full-duplex VoIP communications tests using the *linphone* application. We used a pulse code modulation (PCM) voice codec (64 Kbps). To ensure continuous data transmission, we sent music rather than human voice. At very low speeds, the user noticed no communication gaps. Perceptible gaps arose as user speed increased. Link quality estimation may become more difficult as nodes move faster, which causes an aliasing problem. (Link quality is sampled periodically, at a frequency equal to the hello message rate.) Figure 15.10 illustrates the throughput received during one such experiment. No connectivity gaps occurred due to the LQE available in the OLSR implementation. Similar tests, in which the firefighter was replaced by a Wifibot, produced similar results. In this case, a 256 Kbps MPEG 4 video stream over UDP was generated by a webcam included in the Wifibot.

15.6.4   Multicast Communications

Finally, we evaluated the suitability of the system to emulate firefighter voice group services over our proposed WMN. For this type of application, multicast communications may be a good option for several reasons. First, they are efficient in point-to-multipoint transmissions. Second, in multicast, no ACK is transmitted at the MAC level. Lost

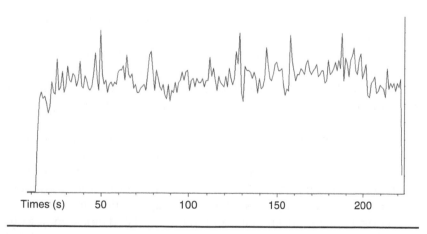

Figure 15.10   Throughput received during a unicast VoIP experiment in the emulated emergency scenario. The vertical axis depicts the normalized instantaneous throughput received.

frames are not retransmitted, thereby minimizing end-to-end delays in the event of losses, which may be preferable for the transmission of voice traffic. We performed an initial analysis to determine the capacity of an IEEE 802.11b link to provide unicast and multicast VoIP communications [56]. The results showed that, with a bit rate of 2 Mbps, a maximum of 7.20 flows can be performed in unicast mode and a maximum of 8.52 can be performed in multicast. The assumed conditions were a PCM voice codec (64 Kbps), 60% use of WLAN capacity, and no RTS/CTS mechanism. Furthermore, silence and activity periods were not detected. We also analyzed the capacity gain obtained with a GSM 6.10 codec (13 Kbps). In this case, the maximum number of flows was higher: 10.45 with unicast and 13.33 with multicast. However, the PCM codec offers higher quality, demands less CPU processing, and presents less degradation in the event of packet losses. These results correspond to the worst-case scenario (with information transmitted at the lowest rate). Higher rates allow more VoIP sessions. In multicast mode, some WLAN cards transmit at the minimum available rate in order to ensure that all destinations receive the message. However, this limitation can usually be overcome by permitting transmissions up to 11 Mbps [56]. Capacity can also be increased by using IEEE 802.11a interfaces. In the proposed WMN system, the number of simultaneous VoIP flows is limited due to the multihop nature of the network. For instance, in a four-hop scenario, the number of multicast VoIP communications using a GSM codec is reduced to 2.14 flows. In emergency operations, a high system capacity for VoIP communications is desirable in order to prevent congestion collapse. However, the use of multiple channels can be harmful since firefighters may find it difficult to manage different channels. This can lead to tragic results [57]. Moreover, group communication systems allow trunking, which results in a more efficient use of channels. Hence, despite capacity limitations, the deployment of multicast VoIP services over WLAN may be a viable approach for the proposed WMN system.

In the second stage, we evaluated the performance of multicast communications using the MAODV protocol. The tests involved transmitting VoIP traffic using a PCM codec and sharing data using a whiteboarding application. The RTS/CTS mechanism was disabled in all tests. We validated the feasibility of both applications in static WMN scenarios. However, from trials held in the emulated emergency scenario shown in Figure 15.8 we observed some problems related to protocol performance and topology changes. We measured the following performance parameters:

- *Tree generation latency* (TGL): The first time a user joins a nonexistent group, a multicast tree must be created. This process is not negligible and takes about 6 s. During this time, if possible, data are stored in an application buffer and delivered to the user once the multicast tree has been built. Otherwise, the user observes a communication gap at the beginning of the transmission.
- *Route change latency*: We measured the RCL in a situation in which all nodes belonged to the multicast tree. In the event of link failure, route recovery takes about 3 s, which affects the continuity of the voice communication.

15.7 CONCLUSIONS

This chapter focused on the suitability of WMNs for overcoming limitations in the functionality and capacity of current firefighter communications equipment. We proposed a WMN system that integrates several technologies based on four main elements: CC, ONs, seeds, and FTs. We listed several general and specific requirements for our WMN system. We performed a comprehensive market study and analyzed the features of current commercial WMN solutions and their suitability for use as a platform for our WMN. We chose the 4G Systems access cube due to its form factor and weight features and because it is an open platform and uses standards. We described the default functionality of the access cubes and the other hardware and software elements we added to them to build our proposed WMN. We then validated the ability of our prototype WMN to offer broadband services such as file transfer and video as well as VoIP. Our tests proved that the performance requirements for end-to-end bandwidth and delay can be fulfilled. In the event of sudden topology changes, the tested AODV implementation reacts faster than the OLSR implementation, although significant connectivity gaps occur in both cases. However, link quality routing mechanisms included in the OLSR implementation are shown to be valuable when smooth topology changes occur, since a route change can be performed almost without connectivity interruptions. We also evaluated multicast communications over our WMN as a way of emulating the voice group communication used by firefighters. We adapted and tested the MAODV routing protocol, which allows group communication but exhibits significant tree generation and route change latencies. During this work, we identified several open issues, which are summarized in Section 15.8.

15.8 OPEN ISSUES

15.8.1 Congestion Control

Firefighters complain about the lack of capacity of current communication systems. Radio channels become congested as soon as a team is deployed in an incident area. In some cases, firefighters switch off their communications equipment to avoid continuous, distracting noise [58]. Even with an increase in capacity, it is impossible to ensure that congestion will be avoided. A more realistic approach would be to differentiate between communications and ensure that the most relevant messages are given priority over others [59,60]. WMNs are as prone to buffer overflow as other networks (e.g., wired ones). In addition, when the number of transmissions increases, the related interference reduces the capacity of the links. The network requires distributed congestion control that can prevent terminals from sending certain low-priority data and discard those packets in the relaying nodes.

15.8.2 Location

Location services are a key facility in fire emergencies. There are a wide range of location solutions on the market, but most of them cannot be applied ad hoc (i.e., without previous calibration). Others, such as GPS devices, are not suitable for indoor environments. Some solutions derive location information using the radio communication system. This information is not accurate enough for fire emergency tasks. Of the WMN products considered, the best reported location performance is 10 m [24], which is nowhere near the location requirement for these scenarios. Emerging ultra wideband (UWB) radio interfaces promise to provide highly accurate location services with subcentimeter precision.

15.8.3 Link Layer Notification for Routing Protocol Connectivity Maintenance

The connectivity of firefighter equipment to the emergency network is an extremely important issue. Even in an incident area with sufficient coverage, network topology changes lead to connectivity gaps, in which vitally important messages may be lost. The time a routing protocol needs to react to a sudden topology change is approximately equal to the time needed to detect the related link failure. Publicly available routing protocol implementations for ad hoc networks mainly use layer-3 approaches to maintain local connectivity and detect when a link has failed. One such example is the hello message mechanism, used by either OLSR (as a protocol component) or AODV (as an accessory mechanism). Feasibility is the main reason for using a layer-3 approach. A routing protocol could use information from layer 2 to quickly detect when a link fails (e.g., by relying on the lack of layer-2 acknowledgments). However, the routing protocol implementation should be able to communicate with the related wireless card driver, which is only feasible if both the routing protocol implementation and the driver are designed for this purpose. Since typical hello interval values are between 1 and 2 s and links fail after the loss of several consecutive hello messages, RCL values typically fall between a few seconds and tens of seconds [10,14]. Custom-developed fire emergency WMN equipment could use layer-2 notification to dramatically reduce RCL values from a few seconds to tens of milliseconds [61].

15.8.4 Quality of Service

Various QoS parameters need to be guaranteed in order to provide fire emergency services. Examples include reliability, end-to-end bandwidth, and end-to-end delay. Loss of relevant information should be avoided, as mentioned earlier. A fire emergency network needs to support data flows with different characteristics. Traffic prioritization mechanisms are needed to fulfill the requirements of each service. In this regard, the IEEE 802.11e standard is a promising technology. However, it currently provides QoS only in infrastructure mode. Several mechanisms have been proposed to provide QoS at several layers in multihop networks, including MAC mechanisms [59] and routing protocol–based approaches [62]. Further research is needed in this area.

15.8.5 Multicast

From the tests performed and the corresponding literature, we conclude that support for multicast in mobile mesh networks has not been solved in terms of communication channel availability [63]. The mobility of the user (i.e., the firefighter) may result in link breaks, and the corresponding ad hoc routing protocol should repair them. Once connectivity has been solved, the corresponding multicast protocol should incorporate the new path into the multicast tree. If the two phases are carried out in different steps, the resulting path-break time may be unacceptable in view of the strict availability requirements. It is clear that multicast tree creation should be related to route discovery. This issue is not yet fully resolved and significant improvements are needed to meet the requirements of emergency applications.

15.8.6 Radio Interface

On the basis of our market analysis, we concluded that, even though IEEE 802.11 interfaces are cheap, most manufacturers propose proprietary radio interfaces. The main reason for this is the poor performance of the MAC protocol in mesh scenarios [64]. This problem has been identified and most IEEE 802 groups related to wireless communication are working to overcome this limitation. One expected result is the IEEE 802.11s [65], but it will take some time for products to reach the market. Fortunately, the groups promoting the IEEE standard have reached a unified proposal. The new standard will use layer-2 addressing, one or more radios, self-configuration, QoS, and security based on IEEE 802.11i. However, an analysis of the performance of IEEE 802.11s and the implications for upper layers is required, which could be the subject of further research.

15.8.7 Multiple Radio Interfaces

The use of a single radio is proposed by some manufacturers due to the simplicity and low cost of this option. Because of robustness, security, and capacity requirements, the need for multiple radios seems evident. The use of several radios makes it possible to reduce the interference produced by the WMN nodes and obtain better performance [66]. Robustness and security (related to DoS attacks) can be improved if different channels can be used. In the event of excessive interference, the system should be able to select a new channel. If the new 4.9 GHz band is used, the number of available IEEE 802.11 channels is limited to just two. However, because it is a licensed band (at least in North America), the only interference present is due to the system itself. ISM bands may have more interference, although this limitation can be overcome by choosing the most suitable channel from among the available ones. With reference to the use of multiple radios, channel selection and notification between nodes of the selected channel are problems that have not yet been fully solved. An advantage is that standard routing protocols like OLSR, AODV, and the dynamic MANET on-demand (DYMO) routing protocol [67] support multiple interfaces.

ACKNOWLEDGMENT

This project was supported by Vodafone Group Research & Development. ES, the I2CAT Foundation, the Spanish government through project TIC2003-01748, the Catalan government's Ministry for Universities, Research and the Information Society (DURSI), the European Regional Development Fund (ERDF), and the European Science Foundation (ESF).

REFERENCES

1. R.D. Paulison. "Working for a Fire Safe America: The United States Fire Administration Challenge," 2002. Available at: http://usfa.fema.gov/about/index.shtm.
2. US Fire Administration (F.E.M.A.) NFIRS (National Fire Incident Reporting System). Available at: http://nfirs.fema.gov/.
3. J. Dunlop, D. Girma, J. Irvine. "*Digital Mobile Communications and the TETRA System,*" Wiley ed., 1999.
4. TSB-102-A (November 1995), "APCO Project 25—Systems and Standards Definition," 1995.
5. The SAFECOM Program. US Department of Homeland Security, "Statement of Requirements for Public Safety Wireless Communication and Interoperability," Version 1.0, March 2004.
6. "FCC Second Report and Order and Further Notice of Proposed Rulemaking in the Matter of the 4.9 GHz Band Transferred from Federal Use," 27 February 2002.
7. T.L. Doumi. "Spectrum Considerations for Public Safety in the United States," *IEEE Communications Magazine*, vol. 44, no. 1, pp. 30–37, January 2006.
8. I.F. Akyildiz, X. Wang, W. Wang. "Wireless Mesh Networks: A Survey," *Elsevier Computer Networks*, vol. 47, pp. 445–487, January 2005.
9. R. Draves, J. Padhye, B. Zill. "Routing in Multi-radio, Multi-hop Wireless Mesh Networks," *ACM MobiCom*, Philadelphia, PA, September 2004.
10. C. Gomez, M. Catalan, X. Mantecon, J. Paradells. "Evaluating Performance of Real Ad-Hoc Networks Using AODV with Hello Message Mechanism for Maintaining Local Connectivity," 16th International Symposium on Personal Indoor and Mobile Radio Communications (PIMRC 2005), Berlin, September 2005.
11. R. Draves, J. Padhye, B. Zill. "Comparisons of Routing Metrics for Static Multi-Hop Wireless Networks," ACM Annual Conference Special Interest Group on Data Communication (SIGCOMM), pp. 133–144, August 2004.
12. D.S.J. De Couto, D. Aguayo, D. Bicket, R. Morris. "A High-Throughput Path Metric for Multi-Hop Wireless Routing," *Proceedings of ACM MOBICOM*, San Diego, CA, September 2003.
13. T. Kitahara, Y. Kishi, Y. Imagawa, K. Tabata, S. Nomoto, A. Idoue. "An Adaptive Load Balancing in Multi-hop Mesh Networks for Broadband Fixed Wireless Access Systems," IEEE Radio Wireless Conference, Atlanta, USA, September 2004.

14. C. Gomez, D. Garcia, J. Paradells. "Improving Performance of a Real Ad-Hoc Network by Tuning OLSR Parameters," 10th IEEE Symposium on Computers and Communications (ISCC'05), Cartagena, June 2005.

15. R. Bruno, M. Conti, E. Gregori. "Mesh Networks: Commodity Multihop Ad Hoc Networks," *IEEE Communications Magazine*, vol. 43, no. 3, pp. 123–131, March 2005.

16. G. Held. "*Wireless Mesh Networks*," Auerbach Publications, 1st ed., 2005.

17. Firetide. http://www.firetide.com.

18. TROPOS Networks. http://www.tropos.com.

19. Nova Engineering. http://www.nova-eng.com.

20. SkyPilot Networks. http://www.skypilot.com.

21. Nortel Networks. http://www.nortelnetworks.com.

22. Strix Systems. http://www.strixsystems.com.

23. 4G Systeme/s. http://www.4g-systems.com.

24. Motorola Networks. http://www.motorola.com.

25. NexGen City. http://www.nexgencity.com.

26. NovaRoam. http://www.novaroam.com.

27. T. Clausen, P. Jacquet. "RFC3626. Optimized Link State Routing Protocol (OLSR)." Status: Experimental, October 2003.

28. C. Perkins, E. Belding-Royer. "RFC3561. Ad-Hoc On-Demand Distance Vector (AODV)," Status: Experimental, July 2003.

29. LocustWorld. http://www.locustworld.com.

30. PacketHop Networking. http://www.packethop.com.

31. R. Ogier, F. Templin, M. Lewis. "RFC3584. Topology Broadcast Based on Reverse-Path Forwarding (TBRPF)," Status: Experimental, February 2004.

32. MeshDynamics. http://www.meshdynamics.com.

33. Microsoft Mesh Networking. http://research.microsoft.com/mesh.

34. D.B. Johnson, D.A. Maltz, Y. Hu. "The Dynamic Source Routing Protocol for Mobile Ad Hoc Networks (DSR)," IETF MANET Working Group Internet-Draft, 19 July 2004 (work in progress).

35. BelAir Networks. http://www.belairnetworks.com.

36. IEEE Std 802.1w-2001. "IEEE Standard for Information Technology—Telecommunications and Information Exchange between Systems—Local and Metropolitan Area Networks—Common Specifications—Part 3: Media Access Control (MAC) Bridges: Amendment 2: Rapid Reconfiguration."

37. J. Moy. "RFC2328. Open Shortest Path First (OSPF) Version 2," April 1998.

38. IEEE Std 802.11e-2005. "IEEE Standard for Information Technology—Part 11: Wireless LAN Medium Access Control (MAC) and Physical Layer (PHY) Specifications: Amendment 8: Medium Access Control (MAC) Quality of Service Enhancements."

39. Atheros AR5004 WLAN Chipset Product Bulletins. Available at: http://www.atheros.com/pt/AR5004XBulletin.htm.

40. J. Rosenberg, H. Schulzrinne, G. Camarillo et al. "RFC3261. SIP: Session Initiation Protocol," June 2002.

41. IEEE Std 802.11i-2004. "IEEE Standard for Information Technology—Telecommunications and Information Exchange between Systems—Local and Metropolitan Area Networks—Specific Requirements—Part 11: Wireless LAN

Medium Access Control (MAC) and Physical Layer (PHY) Specifications: Amendment 6: Medium Access Control (MAC) Security Enhancements."

42. Optimized Link State Routing Protocol (OLSR) Implementation. Version 0.4.9. http://www.olsr.org.

43. The DOT Language. http://www.graphviz.org/doc/info/lang.html.

44. Graphviz Graph Visualization Software. http://www.research.att.com/sw/tools/graphviz/.

45. MANET Implementations. http://www.comnets.uni-bremen.de/koo/manet-impl.html.

46. AODV-UU. AODV implementation from Uppsala University. http://core.it.uu.se/Adhoc/AodvUUImpl.

47. H. Lundgren, E. Nordstrom, C. Tschudin. "Coping with Communication Gray Zones in IEEE 802.11b Based Ad Hoc Networks," *Proceedings of the 5th ACM International Workshop on Wireless Mobile Multimedia*, Atlanta, USA, September 2002.

48. Multicast Extensions of AODV (MAODV) University of Maryland implementation Version 0.6. http://www.isr.umd.edu/CSHCN/research/maodv/MAODV-UMD.html.

49. Simple Multicast OLSR (SMOLSR) and Tree-Based Multicast OLSR (MOLSR). INRIA Implementation. http://hipercom.inria.fr/SMOLSR-MOLSR/index.html.

50. Multiband Atheros Driver for WiFi. MadWifi. http://madwifi.org.

51. Netfilter/iptables: firewalling, NAT, and packet mangling for Linux. http://netfilter.org.

52. P. Gupta, R. Gray, P.R. Kumar. "An Experimental Scaling Law for Ad Hoc Networks," University of Illinois Urbana Champaign, May 2001.

53. C. Tschudin, P. Gunningberg, H. Lundgren, E. Nordstrom. "Lessons from Experimental MANET Research," *Ad Hoc Networks*, vol. 3, no. 2, pp. 221–233, March 2005.

54. Linphone (telephony on Linux). http://www.linphone.org.

55. Wifibot. http://www.wifibot.com/.

56. W. Wang, S.C. Liew, V.O.K. Li. "Solutions to Performance Problems in VoIP over a 802.11 Wireless LAN, *IEEE Transactions on Vehicular Technology*," vol. 54, no. 1, pp. 366–384, January 2005.

57. A. Thiel. "Improving Firefighter Communications. Special Report," Technical Report Series. Federal Emergency Management Agency. United States Fire Administration.

58. H. Stambaugh. "Improving Firefighter Communications. Special Report," United States Fire Administration.

59. M.A. Haq, M. Matsumoto, J.L. Bordim et al. "Admission Control and Simple Class Based QoS Provisioning for Mobile Ad Hoc Network," Vehicular Technology Conference, vol. 4, pp. 2712–2718, September 2004.

60. S.Z. Ozer, S. Zeng, C. Barker. "Congestion Control and Service Differentiation in Multihop Wireless Networks," *IEEE Communications Society Globecom*, pp. 62–67, December 2004.

61. C. Gomez, P. Salvatella, O. Alonso, J. Paradells. "Adapting AODV for IEEE 802.15.4 Mesh Sensor Networks: Theoretical Discussion and Performance Evaluation in a Real Environment," 7th IEEE International Symposium on a

16.1 INTRODUCTION

Reliable and efficient communication is absolutely crucial for public safety in general, and emergency response and disaster recovery operations in particular. Recent events such as 9/11 and Hurricane Katrina have dramatically demonstrated that there exist significant inadequacies in current first responder communications. One of the main problems that plagued rescue teams and emergency services during these disasters was the lack of interoperability between communications equipment used by different public safety agencies and jurisdictions. The 9/11 commission report [1] noted that a patchwork of incompatible technology and the uncoordinated use of frequency bands were the main reasons for nonexisting or poor interagency communication during emergency response and recovery operations. Shouting, waving signs, and runners with handwritten messages often had to be used as a primitive alternative. Another problem of public safety and disaster recovery (PSDR) communication is the strong reliance on terrestrial communications infrastructure such as traditional landline and cellular telephony as well as infrastructure-based land mobile radio (LMR). Hurricane Katrina uprooted hundreds of wireless base stations, disconnected numerous vital communications cables, and flooded central offices. The remaining functional parts of the network were often completely overloaded and unable to provide adequate services in the aftermath of the disaster. First responders were surprised and severely hampered by a near-complete breakdown of the fixed terrestrial communications infrastructure. In a number of recent major disasters, communication systems relying on fixed terrestrial infrastructure have proven to be rather unreliable. A strong dependence on point-to-point communication links and a limited degree of redundancy give these systems an insufficient level of resilience and robustness in disaster scenarios. A further shortcoming of current PSDR communications is the lack of support for broadband data rates. It is widely recognized that data-intensive multimedia applications have a great potential to improve the efficiency of disaster recovery operations. Real-time access to critical data such as high-resolution maps or floor plans can be extremely valuable for frontline first responders. Being able to send a live video stream from the incident site back to the command post would greatly increase the situational awareness and would allow more efficient decision-making. The need for broadband communication capabilities for PSDR agencies is also pointed out in a report by the SAFECOM program of the US Department of Homeland Security [2,28]. The document states that "voice communications are critical, but voice communications requirements are not the only issue . . . public safety agencies are increasingly dependent on sharing of data, images, and video." Unfortunately, current PSDR communication systems do not provide the necessary broadband capabilities for bandwidth-intensive multimedia applications. Given the shortcomings of current PSDR communications mentioned earlier, Wireless mesh networks (WMNs) provide an interesting alternative technology. The key features of WMNs such as broadband support, fault tolerance, and a high level of interoperability provide them with a great potential as a platform for PSDR communication. The aim of this chapter is to discuss the suitability of WMN technology for PSDR applications. Section 16.2 gives a brief overview of WMN technology and highlights its key characteristics and features. Section 16.3 provides a background on PSDR communications and discusses some of the key technologies and standards. It also specifies the key requirements of PSDR communication systems in terms of functionality and performance. In Section 16.4, we investigate to what extent WMNs are able to meet these requirements, and highlight areas of strength as well as those of weakness, in which further research is required. Section 16.5 gives an overview of the key research activities that are underway to address the main limitations and shortcomings of current WMNs. Finally, Section 16.6 concludes the chapter.

16.2 WIRELESS MESH NETWORKS

Today, wireless local area networks (WLANs) are primarily used to provide mobile users access to a fixed network infrastructure. These networks allow users to roam freely throughout the office or any other space within network coverage with untethered broadband network connectivity. This support for mobile broadband connectivity combined with the rapidly decreasing cost of IEEE 802.11-based commodity hardware has resulted in a phenomenal success of wireless networking technology in the past few years. In traditional WLAN deployments, clients are associated with wireless access points that are interconnected by a wired backbone network. In this case, the wireless network constitutes only a single hop of the end-to-end path. Clients therefore need to be within a single-hop range of a wireless access point to gain connectivity. To achieve wide area coverage, a large number of fixed access points need to be deployed and the corresponding wiring for the backbone (or backhaul) network needs to be installed. Deployment of large-scale WLANs is therefore a very costly and time-consuming undertaking. In contrast, WMNs can provide wireless network coverage of large areas without relying on a wired backbone infrastructure or dedicated access points. In WMNs, a collection of wireless *mesh routers*, typically implemented using commodity 802.11 hardware operating in ad hoc mode, provide network access to wireless clients. However, communication between mesh routers is achieved via the wireless network, typically involving multiple wireless hops. One or multiple mesh routers that are connected to the Internet can then serve as gateways and provide Internet connectivity for the entire mesh network. Figure 16.1 and Figure 16.2 illustrate the difference between traditional WLAN deployment and WMN. Figure 16.1 shows a WLAN where clients are associated with wireless access points that are connected to a central switch by a wired backhaul. In the WMN scenario (Figure 16.2), clients can either be directly communicating with a mesh router or can be associated with an access point that is connected to a mesh router. The key difference here is that the wired backbone is replaced by a WMN.

16.2.1 WMN Architecture

We can differentiate between three basic types of WMN architectures: infrastructure mesh, client mesh, and hybrid mesh [9]. Figure 16.2 is an example of an infrastructure WMN. The mesh routers collectively provide a wireless backbone infrastructure. In this architecture, clients have a passive role and do not contribute to the mesh infrastructure. In a client mesh architecture (Figure 16.3), the network is made up of user devices only and no dedicated network infrastructure is involved. Since client devices themselves constitute the network, they need to

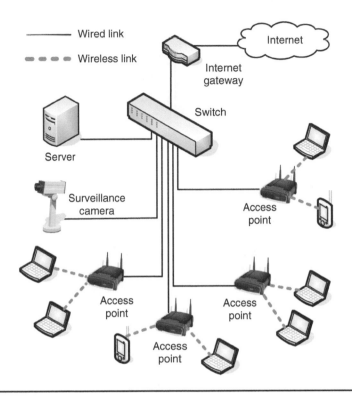

Figure 16.1 Traditional WLAN.

perform functions such as routing and self-configuration themselves. Clients relay packets on behalf of peers and essentially implement the role of mesh routers, in addition to running the user applications. This places a significantly higher demand on the end-user terminals, both in terms of functionality and resource consumption, e.g., computation and power. A client mesh is essentially identical to a traditional ad hoc network [10].

A hybrid mesh architecture combines the concepts of infrastructure and client mesh networks (Figure 16.4). A hybrid WMN consists of mesh routers that form the backbone of the network. In addition, mobile clients can actively participate in the creation of the mesh by providing network functionalities, such as routing and forwarding of data packets. Clients implementing these functionalities can therefore act as a dynamic extension to the more static infrastructure part of the mesh. The hybrid mesh is very flexible and allows combining the benefits of both infrastructure and client mesh architectures.

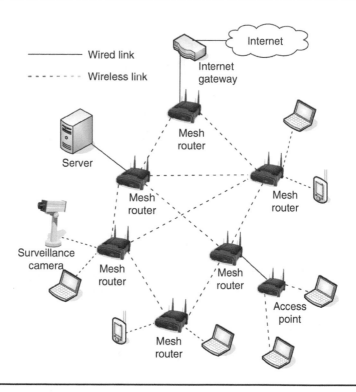

Figure 16.2 Wireless mesh network (infrastructure mesh).

16.2.2 Characteristics of WMN

One of the key features of WMNs is the ability to dynamically self-organize and self-configure. The nodes in a WMN automatically detect neighbor nodes and establish and maintain network connectivity in an ad hoc fashion, typically implemented at the network layer through the use of ad hoc routing protocols [11]. The self-configuring nature of WMNs allows easy and rapid deployment. WMNs also have the ability to dynamically adapt to changing environments and to essentially self-heal in case of node or link failure. Unlike existing point-to-point radio systems used in PSDR communications, mesh networks are inherently redundant and therefore have a high level of fault tolerance and robustness.

The availability of low-cost commodity hardware based on IEEE 802.11 standards has been one of the key drivers behind the recent surge of interest in WMN technology, in terms of both research and product development. Currently, a wireless mesh router can easily be

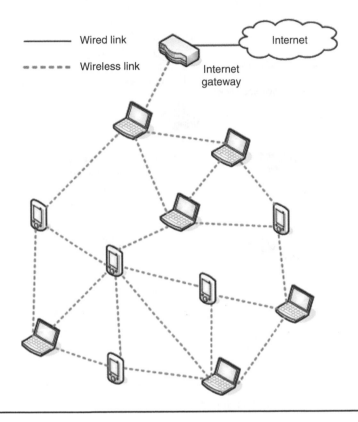

Figure 16.3   Client mesh architecture.

assembled from commercial-off-the-shelf (COTS) hardware components for a few hundred dollars. Finally, since most WMNs are based on
Internet Protocol (IP) standards, integration and interoperation with
other networks are not a problem, and bridges to legacy networks can
easily be supported. The following list summarizes the key features of
WMNs that are relevant to PSDR applications:

- Broadband support (e.g., up to 54 Mbit/s theoretical peak data
  rate for IEEE 802.11a/g)
- Rapid and easy deployment through self-configuration
- Cost-effective wide area coverage through wireless multi-hop
  technology
- Robustness and fault tolerance through redundancy and self-
  healing capabilities
- Low cost through use of commodity hardware
- Interoperability through IP standards

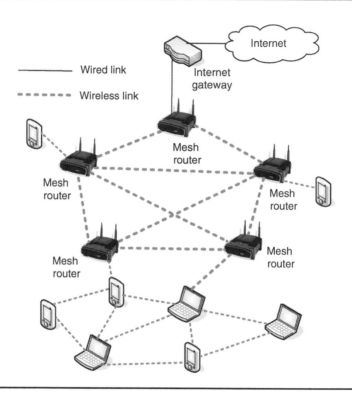

Figure 16.4 Hybrid mesh architecture.

This set of features makes WMN technology a promising platform for PSDR communications, overcoming many of the shortcomings of currently deployed systems.

16.2.3 WMN Systems and Standards

Several companies (e.g., Tropos Networks, Firetide, Motorola, Strix, and PacketHop) have realized the potential of WMNs and have started offering mesh networking products for a range of application scenarios, including PSDR communications. Most of these products are based on commodity IEEE 802.11 hardware to leverage the low cost and high performance of this technology. However, a majority of these commercial systems implement their own proprietary mesh protocols for routing and network configuration. Unfortunately, this makes integration of mesh routers from different vendors into a single WMN very difficult, if not impossible. However, interconnection between mesh networks from different vendors is not a problem due to the use of IP as a common network protocol. Open standards are important because they allow interoperability and reduce cost due to economies of scale. Efforts are underway in several IEEE working groups (e.g., 802.11, 802.15, 802.16, and 802.20) to define mesh standards. IEEE 802.11s [44] is the most relevant emerging standard for WMN technology in the context of PSDR communications. Its aim is to extend the medium access control (MAC) protocol of 802.11 networks to support mesh functionality. This is in contrast to the present WMNs that implement mesh networking at the network layer. The 802.11s standardization effort is in its early stages and hence the approval of a standard is not expected until 2008.

16.3 PUBLIC SAFETY AND DISASTER RECOVERY COMMUNICATIONS

16.3.1 PSDR Command and Communication Structure

To understand the requirements of PSDR communications, it is important to consider the typical command and control structure used in emergency and disaster recovery situations. The Incident Command System (ICS) is one of the predominant emergency response management systems. The ICS provides a generic framework for the coordination and management of emergency response and disaster recovery operations, where various agencies of different jurisdictions and disciplines are involved. One of the key characteristics of the ICS is its ability to scale to incidents of any magnitude. The ICS is used in the United States as part of the National Incident Management System (NIMS) [45] and also employed in other countries, such as Canada and the United Kingdom. Australia has implemented a similar system known as the Australian Inter-service Incident Management System (AIIMS) [46]. A key element of ICS and similar systems is a strictly hierarchical command and control structure. Figure 16.5 shows a generic incident command structure. At the top level of the hierarchy are the five aspects (or sections) of incident management according to ICS: command, operations, planning, logistics, and finance/administration. Figure 16.5 shows how operational incident management is further divided into smaller organizational structures, such as branches, divisions (or groups), down to units and individual resources. A typical example of a resource is a fire truck and its

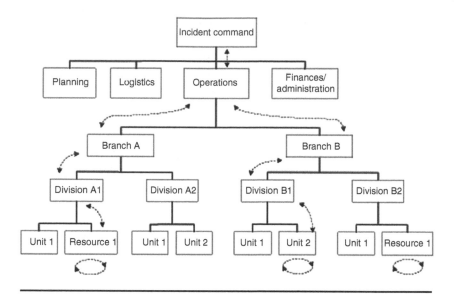

Figure 16.5   Incident command structure.

crew. The number of hierarchy levels can vary depending on the magnitude of an emergency or disaster. The section "operations" of ICS is responsible for the tactical handling of an incident and therefore places the highest demand on the communications infrastructure in terms of traffic volume and priority. Examples of typical traffic flow are indicated in Figure 16.5 with dashed arrows. Traffic can flow directly between first responders on a peer-to-peer basis within the same unit or resource. Communication with other entities flows along the structure of the command hierarchy, i.e., the chain of command. The structure of the incident command management system is an important consideration when designing communication systems for PSDR applications. The structure directly reflects the traffic pattern that can be expected in a disaster event and therefore gives indications regarding the amount of bandwidth that is required in the various parts of the network.

16.3.2   Types of PSDR Communication Networks

As defined in the SAFECOM report [2], we can differentiate between the following types of PSDR communication networks.

■ Personal area networks (PANs) interconnect various devices carried by individual first responders. For example, fire fighters might be equipped with devices to detect hazardous gas or to monitor their vital statistics, geographical location, or oxygen tank status. These devices are connected wirelessly by a PAN.

■ An incident area network (IAN) is a temporary network created for the duration of a specific incident. An IAN is necessary when fixed infrastructure networks are unavailable at the incident scene, because either they have been destroyed or they simply do not exist (e.g., in rural areas or in a subway tunnel). IANs allow first responders to share mission critical data and to coordinate their recovery efforts.

■ A jurisdiction area network (JAN) is the main communication network for first responders for all data and voice traffic that is not handled by the IAN. JANs are permanent networks that are typically installed by municipalities or public safety agencies to provide wide area (e.g., city wide) communications infrastructure for use in emergency and disaster situations.

■ Extended area networks (EANs) provide wide area connectivity between various regional, state, and national public safety networks.

Figure 16.6 shows a conceptual network diagram with these four types of networks and their interrelationship, as defined in the SAFECOM

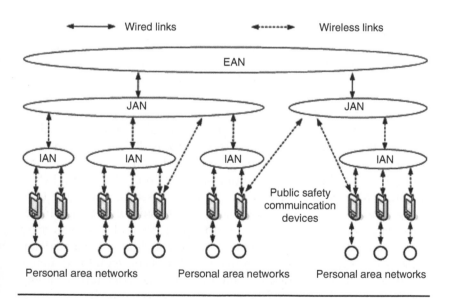

Figure 16.6   PSDR network types.

report [2]. Even though it is possible to apply WMN technology to PANs and EANs, its greatest potential is for IANs and JANs. The remainder of this chapter will therefore focus on these two categories of PSDR networks. The main purpose of IANs and JANs is to support the operational and tactical aspects of incident management. This involves communication between individual first responders and between team leaders and decision-makers at various levels in the command hierarchy (Figure 16.5). The specific role of an IAN is to provide localized communication at the incident site, between individual first responders in a peer-to-peer fashion, and between first responders and their on-site team leaders. The size of IANs can range considerably from very small, such as a single ambulance and its crew, to a size of hundreds of emergency responders in major disaster events. IANs do not rely on any fixed communications infrastructure and are typically created dynamically, in an ad hoc manner by emergency service personnel upon arrival at the incident site. WMNs seem an ideal match for IANs, due to their rapid deployability and self-configuration capabilities. A WMN deployed as an IAN would most likely have a hybrid architecture and would be comprised of a combination of mobile handheld terminals and more static mesh routers. In a disaster scenario, static mesh routers can be deployed at selected locations at the incident site to create an instant wireless hot zone on the fly. The coverage of this network can be further extended by mobile terminals participating in the routing and forwarding of packets.

In contrast to the localized communication of IANs, JANs provide wide area connectivity to remote entities, including team leaders at various levels of the command hierarchy. An important role of a JAN is to relay environmental, personal, and equipment data gathered at the incident site back to the remote command post to support informed decision-making. JANs also allow first responders to access remote databases or the Internet to download crucial information such as maps or floor plans. If IANs and JANs employ different communication technologies, a gateway node equipped with multiple interfaces and with the ability to communicate over both IAN and JAN is required. Alternatively, if IAN and JAN are based on the same technology and use the same air interface, direct communication between nodes is possible. LMR is the technology with which JAN functionality has traditionally been implemented. As we will discuss in more detail, these systems typically suffer from a number of shortcomings, such as limited support for broadband communication, lack of interoperability, and limited resilience to failure. WMN technology does not suffer from these limitations and therefore has a great potential to serve as the basis for JANs. The key features of WMNs that are relevant in this context are their inherent fault tolerance, self-healing capability, and the relatively low cost of wide area deployment. To what extent WMN technology can meet the generic requirement of PSDR communications, for both IANs and JANs, will be discussed later.

16.3.3   Current PSDR Communication Technology

The mainstay of public safety communications has been and still is LMR, also known as professional mobile radio or private mobile radio (PMR). Traditional LMR systems provide analog voice communication for closed user groups over dedicated ultra high frequency (UHF) or very high frequency (VHF) bands. Modern LMR systems are digital and have limited data capabilities. The two most relevant standards are digital and the Association of Public Safety Communications Official (APCO) Project 25 (P25) [3], standardized by Telecommunications Industry Association (TIA), and terrestrial trunked radio (TETRA) [4], developed by the European Telecommunications Standards Institute (ETSI). Most of the versions of LMR systems that are currently being deployed (TETRA release 1 and P25) only support narrowband communication with data rates of 9.6 or 28 Kb/s. Both TIA and ETSI have recently developed new standards supporting data rates of up to 473 or 690 Kb/s, respectively [5,6]. This is still not sufficient for high-quality video and other broadband applications, which require a throughput capacity of multiple Mb/s. Further standardization efforts are underway to develop public safety communication systems with true broadband capabilities [7]. However, these efforts are in their early stages and it is not expected that any standards or products will be available in the short to medium term. We refer the readers to Balachandran et al. [8] for a more detailed overview and discussion of PSDR communications technology.

16.3.4   Functional Requirements

To evaluate the suitability of WMN technology for PSDR applications, we need to consider their specific requirements. The SAFECOM program of the US Department of Homeland Security recently issued a statement of requirements (SoR) for public safety wireless communication [2]. This report defines a comprehensive set of requirements that need to be met by current and future PSDR communication systems. These requirements will form the basis of our evaluation of WMN technology. This section lists the relevant functional requirements and the next section will give an overview of the corresponding performance requirements.

■ *Interoperability*: It is absolutely crucial that public safety personnel and first responders from different agencies (e.g., ambulance, police, and fire) and jurisdictions (e.g., local, regional, state, and federal) are able to communicate and coordinate their efforts. Interoperability capability is a top priority requirement for PSDR communication systems.

■ *Support of voice and data services*: The SoR document identifies voice and data as the two main categories of services that are required for public safety communications. Even though voice could be considered as just another data service, it is treated as a separate category due to its primary role in first responder communication. The SoR report also lists a number of interactive data services that PSDR communication technology should support, such as instant messaging, video conferencing, and database queries. Further requirements are Internet connectivity and support for web-based services. The system should also be capable of supporting real-time transmission of vital statistics of objects or persons, such as the heart rate of a fire fighter or the level of her oxygen tank. Noninteractive data services that need to be supported include e-mail and file transfer.

■ *Support for mobility*: Public safety users must be able to have constant communication while traveling at reasonable speeds. The SoR document defines *reasonable speeds*, as the speed of up to and including small aircraft. The mobility requirement includes the ability to roam between different networks, potentially operated by different jurisdictions. A further requirement is the ability to determine the 3-dimensional geographical location of a user solely based on functionality provided through the communication systems. This is very useful in situations where GPS-based solutions fail to perform adequately, such as in urban areas or most indoor environments. The SoR document mentions a required accuracy of the location system of 1 m.

■ *Standards-based design*: Open standards are important for PSDR communication systems as they allow interoperability in a multivendor environment. Standards further reduce equipment cost due to economies of scale. PSDR communication systems should use products with the broadest possible market base, while meeting all the essential requirements. The SoR document specifically dictates the use of COTS-based equipment whenever possible.

- *Security:* A number of security aspects are crucial for PSDR communication. The key requirements are privacy, data integrity, authentication, access control, and availability. Privacy guarantees that only the intended recipient of the message is able to receive and understand it. Implementation of appropriate privacy mechanisms might also be required for compliance with state or federal policies, such as the Health Insurance Portability and Accountability Act (HIPAA) [29]. Data integrity prevents unauthorized (accidental or intentional) modification or insertion of information. Access control mechanisms provide authentication of users and their devices and limit access of network resources to authorised entities. Furthermore, the communication system needs to provide a high level of availability and should be resistant to a wide range denial of service attacks, including radio jamming.
- *Communication modes*: The SoR document lists a number of communication modes that need to be supported by PSDR communication systems. The required modes are unicast, multicast, and broadcast. Furthermore, the system needs to be able to operate in peer-to-peer mode, where clients are able to directly communicate with each other in scenarios where fixed communications infrastructure is unavailable.
- *Network management*: Under the heading command and control, maintenance, and operation, the SoR report specifies a number of network management requirements for PSDR communication systems. These cover the traditional areas of network management such as performance management, configuration management, fault management, and security management. For example, administrators need to be able to define and manage user groups and assign role-based permissions to users. Further functionality that is required includes the ability to continuously monitor the state of the network to detect anomalies such as faults or attacks.

16.3.5 Performance Requirements

In addition to the aforementioned functional requirements, the SAFECOM SoR report specifies performance requirements for PSDR communication systems in the following areas:

- *Robustness*: PSDR communication systems need to be highly reliable and robust and be able to function in very adverse and hostile environments. Recent disasters have clearly underscored the shortcomings of currently deployed technology in this regard. PSDR communication systems need to be highly redundant and should not have any single point of failure. In case of link and node failures, systems should support self-healing and self-configuration capabilities.

■ *Scalability*: The SoR document specifies two types of scalability requirements. Horizontal scalability refers to the ability of the network to grow efficiently and cost-effectively in terms of geographical coverage. Vertical scalability stands for the ability to efficiently support increasing number of users. PSDR communication systems need to be able to provide both horizontal and vertical scalability, while still maintaining the required quality of service (QoS).

■ *Quality of service*: As a first priority, a PSDR communication system must be able to provide reliable and high-quality voice communication. Image and video data need to be transmitted to allow rendering in acceptable quality. Beyond specifying that the call-setup time for voice communication should not be more than 250 ms, the SoR report does not provide any quantitative QoS requirements. A further requirement is the ability to differentiate between traffic of different levels of priority. High-priority traffic needs to be given precedence over low-priority traffic to guarantee delivery of urgent messages in situations of network congestion.

16.4   WMNs FOR PSDR COMMUNICATIONS

Section 16.3 gave a summary of the functional and performance requirements that are specific for PSDR communication systems. This section will evaluate the potential of WMN technology as a platform for PSDR communication for both IANs and JANs, by discussing its ability to meet these requirements. We will also highlight the areas where WMNs have inadequacies and where further research is required.

16.4.1   Functional Requirements

16.4.1.1   Interoperability

Recent WMN products are based on commodity IEEE 802.11 hardware. However, the majority of the commercial systems implement their own proprietary mesh protocols for routing and network configuration. Unfortunately, this makes integration of mesh routers from different vendors into a single WMN very difficult, if not impossible. The emerging IEEE 802.11s standard with integrated mesh network support at the MAC layer might solve this problem, but given the state of the standardization process, this is not expected to happen soon. Even though integration of mesh routers from different vendors into a single WMN is currently not possible, interoperability between different WMNs, as well as between WMNs and other networks is not a problem. The IP, the lingua franca of data communications, provides the unifying standard that allows devices on different types of networks, including WMNs, to communicate with each other. An IP-based network is therefore the ideal common platform for communication between multiple emergency response services and different jurisdictions. Interoperability to legacy PSDR communication systems can be provided by dedicated bridging devices. For example, a number of vendors provide radio over IP (or LMR over IP) bridges that allow interconnecting traditional LMR radio systems with an IP-based network.

16.4.1.2   Support for Voice and Data Services

Even though current WMN systems differ in terms of their mesh routing protocols, they share IP as the common network protocol. IP-based networks are very versatile and can be utilized for a wide range of applications. All the application types required for PSDR communication specified in Section 16.3.4 are already widely implemented and can easily be supported by WMNs. Voice over IP (VoIP) technology allows the transmission of digitized voice over an IP-based packet networks. Both the ITU-T as well as the Internet Engineering Task Force (IETF) have defined a set of higher layer standards and protocols, and implementations for VoIP services are readily available. Other services such as instant messaging, video conferencing, or file transfer have been deployed on IP networks for many years and are well supported. Clearly, an IP-based network is also the ideal platform to provide access to the Internet and the World Wide Web. The flexibility and the variety of services that can be supported by WMNs is one of the key advantages over traditional PSDR communication systems.

16.4.1.3   Mobility

According to the SAFECOM report [2], PSDR communication technology should support mobile clients traveling with up to the speed of small aircraft. IEEE 802.11 technology has not been designed to handle mobile clients traveling at such high speeds. However, recent experiments have shown that 802.11b can support mobile clients with speeds of up to 180 km/h [31]. Further research is required to explore the limits of 802.11 technology in this regard. Mobile clients cannot be expected to stay within the physical range of a single access point or mesh router. Current deployments of traditional WLANs support roaming of clients between access points. This type of mobility is handled at the MAC layer and is transparent to higher layers. An inter-access point protocol (IAPP) is used to coordinate the handover between access points. Unless mesh networking is supported at the MAC layer, possibly with the upcoming mesh standard IEEE 802.11s, roaming of clients between mesh routers needs to be implemented at a higher layer. In typical WMNs, where mesh functionality is implemented at the network layer, ad hoc routing protocols handle the mobility management. New routes to the destination are automatically detected, either proactively or reactively, when a client roams between mesh routers. This method of mobility management assumes that roaming occurs within a single IP subnet and mobile clients do not change their IP address. If roaming between different networks (i.e., IP subnets) is to be supported, mechanisms such as mobile IP [30] are required. One of the problems with mobile IP is that it relies on centralized entities, which is a bad match for highly distributed WMNs. A further requirement for PSDR communication systems in the context of mobility is the ability to determine the location of a mobile device. The communication system needs to be able to provide the location of a device or user with an accuracy of 1 m. Even though there have been efforts to implement location services on 802.11 networks using triangulation methods [15], the results are far from meeting PSDR requirements. Multipath signal propagation effects represent the major obstacle to achieve exact location information.

16.4.1.4 Standards-Based Design

IEEE 802.11 technology has experienced a huge success in the market place and is evolving rapidly. In comparison, specialized PSDR communication technology represents a relatively niche market, lacking the resources for research and development of commodity systems. It is not surprising that the functionality and performance of dedicated PSDR communication systems generally lag behind that of commodity systems. The use of COTS hardware based on the IEEE 802.11 standards provides WMNs with high performance and functionality at a very low cost. It is widely recognized that cost is a crucial factor in PSDR communications and is one of the main reason why many outdated systems have not been upgraded or replaced yet. WMNs also meet the standards-based design requirement at the network layer by using the IP. The only area where current WMN systems are lacking common standards is the mesh protocol that provides routing and self-configuration functionality. Current commercial WMN systems employ a variety of proprietary and incompatible mesh protocols and mechanisms, which prevent the deployment of mixed-vendor mesh networks. The efforts behind the upcoming IEEE 802.11s standard are aimed at addressing this problem by defining a wireless mesh networking standard. The corresponding IEEE Task Group has issued a call for proposals and is currently evaluating submissions.

16.4.1.5 Security

Wireless networks are generally more vulnerable to attacks than their wired counterparts due to the broadcast nature of the wireless medium and the lack of any firm physical boundaries. Wired Equivalent Privacy (WEP), defined in [32], was a flawed attempt to provide security for 802.11-based WLANs. Weaknesses in the protocol specification and implementations allowed attackers to compromise a network relatively easily [33]. IEEE 802.11i is a more recent security standard that addresses the flaws and weaknesses of WEP, and provides privacy, integrity, and authentication. IEEE 802.11i is considered secure and so far it has successfully withstood careful scrutiny and analysis. IEEE 802.1x is a related standard that provides port-based access control for wired and wireless networks. Security in wireless networks can also be implemented at higher layers by protocols such as IPSec or TLS. However, these protocols assume either a preexisting trust relationship between participating nodes (e.g., by shared secret keys), or the availability of a trusted third party such as a key distribution server or a certification authority. These assumptions are valid for relatively static networks with centralized infrastructure, but not necessarily for WMNs that are highly distributed and can potentially be very dynamic. Even though it is possible to preestablish keys within closed user groups such as individual emergency service units or agencies, it is an extremely challenging task to do this for all the emergency service and disaster recovery units from various jurisdictions that potentially need to cooperate and communicate during a major disaster. Incompatible security parameter settings and lack of the required keys will make communication between the different first responder and disaster recovery units impossible, even if their communication equipment is otherwise interoperable. One of the research challenges in this context is to define mechanisms for key distribution and management for such dynamic environments and in the absence of centralized infrastructure. This and similar problems have been addressed by the ad hoc network research community for some time [12]. The design space of key management and other security mechanisms needs to be further explored to find viable trade-offs between security and interoperability for different PSDR scenarios.

Availability is a crucial security requirement for PSDR communications and it cannot be guaranteed by cryptographic primitives. Attackers have a range of options to launch denial of service attacks and disrupt availability of communications services in WMNs. Attacks are possible at the various layers of the protocol stack. At the physical layer, an attacker can disrupt communication by simply jamming the radio channels in which WMNs operate. By saturating the frequency band with noise, the signal-to-noise ratio at the receiver can be decreased to a level where successful reception of data becomes impossible. It is relatively easy for an attacker to implement localized jamming of a small area using high-power commodity transmitters and directional antennas. However, wide area jamming of a large number of mesh nodes is very resource intensive and extremely difficult to accomplish for an attacker. Due to their high level of redundancy and self-healing capabilities, WMNs have great resilience to localized jamming.

Wireless LANs and WMNs based on IEEE 802.11 technology are also susceptible to a number of attacks at the MAC layer. The correct operation of the MAC functionality relies on the assumption that all nodes collaborate and strictly adhere to the standard. There are a number of ways in which individual nodes can misbehave to either gain an unfair advantage in terms of resource allocation or to simply deny service to other users [34]. For example, the virtual carrier sense mechanism of 802.11 allows cheating nodes to reserve an almost arbitrary large fraction of the available bandwidth. Another method by which misbehaving nodes can deny other users their fair access to the shared medium is by not respecting the inter-frame spacing (IFS) periods. Finally, a malicious node can simply continually transmit data and deliberately cause collisions by completely ignoring the exponential backoff mechanism of carrier sense multiple access with collision avoidance (CSMA/CA). The result is the disruption of any data transmission within the attacker's radio range. These types of attacks are a fundamental problem of the 802.11 MAC protocol and cannot be prevented with cryptography. Fortunately, considerable efforts are required to implement these kinds of attacks, as they require firmware modifications or other fairly sophisticated techniques [35]. Similar types of attacks are possible at the network layer. Malicious nodes can disrupt the routing and forwarding of packets in a number of ways. By not following the specifications of the routing protocols malicious node can simply refuse to forward any packets on behalf of other nodes (black hole attack) or they can disrupt the routing protocol by advertise false routes. A number of secure ad hoc routing protocols have been proposed to prevent or mitigate such attacks [36,37]. These protocols rely on cryptographic primitives for the implementation of authentication and data integrity mechanisms and therefore require a method for key distribution. As mentioned above, key management for WMNs in large-scale PSDR applications is an open research problem.

16.4.1.6 Communication Modes

Multicast (one-to-many) and broadcast (one-to-all) are important service primitives required for PSDR communication systems, allowing more efficient use of the network resources when identical data needs to be sent to multiple recipients. IEEE 802.11 wireless networks provide support for these modes of communication at the MAC layer, which is relatively straight forward due to the broadcast nature of the wireless medium. However, broadcast/multicast support of the current 802.11 MAC layer is limited to a single wireless cell and does not extend to a multi-hop network. This limitation will be addressed in the future by the emerging wireless mesh standard (IEEE 802.11s), which aims to implement efficient multicast and broadcast primitives for multi-hop networks at the MAC layer. Without MAC layer support, multicast and broadcast need to be implemented at a higher layer. A significant amount of research has recently been done in the area of multicast routing protocols for ad hoc networks. However, it has been shown that these protocols have some significant limitations in terms of scalability and efficiency [38]. These protocols have been designed for pure ad hoc networks where the entire network consists of mobile client nodes. More research is required to design efficient multicast routing protocols that are suitable for the more heterogeneous WMNs with a mix of relatively static mesh routers and mobile clients. Finally, WMNs in both client and hybrid mesh architecture support the required peer-to-peer mode of operation where client devices are able to communicate directly with each other without the support of any communication infrastructure.

16.4.1.7 Network Management

Management and monitoring of multi-hop WMNs is more difficult than for wired or single-hop wireless networks. The reasons are the lack of a hierarchical and static network topology, especially for client and hybrid WMNs, node mobility, and the dynamic nature of ad hoc routing protocols. Traditional network management tools based on the Simple Network Management Protocol (SNMP) provide most of the functionality required for PSDR communication systems stipulated in the SAFECOM report [2]. However, due to their strictly centralized or hierarchical mode of operation, they are a poor match for the distributed and dynamic nature of WMNs. Ramachandran et al. [47] present a basic monitoring system for multi-hop wireless networks using a completely distributed architecture. Key features include autodiscovery of sinks for monitoring information and resilience to individual node failure. The feasibility of the approach is demonstrated with an implementation for a WMN running the ad hoc on-demand vector (AODV) [49] routing protocol. This work presents an important step toward a distributed and efficient solution to network monitoring and management for WMNs, but a lot more work is required in this area.

16.4.2   Performance Requirements

In addition to the above mentioned functional requirements, the SoR document also lists performance requirements that need to be met by PSDR communication systems. These requirements are in regard to scalability, robustness, and QoS.

16.4.2.1   Robustness

Robustness is clearly one of the strengths of WMNs. One of the key features of WMNs is the inherent redundancy of the mesh topology with multiple redundant paths between communication endpoints. The lack of a single point of failure guarantees connectivity even in the event of individual link or node failures. The ability to self-heal and dynamically adapt to a changing environment is another crucial characteristic of WMNs. Mesh routing protocols automatically establish the best possible path between source and destination under the given conditions. WMNs, therefore, easily meet the requirements of PSDR communication systems in terms of reliability, survivability, and restorability as defined in the SAFECOM report [2].

16.4.2.2   Scalability

The coverage area of a WMN (horizontal scalability) can easily be increased by simply deploying additional mesh routers. However, this results in an increased average path length and it has been shown that the throughput of a WMN degrades rapidly with the number of hops involved in the end-to-end path [13,14]. This severely limits the scalability of WMNs in terms of network size and diameter. One of the main problems is co-channel interference, resulting in collisions, reduced throughput, and increased communication delays. In contrast to WLANs, the spectrum in WMNs is typically shared not only by the traffic from clients to the access points, but also with the backbone traffic between mesh routers. Furthermore, the 802.11 MAC mechanism was designed for single-hop networks and does not perform well in the multi-hop environment of WMNs.

The issue of vertical scalability is also very much an open research problem for WMNs. It is difficult to determine exactly how many users can be supported by a WMN, since this depends on a variety of parameters such as network topology, terrain, and type of applications. The fact that most WMNs operate in unlicensed Industrial Scientific and Medical (ISM) frequency bands and therefore have to share the spectrum with other WLANs as well as a range of other wireless devices further aggravates the scalability problem. A lot of public safety and emergency response practitioners are very skeptical to use any unlicensed frequency bands for their mission critical communications. In this context, it is interesting to note that the US Federal Communications Commission has recently made the licensed 4.9 GHz band available for public safety and homeland security applications. Only minor changes are needed to 802.11a equipment operating in the neighboring 5 GHz band to be used for the licensed 4.9 GHz band. This allows manufacturers of PSDR communication systems to leverage the high performance and low cost of such commodity hardware, while operating the equipment in dedicated public safety bands. First commercial WMN products using the 4.9 GHz band have already been announced by companies such as Firetide and Proxim. Spectrum allocation is obviously a key issue in this context and it can vary greatly from country to country. For example, the 4.9 GHz band reserved for public safety applications in the United States is an unlicensed ISM band in Japan and the IEEE 802.11j standard has specifically been defined for wireless LANs in this band.

16.4.2.3 *Quality of Service*

Current WMNs based on commodity hardware fail to provide QoS guarantees. The MAC mechanism of 802.11 networks is based on a randomized algorithm (CSMA/CA), which makes it very difficult to give any guarantees regarding performance parameters such as delay, throughput, or jitter. Furthermore, current 802.11 networks do not allow differentiation of traffic with different levels of priority. This is a crucial requirement for mission critical PSDR communication. In a disaster scenario where the network is likely to be congested, it is crucial to be able to give precedence to high-priority messages. IEEE 802.11e [16] is a recent extension to the 802.11 standard that supports QoS and differentiation of traffic classes for single-hop wireless networks. However, the issue of QoS in multi-hop WMNs is still very much an open research problem. Limited scalability and capacity, combined with the lack of QoS guarantees are currently the most significant shortcomings of WMNs in general. These are also the areas where WMNs fall short of the requirements of PSDR communication systems. Industrial and academic research on WMNs is now trying to address these problems. Section 16.5 highlights the key research activities that are currently underway in this area.

16.5   KEY RESEARCH ACTIVITIES

As mentioned earlier, the main weaknesses of current WMN technology are in regard to capacity, scalability, and QoS. This section gives an overview of the main research activities addressing these issues. These research efforts can be roughly classified according to the layer of the protocol stack that they are focusing on.

16.5.1   Physical Layer

The physical layer of a wireless communication system is concerned with the transmission of raw bits from a sender to a receiver and addresses issues such as modulation, coding, and antenna design. Research in this area is applicable to a wide range of wireless systems and is not necessarily specific to WMNs. Researchers have been exploring concepts such as beam-forming antennas, multiple-input multiple-output (MIMO) systems, and cognitive radios to increase the capacity of wireless communication systems. Beam-forming antennas are able to dynamically concentrate the signal energy and receiver gain in the desired direction, resulting in reduced interference and thereby increased capacity. MIMO systems use multiple antennas and sophisticated signal processing to exploit phenomena such as multipath propagation to reduce error rates and increase throughput of the wireless link. These ideas have been studied for years but high cost and complexity have prevented their widespread adoption into commodity hardware so far. Recently, researchers have also started investigating the idea of the so-called cognitive radios. The underlying idea is a software-based radio with dynamically reconfigurable system parameters, allowing dynamic and optimal adaptation to the operating environment or user demands. For example, a cognitive radio system could measure the utilization of a wide range of the radio spectrum and dynamically switch to more underutilized bands. Even though cognitive radios have a big potential to improve the spectrum efficiency and capacity of wireless systems, the technology is relatively new and it will be a long time before it will make its way into commodity wireless products.

The use of multiple commodity wireless network cards (radios) per node has recently been proposed as a method for increasing the capacity of WMNs [17]. With multiple radios operating on different orthogonal channels, a node can simultaneously transmit or receive data with no (or very little) interference between the channels. The potential of this approach has been demonstrated [39], where it was shown that by adding a second radio to the nodes of a WMN the average throughput could be improved by a factor of 6 to 7. The key benefit of the multiradio solution is that it works with currently available COTS hardware, and is therefore a very cost-effective way to increase the capacity and scalability of WMNs. An interesting problem that is currently being investigated is how to optimally assign channels to the various wireless interfaces of the nodes to minimize interference and maximize throughput [18].

16.5.2 Media Access Control Layer

The MAC layer of communication system coordinates the access to a shared transmission medium by different users. Wireless networks based on the IEEE 802.11 standard use a CSMA/CA scheme that uses a randomized back off mechanism in case of collisions. The IEEE 802.11 MAC protocol has been designed for single-hop wireless networks and has significant limitations in the multi-hop environment of WMNs, where traffic from multiple hops of the data path contends for the shared medium [19]. Even though there have been a number of proposals for new MAC protocols for wireless multi-hop networks [20,21], they have not found their way into commodity systems. The key problems of MAC-based solutions is that they are incompatible with currently available hardware. This represents a much higher obstacle to widespread acceptance compared to solutions that can be implemented with COTS hardware, such the multiradio approach.

16.5.3 Network Layer

The network layer of a WMN is responsible for discovering optimal paths from a source to a destination, considering current network conditions. Traditional routing protocols only consider the path length as the routing metric. Finding the optimal path corresponds to finding the path with the least number of hops, i.e., the shortest path. It has been shown that for WMNs, the shortest path is not always best in terms of throughput, delay, and error rate [22]. For example, it is better to use a 2-hop path consisting of two high-quality wireless links than a single-hop path that has low throughput and high error rates. A number of researchers have been working on defining new routing metrics that are more suitable for WMN environments. These metrics specifically consider the quality of the individual wireless links. Expected Transmission Count (ETX), the metric proposed by De Couto et al. [23], measures the expected number of transmissions needed to send a unicast packet across a link. Since this metric includes retransmissions due to poor signal quality or collisions, it gives a good indication of the link quality. Routing based on the ETX metrics performs significantly better than traditional shortest path protocols in WMNs [22]. A new routing metric specifically designed for WMNs with multiple radios per node has been introduced by Draves et al. [24]. The metric called Weighted Cumulative Expected Transmission Time (WCETT) takes into account both path length as well as link quality, and provides a trade-off between delay and throughput.

Current routing protocols for WMNs can be roughly grouped into two categories. The first consists of protocols that are based on traditional routing protocols for wired networks such as Routing Information Protocol (RIP) or Open Shortest Path First (OSPF). Since these protocols are not able to handle node mobility, their application is restricted to relatively static infrastructure WMNs. The second category consists of protocols that are based on ad hoc routing protocols. Tremendous research efforts have been made in this area over the last few years and an impressive number of ad hoc routing protocols have been proposed. These protocols were designed for networks with highly mobile and typically power con-strained devices. As a consequence, these protocols are able to handle node mobility and the generally dynamic nature of WMNs.

To the best of our knowledge, there currently exist no routing protocols that are specially designed for hybrid WMNs, consisting of a mix of mobile nodes and relatively static mesh routers. However, this is the most likely WMN architecture to be employed for PSDR communications, especially in the case of IANs. Current WMN routing protocols do not differentiate between the type of nodes in the network and are therefore unable to take advantage of the high degree of heterogeneity. Mesh routers and mesh clients can differ greatly in a number of aspects. Mesh clients are generally resource constrained devices with limited battery power, equipped with only a single radio. In contrast, mesh routers are much less resource constrained and are either powered with high-capacity batteries or have access to mains power. In addition to being more static, mesh routers are also likely to be equipped with higher gain antennas and multiple radio interfaces, resulting in a significantly increased capacity compared to mesh clients. In an emergency response scenario where an IAN is implemented via a hybrid WMN, it is crucial to route traffic preferably via mesh routers and only use mesh clients as routers if necessary. This not only maximizes the battery life of the mobile terminals used by first responders, but also results in increased throughput and QoS.

We are currently investigating extensions to AODV that implement such a mesh router preferential routing scheme. The key idea is to replace hop count as AODV's standard routing metric with a metric that not only considers the number of hops but also the type of the individual nodes and their available resources. Simulations of hybrid WMN networks with a topology and mobility pattern that are typical for PSDR scenarios show significant improvements of this simple scheme over standard AODV in terms of latency and packet delivery ratio [48]. However, these are early results and a lot more work is required to develop routing protocols that perform optimally for PSDR applications and are able to provide the required QoS in emergency response situations.

Some researchers also propose an interaction or integration of the routing protocols with the MAC layer and/or the physical layer [25–27]. Such cross-layer mechanisms have a great potential to improve protocol performance. For example, the routing protocol could benefit from having access to information about the link quality from the physical or MAC layer. However, the cross-layer optimization approach in protocol design is somewhat controversial, since it violates the paradigm of clear separation of protocol layers. The potential consequences are incompatibility with current protocols, reduced modularity, and increased maintenance and management costs. It remains to be seen to what extent cross-layer optimization will be incorporated into future WMN designs.

16.5.4 Higher Layers

Due to the physical characteristics of the transmission medium, wireless networks in general and WMNs in particular have higher transmission error rates than wired networks. It is a well-known fact that TCP, the most predominant transport protocol, does not perform well in wireless environments. This is mainly due to the fact that TCP assumes a reliable physical layer and associates packet loss with network congestion. Over the last few years, a number of solutions have been proposed to improve the performance of TCP over wireless networks [40,41]. Some researchers have also proposed completely new transport protocols to replace TCP [42]. Given the installed base of TCP and the dependency of many applications on it, it is unlikely that new and incompatible protocols will have a good chance of adoption in the marketplace. A further proposal to deal with the dynamic nature of wireless networks is the concept of application adaptivity [43]. In this approach, applications are aware of the characteristics and resources of the network and can adapt accordingly. For example, a video conferencing application could react to a drop in available bandwidth either by reducing the quality of the video, or even by entirely removing the video frames to guarantee that the most important voice packets are transmitted timely and reliably. Adaptive applications are a promising method to deal with the dynamic nature of WMNs, but a lot more work is required to explore their potential.

16.6 CONCLUSIONS

WMNs based on commodity hardware have a great potential for PSDR applications. Economies of scale and a tremendous investment in research and development resulted in the availability of high-performance wireless equipment at a very low cost. WMNs leverage commodity hardware to provide wide area broadband connectivity cost-effectively. As discussed in Section 16.2, further characteristics of WMNs are interoperability, robustness, self-configuration, and self-healing capabilities, making them a promising platform for PSDR communications. Reliable and efficient communication is absolutely mission-critical for PSDR operations. Communication technologies used in this context need to meet a set of application specific functional and performance requirements, as outlined in Section 16.3. Even though WMNs are able to meet most of the functional requirements, they currently suffer from significant limitations in terms of scalability and QoS, which are crucial performance requirements in the PSDR context. Given the current research efforts that are underway in this area, we are quite optimistic that these limitations can be overcome and that WMN technology will play a vital role in future PSDR communications.

ACKNOWLEDGMENTS

National ICT Australia is funded by the Australian Government's Department of Communications, Information Technology, and the Arts, the Australian Research Council through Backing Australia's ability, the ICT Research Centre of Excellence programs, and the Queensland Government.

REFERENCES

1. Kean, T.H., et al., "The 9–11 Commission Report," http://www.9-11commission. gov/report/911Report.pdf (accessed January 2006).
2. SAFECOM Program, US Department of Homeland Security, "Statement of Requirements for Public Safety Wireless Communication and Interoperability," Version 1.0, March 2004, http://www.safecomprogram.gov/SAFECOM/ (accessed January 2006).
3. Telecommunications Industry Association (TIA), "APCO Project 25 System and Standards Definition," TIA/EIA Telecommunication Systems Bulletin TSB102-A, November 1995.
4. European Telecommunications Standards Institute (ETSI), "Terrestrial Trunked Radio (TETRA); Voice plus Data (V+D); Designers Guide; Part 1: Overview, Technical Description and Radio," ETR 300–1, May 1997.
5. Telecommunications Industry Association (TIA), "Wideband Air Interface (SAM) Radio Channel Coding Specification Public Safety Wideband Standards Project Digital Radio Technical Standards," TIA-902 BAAD, September 2002.
6. European Telecommunications Standards Institute (ETSI), "Terrestrial Trunked Radio (TETRA): TETRA Advanced Packet Service (TAPS)," ES 201–962, September 2001.
7. Mobility for Emergency and Safety Applications (MESA), http://www.porject mesa.org (accessed January 2006).
8. Balachandran, K., et al., "Mobile Responder Communication Networks for Public Safety," IEEE Communications Magazine, vol. 44, no. 1, January 2006.
9. Akyildiz, I.F., and Wang, X., "A Survey on Wireless Mesh Networks," IEEE Communications Magazine, vol. 43, no. 9, September 2005.
10. Perkins, C.E., "Ad-Hoc Networking," Addison Wesley Professional, Reading, MA, 2001.

11. Royer, E.M., and Toh, C.K., "A Review of Current Routing Protocols for Ad Hoc Mobile Wireless Networks," *IEEE Personal Communications Magazine*, April 1999.

12. Hubaux, J.-P., Buttyan, L., and Capkun, S., "Self-Organized Public-Key Management for Mobile Ad Hoc Networks," *Transaction on Mobile Computing*, vol. 2 no. 1, pp. 52–64, January to March 2003.

13. Gupta, P., and Kumar, P.R., "Capacity of wireless networks," Technical Report, University of Illinois, Urbana-Champaign, 1999.

14. Li, J., Blake, C., De Couto, D.S.J., Lee, H.I., and Morris, R., "Capacity of Ad Hoc Wireless Networks," in Proceedings of ACM/IEEE MOBICOM, July 2001.

15. Bahl, P., and Padmanabhan, V.N., "Radar: An In-Building RF Based User Location and Tracking System," in Proceedings of IEEE INFOCOM 2000, pp. 775–784, March 2000.

16. IEEE WG, "IEEE 802.11e/D13.0, Draft Supplement to Part 11: Wireless Medium Access Control (MAC) and Physical Layer (PHY) Specifications: Medium Access Control (MAC) Enhancements for Quality of Service (QoS)," January 2005.

17. Bahl, P., Adya, A., Padhye, J., and Wolman, A., "Reconsidering Wireless Systems with Multiple Radios," ACM Computer Communications Review, July 2004.

18. Raniwala, A., Gopalan, K., and Chiueh, T., "Centralized Channel Assignment and Routing Algorithms for Multi-Channel Wireless Mesh Networks," *SIGMOBILE Mobile Computer Communications Review*, vol. 8, no. 2, pp. 50–65, April 2004.

19. Xu, S., and Saadwi, T., "Does the IEEE 802.11 MAC Protocol Work Well in Multi-Hop Wireless Ad Hoc Networks?" *IEEE Communications Magazine*, vol. 39, no. 6, June 2001.

20. Tzamaloukas, A., and Garcia-Luna-Aceves, J.J., "A Receiver-Initiated Collision-Avoidance Protocol for Multi-Channel Networks," in Proceedings of IEEE INFOCOM, 2001.

21. Lin, C.R., and Gerla, M., "MACA/PR: An Asynchronous Multimedia Multi-Hop Wireless Network," in Proceedings of IEEE INFOCOM, 1997.

22. Draves, R., Padhye, J., and Zill, B., "Comparison of Routing Metrics for Static Multi-Hop Wireless Networks," in Proceedings of the 2004 Conference on Applications, Technologies, Architectures, and Protocols for Computer Communications, SIGCOMM'04. ACM Press, New York, pp. 133–144, 2004.

23. De Couto, D., Aguayo, D., Bicket, J., and Morris, R., "High-Throughput Path Metric for Multi-Hop Wireless Routing," in Proceedings of IEEE//ACM MOBICOM, September 2003.

24. Draves, R., Padhye, J., and Zill, B., "Routing in Multi-Radio, Multi-Hop Wireless Mesh Networks," in Proceedings of IEEE//ACM MOBICOM, September 2004.

25. Bhatia, R., and Kodialam, M., "On Power Efficient Communication over Multi-Hop Wireless Networks: Joint Routing, Scheduling, and Power Control," in Proceedings of IEEE INFOCOM, 2004.

26. Chiang, M., "To Layer or Not to Layer: Balancing Transport and Physical Layers in Wireless Multi-Hop Networks," in Proceedings of IEEE INFOCOM, 2004.

27. Kozat, U.C., Koutsopoulos, I., and Tassiulas, L, "A Framework for Cross-Layer Design of Energy-Efficient Communication with QoS Provisioning in Multi-Hop wireless Networks," in Proceedings of INFOCOM, 2004.
28. Miller, L.E. "Wireless Technologies and the SAFECOM SoR for Public Safety Communications," NIST Technical Report, http://www.antd.nist.gov/wctg/manet/docs/WirelessAndSoR060206.pdf (accessed February 2006).
29. Standard for privacy of individually identifiable health information, Federal Register, vol. 66, no. 40, February 2001.
30. Perkins, C., "IP Mobility Support for IPv4," RFC 3344, IETF, August 2002.
31. Ott, J., and Kutscher, D., "Drive-thru Internet: IEEE 802.11b for Automobile Users," In Proceedings of INFOCOM 2004.
32. LMSC of the IEEE Computer Society, "Wireless LAN medium access control (MAC) and physical layer (PHY) specifications," IEEE Standard 802.11, 1999.
33. Borisov, N., Goldberg, I., and Wagner, D., "Intercepting Mobile Communications: The Insecurity of 802.11," in Proceedings of MOBICOM'01, ACM Press, pp. 180–189, July 2001.
34. Gupta, V., Krishnamurthy, S., and Faloutsos, M., "Denial of Service Attacks at the MAC Layer in Wireless Ad Hoc Networks," in Proceedings of IEEE Military Communication Conference (MILCOM), pp. 1118–1123, 2002.
35. Bellardo, J., and Savage, S., "802.11 Denial-of-Service Attacks: Real Vulnerabilities and Practical Solutions," in USENIX Security Symposium, p. 1528, 2003.
36. Hu, Y.-C., Johnson, D.B., and Perrig, A., "SEAD: Secure Efficient Distance Vector Routing in Mobile Wireless Ad Hoc Networks," in Proceedings of the Fourth IEEE Workshop on Mobile Computing Systems and Applications (WMCSA02), 2002.
37. Hu, Y.-C., Perrig, A., and Johnson, D.B., "Ariadne: A Secure Ondemand Routing Protocol for Ad Hoc Networks," in Proceedings of MOBICOM, 2002.
38. Lee, S.J., Su, W., Hsu, J., Gerla, M., and Bagrodia, R., "A Performance Comparison Study of Ad Hoc Wireless Multicast Protocols," in Proceedings of IEEE INFOCOM, 2000.
39. Raniwala, A., and Chiueh, T., "Architecture and Algorithms for an IEEE 802.11-Based Multi-Channel Wireless Mesh Network," in Proceedings of INFOCOM, 2005.
40. Bakre, A., and Badrinath, B., "I-TCP, Indirect TCP for Mobile Hosts," in Proceedings of 15th International Conference on Distributed Computing Systems (ICDCS), 1995.
41. Balakrishnan, H., Seshan, S., Amir, E., and Katz, R.H., "Improving TCP/IP Performance over Wireless Networks," in Proceedings of MOBICOM, 1995.
42. Sundaresan, K., Anantharaman, V., Hung-Yun Hsieh, and Sivakumar, A.R., "ATP: A reliable Transport Protocol for Ad Hoc Networks," IEEE Transactions on Mobile Computing, vol. 4, no. 6, pp. 588–603, November to December 2005.
43. De Silva, R., Landfeldt, B., Seneviratne, A., and Diot, C., "Managing Application Level QoS Through TOMTEN," Computer Networks Special Issue on High Performance Protocol Architecture, vol. 31, no. 7, 1999.
44. Hauser, J., Draft PAR for IEEE 802.11 ESS Mesh, IEEE Document Number: IEEE 802.11-03/759r2.

45. Department of Homeland Security, "National Incident Management System," http://www.fema.gov/emergency/nims/index.shtm (accessed May 2006).

46. Australasian Fire Authority Council, "The Australasian Inter-Service Incident Management System," http://www.afac.com.au/awsv2/publications/aiims. htm (accessed May 2006).

47. Ramachandran, K.N., Belding-Royer, E.M., and Almeroth, K.C., "DAMON: A Distributed Architecture for Monitoring Multi-Hop Mobile Networks," in Proceedings of IEEE SECON, October 2004.

48. Pirzada, A.A., Portmann, M., and Indulska, J., "Mesh-Aware Ad-Hoc On-demand Distance Vector Routing," submitted to ICNP 2006.

49. Perkins, C.E., and Royer, E.M., "Ad-Hoc On-Demand Distance Vector Routing," in Proceedings of WMCSA, 1999.

INDEX